Fisheries and Oil Development on the Continental Shelf

Cover photo by Carl R. Sullivan

Support for the Outer Continental Shelf Fisheries Management Conflicts Symposium and publication of this book was provided by

Minerals Management Service,
U.S. Department of the Interior

Fisheries and Oil Development on the Continental Shelf

Edited by

Colleen S. Benner

Robert W. Middleton

American Fisheries Society Symposium 11

Bethesda, Maryland
1991

The American Fisheries Society Symposium series is a registered serial. Suggested citation formats follow.

Entire book

Benner, C. S., and R. W. Middleton, editors. 1991. Fisheries and oil development on the continental shelf. American Fisheries Society Symposium 11.

Article within the book

Chamberlain, D. W. 1991. Effects of nonexplosive seismic energy releases on fish. American Fisheries Society Symposium 11:22–25.

Library of Congress Catalog Card Number: 91-71563
ISBN 0-913235-73-3 ISSN 0892-2284

Address orders to

American Fisheries Society
5410 Grosvenor Lane, Suite 110
Bethesda, Maryland 20814, USA

Contents

Symbols and Abbreviations

The following symbols and abbreviations are used in this book without definition.

Prefixes

kilo	(10^3)	k
centi	(10^{-2})	c
milli	(10^{-3})	m
micro	(10^{-6})	µ

Time and Temperature

day	d
degrees Celsius	°C
hour	h
minute	min
second	s

Weights and Measures

foot	ft
gram	g
liter	L
meter	m
1,000 kg	tonne
2,000 pounds	ton

Mathematics and Statistics

all standard mathematical signs, symbols, and abbreviations	
base of natural logarithm	e
common test statistics (F, t, etc.)	
correlation or regression coefficient	r
degree (angular)	°
degrees of freedom	df
logarithm	log
percent	%
probability	P
sample size	N
standard deviation	SD
standard error	SE

Physics and Chemistry

all atomic symbols	
decibel	dB
per mille (per thousand)	‰

General

Company	Co.
compass directions	
east	E
north	N
south	S
west	W
et alii	et al.
et cetera	etc.
for example	e.g.
Incorporated	Inc.
Limited	Ltd.
species (singular)	sp.
species (plural)	spp.
that is	i.e.
United States (adjective)	U.S.
United States of America (noun)	USA
United States dollar	$

Preface

The coastal and offshore waters of the United States are a source of abundant natural resources. This abundance, however, does not preclude conflict among those who use these waters. Concerns have been raised that activities such as fisheries, ocean transportation, recreation and tourism, energy development, and subsistence activities are not compatible and may be mutually exclusive. Conflict can develop from real, potential, or perceived effects of any one coastal or marine activity on the others; thus, a clear understanding of current information and research is needed to discriminate among these sources of conflict.

Two uses of coastal and marine waters are fisheries and oil and gas development. To further understanding of the interactions between these two activities, the Minerals Management Service, U.S. Department of the Interior, sponsored the symposium "Outer Continental Shelf Fisheries Management Conflicts" on September 7–8, 1989, at the 119th annual meeting of the American Fisheries Society in Anchorage, Alaska. This proceedings includes many of the papers presented there.

It was clear at the outset of this effort, and remains clear, that the issues related to multiple use of coastal and marine waters are not easily addressed and that many of the data are still open to interpretation. Nonetheless, it is intended that the information and discussion generated by this symposium will move us closer to comprehending the interactions among and the concerns of each coastal and marine enterprise.

We acknowledge the guidance and assistance of many who contributed to this undertaking. First and foremost, we appreciate the support of the late Carl Sullivan, former Executive Director of the American Fisheries Society, for this project in its early stages. Second, we recognize Randy Bailey and the other American Fisheries Society members who organized and staffed the meetings in Anchorage. The editorial help and direction provided by Catherine Richardson and Robert Kendall of the American Fisheries Society smoothed the course that this published proceedings traveled. Special thanks are reserved for Rosalind Cohen of the Minerals Management Service who provided the initial impetus for this project and the inertia that carried us through to the end. And finally, we are most grateful to the authors for their many hours of work that made this proceedings possible.

Colleen S. Benner
Robert W. Middleton
Minerals Management Service
Herndon, Virginia

American Fisheries Society Symposium 11:1–8, 1991

Assessing the Ecological Effects of Oil Pollution from Outer Continental Shelf Oil Development

ROBERT W. HOWARTH

Section of Ecology and Systematics, Division of Biological Sciences
Cornell University, Ithaca, New York 14853, USA

Abstract.—Despite intensive research, it is still not possible to determine the extent of the threat that offshore oil development poses to marine ecosystems and fisheries. However, most analyses appear to underestimate the potential for harm. In this paper, I briefly review important factors that are commonly underemphasized or ignored. These include the areal extent of oil spill slicks, the persistence of oil in the water column after spills, sublethal effects of oil, effects of chronic discharges and small spills, and the cumulative effects of oil on organisms in lower trophic levels. Major controversies over the ecological effects of offshore oil development will persist until these issues are adequately addressed.

The past 20 years have seen a tremendous burst of research on the effects of oil pollution in marine ecosystems, much of it directed toward understanding the effects of offshore oil development. A great deal has been learned, yet the scientific community is still unable to adequately assess the impacts, potential or extant, of oil development on fisheries. In part, this is a result of the high variability in fishery recruitment and our poor measures of it. Traditional stock assessment methods are able at best to detect only very large decreases (an order of magnitude) in recruitment for a given year, severely limiting our knowledge of the effects of oil spills (Longhurst 1982; Sinclair 1982; Ware 1982; Wiseman et al. 1982). Thus, although there is some evidence that nearshore oil spills have adversely affected recruitment (NAS 1975; Laubier 1980; Desaunay 1981), there is no conclusive evidence that this is true for offshore spills (Royal Society of London 1980; Royal Commission on Environmental Pollution 1981; McIntyre 1982; Teal and Howarth 1984; Howarth 1989). We need to clearly distinguish between observing "no statistically detectable effect" and concluding that there is "no effect" when our methods of detection are so poor (Longhurst 1982; Wiseman et al. 1982).

With regard to potential effects of chronic pollution on fishery resources in offshore oil fields, catches per unit effort have been decreasing in both the Gulf of Mexico and in the North Sea since the start of oil development (NAS 1975; Manners 1982; Reid 1987). It has not proven possible to determine if these trends are caused by oil pollution, by other pollutants, or by overfishing, or if they are natural trends or fluctuations (NAS 1975, 1985; Manners 1982; Reid 1987). Catches per unit effort have been decreasing in many other fisheries of the world during these same time periods in areas without oil development.

In addition to data limitations, our understanding of the effects of oil on fisheries is limited by our incomplete knowledge of the relationship between the structure and function of marine ecosystems and by our incomplete understanding of the factors controlling fishery recruitment. Despite these limitations, evidence is accumulating to support the conclusion that oil pollution, even at very low levels, may adversely affect fishery resources (NAS 1985; Howarth 1989). By too often ignoring this evidence, and by making overly optimistic assumptions, many impact statements and other analyses have underestimated potential damage. In this paper, I briefly review a series of factors that commonly are ignored, underemphasized, or misused in assessing the potential effects of offshore oil development on fisheries.

Areal Extent of Oil Spills

At least some impact assessments for offshore oil development prepared by the U.S. Minerals Management Service have incorrectly predicted the adverse effects of oil spills by greatly underestimating the area covered by potential spills. Relying on an unverified model, the impact statement for lease sale 92 (North Aleutian Basin, Alaska) predicted that a "worst-case" spill of 15,000 tonnes would cover an area of 200 km^2, or 0.013 km^2/tonne of spilled oil (MMS 1985). Consequently, although this impact statement admitted that oil spills could harm fish and other organisms, it concluded that the extent of damage would be quite small.

This conclusion is not justified by observations of actual spills. Ford (1985) compiled data for the Minerals Management Service on spill volumes and actual reported sizes of oil slicks from 21 spills. Spill volumes ranged from 0.7 to 55,000 tonnes; the size of the slicks (reported 1–24 d after the accidents) per tonne of oil spilled ranged from 0.005 to 9.5 km^2/tonne, with a mean value of 0.70 km^2/tonne and a median value of 0.20 km^2/tonne (calculated from Ford 1985). Only 2 of the 21 spills had slick areas per volume of oil spilled that were less than the "worst-case" estimate of 0.013 km^2/tonne in the Mineral Management Service's impact statement for lease sale 92. Data on slick size for both of these spills were reported for only the first day after the spill; because slick size increases markedly with increasing time (at least up to 10 d; Ford 1985), the slicks from these spills almost certainly covered areas much larger than those initially reported. Thus, the estimate of slick area in the impact statement (MMS 1985) is certainly optimistic, and perhaps understates the area likely to be covered by an oil spill by factors of 15 to 54 (estimated from the median and mean values for the data calculated from Ford 1985). That is, a "worst-case" spill of 15,000 tonnes would likely cover an area of at least 3,000 km^2 (median slick area) to 10,500 km^2 (mean slick area) or even more, not 200 km^2 as predicted in the impact assessment for lease sale 92.

Persistence of Oil in the Water Column

Authors of some environmental impact statements for offshore oil development have assumed that toxic concentrations of oil remain in the water column for at most a few days after a spill (MMS 1983, 1985). An assessment by the Royal Society of London (1980) was even more optimistic and assumed that the toxicity of spilled oil becomes negligible in less than a day due to evaporation of toxic components. These assumptions are based on models of oil dissolution and evaporation that have been poorly tested in the field. Field data from actual spills show these assumptions sometimes are wrong, perhaps because evaporation of toxic components is greatly slowed once oil is mixed into the water column (Vandermeulen 1982), or because cold water temperatures slow evaporation and favor dissolution of aromatic compounds (Howarth 1987).

Following the Argo Merchant spill on Georges Bank in December 1976, toxic aromatic hydrocarbons and other dissolved constituents of the spilled oil persisted at elevated concentrations throughout the water column over large areas for at least 5 months (Boehm et al. 1978; Farrington and Boehm 1987; Howarth 1987). Concentrations on Georges Bank ranged from 10 to 100 μg/L (average, 44) 2 months after the spill and from 1 to 50 μg/L (average, 11) 5 months after the spill (Boehm et al. 1978). These concentrations may well have been high enough to have ecological consequences, as discussed below. The concentrations of dissolved oil hydrocarbons did decrease exponentially over time, but nowhere near as rapidly as predicted by the Minerals Management Service (MMS 1983, 1985) or the Royal Society of London (1980). Elevated concentrations of oil also persisted in the water column for at least a month after the Amoco Cadiz spill in France in 1978 (Law, as reported in Olsen et al. 1982).

Oil also regularly reaches bottom sediments following spills (Teal and Howarth 1984; NAS 1985). Once in fine-grained, anoxic sediments, toxic oil hydrocarbons can persist for decades (Teal and Howarth 1984) and produce long-lasting ecological effects, as discussed below.

Toxicity of Oil and Sublethal Effects

In at least some environmental impact statements, the Minerals Management Service has relied on LC50 data to estimate the toxicity of oil (MMS 1983, 1985). This measure is obtained by exposing test organisms to different concentrations of oil in laboratory studies for 96 h and estimating the concentration lethal to 50% of these organisms in that time period. The National Academy of Sciences (1985) and others (Clark 1987; Howarth 1987, 1989) have argued against using this approach because it can greatly underestimate the potential for harm. One major problem is that the test organisms tend to be pollution-tolerant, opportunistic species. These species are also more tolerant of other forms of stress and are easier to grow under laboratory conditions than are more pollution-sensitive species; thus, toxicity tests with pollution-tolerant species are inadequate for assessing toxicity to the sensitive species (Howarth 1989). When more sensitive species are used, even the control organisms exposed to no pollutant often suffer appreciable mortality rates in these laboratory studies, reducing the power of the test. Other problems with LC50 tests, such as loss of oil through volatilization and biodegradation and underestimation of the increased toxicity that may result from photooxidation, are discussed elsewhere (Howarth 1989).

When using LC50 toxicity data to predict deleterious effects in the field, the Minerals Management Service has applied a 10-fold "safety factor." For dissolved oil hydrocarbons, LC50 values of 1,000–3,000 μg/L are typically estimated (MMS 1985). The Minerals Management Service therefore estimated that oil concentrations lower than 100–300 μg/L would not have deleterious effects (MMS 1985). I know of no scientific basis for choosing a 10-fold safety factor, and the resulting safety level concentration of 100–300 μg/L is permissive (Howarth 1989). Vandermeulen and Capuzzo (1983) concluded that markedly lower dissolved oil concentrations, 2–10 μg/L, can decrease the viability of larval fish, and sublethal effects of oil have been found at concentrations as low as 0.2 μg/L (Jacobson and Boylan 1973; Johnson 1977; Steele 1977; NAS 1985).

Sublethal effects can cause population declines in the real world, for instance as a result of increased mortality from predation due to slower avoidance behaviors (Krebs and Burns 1977; Linden et al. 1980; Pearson et al. 1981). Dissolved oil hydrocarbon concentrations of 1–10 μg/L can interfere with salmon migration (D. Martin, Dames and Moore, unpublished). Kuhnhold et al. (1978) reported that when adult flounder were exposed to average oil concentrations of 10 μg/L for 10 d during gonad maturation, the growth rate of the resulting larvae was significantly slowed. Because eggs and larvae are particularly susceptible to predation (Ellertsen et al. 1981; Oiestad 1982), a slower larval growth rate may greatly decrease recruitment (Cushing 1976). Thus, low levels of oil pollution have the potential to decrease fishery recruitment through sublethal effects and complex ecological interactions (Howarth 1987, 1989).

Effects of Chronic Pollution

Impact assessments tend to downplay possible effects from long-term, chronic release of oil and instead concentrate on oil spills, which are much rarer events (MMS 1983, 1985). In fact, our ability to predict the effects of chronic pollution is severely limited by inadequate data. One of the few valid studies of the chronic effects of pollution in an offshore oil field was that of Addy et al. (1978) in the North Sea. They examined numbers of animals and animal species in the benthos along several transects moving away from an underwater oil storage platform. After the platform had been in operation for 4 years, total numbers of animals and numbers of animal species had decreased within 5 km of the platform (Addy et al. 1978). The platform was so designed that it discharged oil-contaminated water every time it was filled. Currently, most oil produced in offshore fields in U.S. waters is piped ashore rather than shipped by tanker, so offshore storage platforms are rare (there are apparently no storage platforms in U.S. waters of the type studied by Addy et al. in the North Sea; personal communication from two anonymous reviewers). Consequently, one should be careful in extrapolating the findings of Addy et al. (1978) to other oil fields, although other sources of chronic pollution such as formation waters (oil-contaminated brines brought up with the oil and typically discharged in the environment) and small spills from connecting pipelines probably are of comparable magnitude to the release of oil from storage platforms (Read 1980; Howarth 1987).

Few other studies have examined the effects of oil pollution on the benthos in offshore fields, and of these, most are seriously flawed. For instance, the three large studies in the Gulf of Mexico (GURC 1974; Bedinger et al. 1981; Middleditch 1982) lacked adequate controls (Sanders and Jones 1981; Carney 1987; Howarth 1989).

The possible effects of chronic pollution on planktonic organisms and ecosystems have received even less attention. Impact assessments seldom discuss such effects, and monitoring programs have seldom examined planktonic effects or even measured concentrations of dissolved oil hydrocarbons in the waters of offshore oil fields. In the Buccaneer field in the Gulf of Mexico, Wiesenburg et al. (1981) reported mean concentrations of dissolved volatile liquid hydrocarbons (principally benzene and naphthalenes) of approximately 1.5 μg/L. Although this concentration may seem low, it is certainly above background levels (generally less than 0.06 μg/L; NAS 1985) and may be high enough to have some biological effects. I know of no other reliable data on dissolved hydrocarbon concentrations resulting from chronic oil pollution for oil fields in U.S. waters.

Information from the North Sea has been obtained only using techniques such as fluorometry that are difficult to calibrate and subject to interference. These data indicate that concentrations of dissolved aromatic hydrocarbons (probably the most toxic components of oil) are higher near oil fields than in more pristine areas (Gunkel et al. 1980) and tend to be higher near oil rigs (Ward et al. 1980). Reported concentrations of aromatic hydrocarbons range from 0.35 to 88 μg/L in the North Sea and average approximately 5 μg/L (Gunkel et al. 1980; Ward et al. 1980; Massie et al. 1985). These concentrations actually may not be higher than those in the Buccaneer field in the Gulf of Mexico. The apparently higher values in the North Sea may be a result of a different, and probably less accurate, measurement technique. On the other hand, actual concentrations may be higher in the North Sea, either because of greater inputs or because of longer persistence of dissolved aromatic hydrocarbons in the colder waters. The reported concentrations in the North Sea, if accurately measured, are probably high enough to have ecological consequences (Howarth 1989).

Cumulative and Ecosystem-Level Effects

Impact assessments tend to ignore effects on fisheries that might result indirectly from oil-induced changes elsewhere within the ecosystem. For instance, oil pollution is well known to cause changes in benthic community structure. Generally, sensitive species are lost and are replaced by a smaller number of opportunistic species (d'Ozouville et al. 1979; Linden et al. 1980; Sanders et al. 1980; Boucher 1981; Cabioch et al. 1981; Glemarec and Hussenot 1981; Grassle et al. 1981; Elmgren and Frithsen 1982; Teal and Howarth 1984; NAS 1985; Howarth 1989). Such changes have been observed in the vicinity of an oil platform in the North Sea (Addy et al. 1978). These effects are of particular concern because oil can persist for decades in fine-grained sediments, continuing to affect the benthic community (Thomas 1977; Gilfillan and Vandermeulen 1978; Teal et al. 1978; Sanders et al. 1980; Teal and Howarth 1984). There have been virtually no studies of how changes in benthic community composition may affect fishery dynamics. An effect seems possible, because ampeliscid amphipods, a favored food for many fish, are quite sensitive to oil pollution and usually disappear from oiled sediments (Elmgren and Frithsen 1982; Teal and Howarth 1984; Howarth 1989). These and other such indirect considerations have not been discussed in impact statements for offshore oil development.

Oil pollution also potentially can cause changes in planktonic community structure (Parsons et al. 1976; Lee et al. 1977; Lee and Takahashi 1977; Davies et al. 1980; NAS 1985; Howarth 1987, 1989, 1991). More than 10 years ago, Greve and Parsons (1977) hypothesized that oil-induced changes in phytoplankton community composition might adversely affect fishery resources. They speculated that oil could cause the replacement of a phytoplankton community dominated by large diatoms with a community composed of smaller species, and that this in turn could cause a replacement of large zooplankton by smaller forms, which are less favorable foods for juvenile fish. Their hypothesis has yet to be adequately tested (Howarth 1991). However, the replacement of large phytoplankton species by smaller species can be induced by low-level experimental oiling in mesocosms (Parsons et al. 1976; Lee et al. 1977). A similar phytoplankton replacement induced by arsenic additions to mesocosms caused a shift in the zooplankton toward species less suitable as food for juvenile fish, that is, a loss of the dominant copepods *Acartia tonsa* and *Eurytemora affinis* (Sanders 1986; Sanders and Cibik 1988). As with the indirect effects of benthic changes on fishery resources, the possible significance of changes in planktonic community structure has not been addressed in impact statements.

Conclusions

I have discussed several reasons why many impact statements and other assessments of the effects of offshore oil development may have seriously underestimated the adverse affects on fishery resources. These include (1) underestimating the area of oil slicks following accidents, (2) underestimating the persistence of toxic concentrations of oil, (3) underestimating the toxicity of oil and paying insufficient attention to the sublethal effects of oil, (4) ignoring the ecological consequences of chronic oil pollution, and (5) not assessing the effects of oil on the structure and function of marine ecosystems that may then affect fisheries.

The limited ecological analysis in these environmental assessments and impact statements is surprising given the decline of major commercial fishery resources in both the Gulf of Mexico and the North Sea since the start of oil development (NAS 1975; Manners 1982; Reid 1987). The decline in the North Sea has been particularly dramatic, with decreases in catches per unit effort and total catches (Reid 1987). Fishery resources have declined in many other regions of the world as well, and oil development obviously is not the cause in most cases. Nonetheless, the scientific community at present has insufficient evidence to determine if the troubling trends in the North Sea and Gulf of Mexico are or are not linked to offshore oil development.

Even if we knew that offshore oil development has not harmed fishery resources in the Gulf of Mexico and the North Sea, we should not complacently extrapolate this finding to all new oil fields. In both the Gulf of Mexico and the North Sea, most produced oil is piped to shore, whereas it has been suggested that oil would be taken to shore in tankers in some of the proposed new fields (BLM 1977; MMS 1986). The chronic oil pollution associated with storing and shipping oil is much greater than that from piping oil (Howarth 1987).

Of even greater significance is the fact that ecosystems can vary in their sensitivity to pollution, and a given level of oil pollution will have greater consequences in some ecosystems than in others. Consideration of both the openness of an ecosystem and its history of stress may be necessary to determine the relative sensitivity of an ecosystem (Howarth 1991). Ecosystems that previously have received a large amount of stress, either from natural forces or from other toxic chemicals, are probably more resistant to further stress from oil pollution (Howarth 1991). And there is at least some reason to believe that ecosystems that are open with regard to their element cycles are more resistant to toxic chemical stress (Howarth 1991). On both counts, it would seem that the open, highly stressed (Carney 1987) areas of the Gulf of Mexico where most oil development has occurred are much more resistant to the harmful effects of oil pollution than are less anthropogenically influenced ecosystems such as Georges Bank or many of the proposed Alaska offshore fields. Further factors such as hydrologic patterns and locations of fish nursery grounds may also make systems such as Georges Bank more sensitive to oil pollution (Howarth 1987).

Acknowledgments

I thank Diane Sherman and Roxanne Marino for assistance in preparing this manuscript and Robert Middleton for his flexibility in allowing the inclusion of the manuscript in these proceedings. Travel costs for my attendance at the symposium were partially supported by Greenpeace; I thank Dorrie Smith for arranging this support. This manuscript represents my personal views and not positions of Cornell University or any other institution.

References

Addy, J. M., D. Lovell, and J. P. Hartley. 1978. Biological monitoring of sediments in Ekofisk oilfield. Pages 515–539 *in* Proceedings, conference on assessment of ecological impacts of oil spills. American Institute of Biological Sciences, Washington, D.C.

Bedinger, C. A., R. E. Childers, J. W. Cooper, K. T. Kimball, and A. Kwok. 1981. Central Gulf platform study. Part 1. Background, program organization and study plan. Pages 1–53 *in* C. A. Bedinger, Jr., editor. Ecological investigations of petroleum production platforms in the central Gulf of Mexico. Volume 1. Pollutant fate and effect studies. Report of Southwest Research Institute, contract AA551-CT8-17, to Bureau of Land Management, Washington, D.C.

BLM (Bureau of Land Management). 1977. Final environmental impact statement, proposed 1977 outer continental shelf oil and gas lease sale offshore the north Atlantic States, OCS sale 42. BLM, Washington, D.C.

Boehm, P. D., G. Perry, and D. L. Fiest. 1978. Hydrocarbon chemistry of the water column of Georges Bank and Nantucket Shoals, February–November 1977. Pages 58–64 *in* In the wake of the Argo Merchant. Center for Ocean Management Studies, University of Rhode Island, Kingston.

Boucher, G. 1981. Effets à long terme des hydrocarbures de l'Amoco Cadiz sur la structure des communautes de nematodes libres des sables fins sublittoraux. Pages 539–549 *in* Amoco Cadiz, fates and effects of the oil spill. Centre National pour l'Exploitation des Oceans, Paris.

Cabioch, L., J. C. Dauvin, F. Gentil, C. Retiere, and V. Rivain. 1981. Perturbations induites dans la composition et le fonctionnement des peuplemants benthiques sublittoraux, sous l'effet des hydrocarbures de l'Amoco Cadiz. Pages 513–526 *in* Amoco Cadiz, fates and effects of the oil spill. Centre National pour l'Exploitation des Oceans, Paris.

Carney, R. S. 1987. A review of study designs for the detection of long-term environmental effects of offshore petroleum activities. Pages 651–696 *in* D. F. Boesch and N. Rabalais, editors. Long-term environmental effects of offshore oil and gas development. Elsevier, New York.

Clark, R. B. 1987. Summary and conclusions: environmental effects of North Sea oil and gas developments. Philosophical Transactions of the Royal Society of London, B: Biological Sciences 316:669–677.

Cushing, D. H. 1976. Biology of fishes in the pelagic community. Pages 317–340 *in* D. H. Cushing and J. J. Walsh, editors. The ecology of the seas. Saunders, Philadelphia.

Davies, J. M., I. E. Daird, L. H. Massie, S. J. Fay, and A. P. Ward. 1980. Some effects of oil-derived hydrocarbons in a pelagic food web from observations in an enclosed ecosystem and a consideration of their implications for monitoring. Rapports et

Procès-Verbaux des Réunions, Conseil International pour l'Exploration de la Mer 179:201–211.

Desaunay, Y. 1981. Evolution des stocks de poissons plats dans la zone contaminée par l'Amoco Cadiz. Pages 727–735 *in* Amoco Cadiz, fates and effects of the oil spill. Centre National pour l'Exploitation des Oceans, Paris.

d'Ozouville, L., M. O. Hayes, E. R. Gundlach, W. J. Sexton, and J. Michel. 1979. Occurrence of oil in offshore bottom sediments at the Amoco Cadiz oil spill site. Pages 187–192 *in* J. O. Ludwigson, editor. Proceedings of the 1979 oil spill conference. American Petroleum Institute, Publication 4308, New York.

Ellertsen, V., E. Noksness, P. Solemkal, T. Westgaard, and V. Oiestad. 1981. Growth and survival of cod larvae in an enclosure, experiments and a mathematical model. Rapports et Procès-Verbaux des Réunions, Conseil International pour l'Exploration de la Mer 178:45–57.

Elmgren, R., and J. B. Frithsen. 1982. The use of experimental ecosystems for evaluating the environmental impacts of pollutants: a comparison of an oil spill in the Baltic Sea and two long-term, low level oil addition experiments in mesocosms. Pages 153–165 *in* G. D. Grice and M. D. Reeve, editors. Marine mesocosms: biological and chemical research in experimental ecosystems. Springer-Verlag, New York.

Farrington, J. W., and P. D. Boehm. 1987. Natural and pollutant organic compounds. Pages 195–207 *in* R. Backus, editor. An atlas of Georges Bank. MIT Press, Cambridge, Massachusetts.

Ford, R. G. 1985. Oil slick sizes and length of coastline affected: a literature survey and statistical analysis. Final report of R. Glenn Ford Consulting to U.S. Department of the Interior, Minerals Management Service, Washington, D.C.

Gilfillan, E. S., and J. W. Vandermeulen. 1978. Alterations in growth and physiology of soft-shell clams, *Mya arenaria*, chronically oiled with Bunker C from Chedabucto Bay, Nova Scotia, 1970–1976. Journal of the Fisheries Research Board of Canada 35:630–636.

Glemarec, K., and E. Hussenot. 1981. Définition d'une succession écologique en milieu meuble anormalement enrichi en matières organiques à la suite de la catostrophe de l'Amoco Cadiz. Pages 499–512 *in* Amoco Cadiz, fates and effects of the oil spill. Centre National pour l'Exploitation des Oceans, Paris.

Grassle, J. F., R. Elmgren, and J. P. Grassle. 1981. Response of benthic communities in Marine Ecology Research Laboratory experimental ecosystems to low level chronic additions of no. 2 fuel oil. Marine Environmental Research 4:279–297.

Greve, W., and T. R. Parsons. 1977. Photosynthesis and fish production: hypothetical effects of climatic change and pollution. Helgolaender Wissenschaftliche Meeresuntersuchungen 30:666–672.

Gunkel, W., G. Gassmann, C. H. Oppenheimer, and I. Dundas. 1980. Preliminary results of baseline studies of hydrocarbons and bacteria in the North Sea: 1975, 1976, 1977. Pages 223–247 *in* Resistencia a los antibioticos u microbiologia marina. VI Congreso Nacional Microbiologia, 1977, Santiago de Compostela, Spain.

GURC (Gulf Universities Research Consortium). 1974. The offshore ecology investigation, final project planning council consensus report. Gulf Universities Research Consortium, Report 138, Galveston, Texas.

Howarth, R. W. 1987. The potential effects of petroleum on marine organisms on Georges Bank. Pages 540–551 *in* R. Backus, editor. An atlas of Georges Bank. MIT Press, Cambridge, Massachusetts.

Howarth, R. W. 1989. Determining the effects of oil pollution in marine ecosystems. Pages 60–87 *in* S. A. Levin, M. A. Harwell, J. R. Kelly, and K. D. Kimball, editors. Ecotoxicology: problems and approaches. Springer-Verlag, New York.

Howarth, R. W. 1991. Comparative responses of aquatic ecosystems to toxic chemical stress. Pages 169–195 *in* J. J. Cole, G. Lovett, and J. S. Findlay, editors. Comparative analysis of ecosystems: patterns, mechanisms, and theories. Springer-Verlag, New York.

Jacobson, S. M., and D. B. Boylan. 1973. Effect of seawater soluble fraction of kerosene on chemotaxis in a marine snail, *Nassarius obsoletus*. Nature (London) 241:213–215.

Johnson, F. G. 1977. Sublethal biological effects of petroleum hydrocarbon exposure: bacteria, algae, and invertebrates. Pages 271–219 *in* D. C. Malins, editor. Effects of petroleum on arctic and subarctic marine environments and organisms, volume 2. Academic Press, New York.

Krebs, C. T., and K. A. Burns. 1977. Long-term effects of an oil spill on populations of the salt marsh crab *Uca pugnax*. Science (Washington, D.C.) 197:484–487.

Kuhnhold, W. W., D. Everich, J. J. Stegeman, J. Lake, and R. E. Wolke. 1978. Effects of low levels of hydrocarbons on embryonic, larval, and adult winter flounder (*Pseudopleuronectes americanus*). Pages 677–711 *in* Proceedings, conference on assessment of ecological impacts of oil spills. American Institute of Biological Sciences, Washington, D.C.

Laubier, L. 1980. The Amoco Cadiz oil spill: an ecological impact study. Ambio 9:268–276.

Lee, R. F., and six coauthors. 1977. Controlled ecosystems: their use in the study of the effects of petroleum hydrocarbons on plankton. Pages 323–342 *in* F. J. Vernberg, A. Calabrese, F. P. Thurberg, and W. B. Vernberg, editors. Physiological responses of marine biota to pollutants. Academic Press, New York.

Lee, R. F., and M. Takahashi. 1977. The fate and effect of petroleum in controlled ecosystem enclosures. Rapports et Procès-Verbaux des Réunions, Conseil International pour l'Exploration de la Mer 171:150–156.

Linden, O., R. Elmgren, L. Westin, and J. Kineman.

1980. Scientific summary and general discussion. Pages 43–58 *in* J. J. Kinemann, R. Elmgren, and S. Hansson, editors. The Tsesis oil spill. National Oceanic and Atmospheric Administration, Boulder, Colorado.

Longhurst, A. 1982. Executive summary. Canadian Technical Report of Fisheries and Aquatic Sciences 1096:7–11.

Manners, I. R. 1982. North Sea oil and environmental planning: the United Kingdom experience. University of Texas Press, Austin.

Massie, L. C., A. P. Ward, J. M. Davies, and P. R. Mackie. 1985. The effects of oil exploration and production in the northern North Sea. Part 1: the levels of hydrocarbons in water and sediments in selected areas, 1978–1981. Marine Environmental Research 15:165–213.

McIntyre, A. D. 1982. Oil pollution and fisheries. Philosophical Transactions of the Royal Society of London, B: Biological Sciences 297:401–411.

Middleditch, B. S. 1982. Environmental effects of offshore oil production. The Buccaneer gas and oil field study. Plenum, New York.

MMS (Minerals Management Service). 1983. Final environmental impact statement, North Atlantic outer continental shelf oil and gas lease offering, outer continental shelf oil and gas sale no. 52. U.S. Department of the Interior, Washington, D.C.

MMS (Minerals Management Service). 1985. Final environmental impact statement, North Aleutian Basin lease sale no. 92. U.S. Department of the Interior, Washington, D.C.

MMS (Minerals Management Service). 1986. Norton Basin sale 100, final environmental impact statement. U.S. Department of the Interior, Washington, D.C.

NAS (National Academy of Sciences). 1975. Petroleum in the marine environment. National Academy Press, Washington, D.C.

NAS (National Academy of Sciences). 1985. Oil in the sea, inputs, fates, and effects. National Academy Press, Washington, D.C.

Oiestad, V. 1982. Application of enclosures to studies on the early life history of fishes. Pages 49–62 *in* G. D. Grice and M. R. Reeve, editors. Marine mesocosms: biological and chemical research in experimental ecosystems. Springer-Verlag, New York.

Olsen, S., M. E. Q. Pilson, C. Oviatt, and J. N. Gearing. 1982. Ecological consequences of low, sustained concentrations of petroleum hydrocarbons in temperate estuaries. Technical Report, Marine Ecosystems Research Laboratory, Graduate School of Oceanography, University of Rhode Island, Narragansett.

Parsons, T. R., W. K. W. Li, and R. Waters. 1976. Some preliminary observations on the enhancement of phytoplankton growth by low levels of mineral hydrocarbons. Hydrobiologia 51:85–89.

Pearson, W. H., D. L. Woodruff, P. C. Suganman, and B. L. Olla. 1981. Effects of oiled sediment on the littleneck clam, *Protothaca staminea*, by the Dungeness crab, *Cancer magister*. Estuarine, Coastal and Shelf Science 13:445–454.

Read, A. D. 1980. Treatment of oily water at North Sea installations—a progress report. Pages 127–136 *in* C. S. Johnston and R. J. Morris, editors. Oily water discharges. Applied Science Publishers, London.

Reid, P. C. 1987. The importance of the planktonic ecosystem of the North Sea in the context of oil and gas development. Philosophical Transactions of the Royal Society of London, B: Biological Sciences 316:587–602.

Royal Commission on Environmental Pollution. 1981. Eighth report. Oil pollution of the sea. Her Majesty's Stationery Office, London.

Royal Society of London. 1980. The effects of oil pollution: some research needs. London.

Sanders, H. L., J. F. Grassle, G. R. Hampson, L. S. Morse, S. Garner-Price, and C. C. Jones. 1980. Anatomy of an oil spill: long-term effects from the grounding of the barge Florida off West Falmouth, Massachusetts. Journal of Marine Research 38:265–380.

Sanders, H. L., and C. Jones. 1981. Oil, science, and public policy. Pages 73–94 *in* T. C. Jackson and D. Reische, editors. Coast alert: scientists speak out. Friends of the Earth, San Francisco.

Sanders, J. G. 1986. Direct and indirect effects of arsenic on the survival and fecundity of estuarine zooplankton. Canadian Journal of Fisheries and Aquatic Sciences 43:694–699.

Sanders, J. G., and S. J. Cibik. 1988. Response of Chesapeake Bay phytoplankton communities to low levels of toxic substances. Marine Pollution Bulletin 19:439–444.

Sinclair, M. 1982. What kind of observational programs would be required to detect the effects on biota? Canadian Technical Report of Fisheries and Aquatic Sciences 1096:54–63.

Steele, R. L. 1977. Effects of certain petroleum products on reproduction and growth of zygotes and juvenile stages of the alga *Fucus edentatus* De la Pyl (Phaeophyceae: Fucales). Pages 115–128 *in* D. A. Wolfe, editor. Fate and effects of petroleum hydrocarbons in marine ecosystems and organisms. Pergamon Press, New York.

Teal, J. M., K. Burns, and J. Farrington. 1978. Analysis of aromatic hydrocarbons in intertidal sediments resulting from two spills of no. 2 fuel oil in Buzzards Bay, Massachusetts. Journal of the Fisheries Research Board of Canada 35:510–520.

Teal, J. M., and R. W. Howarth. 1984. Oil spill studies: a review of ecological effects. Environmental Management 8:27–44.

Thomas, J. L. H. 1977. Long term biological effects of bunker C oil in the intertidal zone. Pages 238–245 *in* D. A. Wolfe, editor. Fate and effects of petroleum hydrocarbons in marine organisms and ecosystems. Pergamon Press, New York.

Vandermeulen, J. 1982. What levels of oil contamination may be expected in water, sediments, and what would be the physiological consequences for biota?

Canadian Technical Report of Fisheries and Aquatic Sciences 1096:22–53.

Vandermeulen, J., and J. M. Capuzzo. 1983. Understanding sublethal pollutant effects in the marine environment. Paper 9 *in* M. A. Champ and M. Trainor, editors. Ocean waste management policy and strategies. University of Rhode Island, Kingston.

Ward, A. P., L. C. Massie, and J. M. Davies. 1980. A survey of the levels of hydrocarbons in the water and sediments in the areas of the North Sea. International Council for the Exploration of the Sea, CM 1980/E:48, Copenhagen.

Ware, D. 1982. What is the liklihood that such effects would impair recruitment and that such impaired recruitment would be separable from natural variation? Canadian Technical Report of Fisheries and Aquatic Sciences 1046:64–73.

Wiesenburg, D. A., G. Bodennec, and J. M. Brooks. 1981. Volatile hydrocarbons around a production platform in the northwest Gulf of Mexico. Bulletin of Environmental Contamination and Toxicology 27:167–174.

Wiseman, R. J., J. Payne, and S. Akengead. 1982. Rapporteurs' comments. Canadian Technical Report of Fisheries and Aquatic Sciences 1046:85–94.

American Fisheries Society Symposium 11:9–17, 1991

Rigs to Reefs: Fuel for Fisheries Enhancement through Cooperation

VILLERE C. REGGIO, JR.
U.S. Department of the Interior, Minerals Management Service
1201 Elmwood Park Boulevard, New Orleans, Louisiana 70123, USA

RICK KASPRZAK[1]
Louisiana State University, Center for Wetland Resources
Baton Rouge, Louisiana 70803, USA

Abstract.—There are in excess of 4,500 offshore oil and gas structures in the USA. More than 95% of these structures are located in the Gulf of Mexico off the coasts of Louisiana and Texas. The cumulative cost of onshore disposal of these petroleum structures is estimated to exceed $1 billion by the year 2000 and $7 billion by the year 2020. Disposal of obsolete offshore oil and gas structures not only is a net financial liability for private industry, but can also be a public loss of productive marine habitat. Changes in attitudes, policies, and visions, along with aggressive leadership by Gulf Coast states, are leading to the conversion of nonproductive petroleum structures into productive implements of marine research, development, management, and conservation. The National Fishing Enhancement Act of 1984 and the recently promulgated National Recreational Fisheries Policy have already stimulated increased national interest in planning and developing artificial reefs, as exemplified by Louisiana's artificial reef program. Offshore oil and gas development in the Gulf of Mexico region has and will continue to contribute to the Gulf's position as the nation's most productive and popular offshore marine fishing zone.

In the Gulf of Mexico, where offshore fishing and petroleum development have coexisted for over 40 years, technological innovations permitting the expansion of the petroleum industry have contributed to the expansion and success of offshore recreational and commercial fishing and scuba diving. This does not mean that an offshore workforce of over 30,000 petroleum workers, who use a fleet of around 1,000 vessels and 500 helicopters in moving personnel, equipment, and supplies to over 4,500 offshore work stations, never experience direct conflict with fishermen. Some conflict is inevitable. However, communication, regulation, mitigation, and the involvement of many people with direct interests in both fishing and petroleum operations have helped to minimize conflict and encourage innovation in seeking solutions of mutual benefit. This paper focuses on one component of offshore petroleum development, the production platform, and attempts to demonstrate how knowledge, understanding, and cooperation are benefiting fish, fishermen, and the petroleum industry.

[1]Present address: Louisiana Department of Wildlife and Fisheries, Fisheries Research Division, Post Office Box 98000, Baton Rouge, Louisiana 70898, USA.

Production Platforms—A Fishing Resource

Offshore recreational fishermen and scuba divers in Louisiana and Texas were quick to learn in the 1950s and 1960s that oil and gas production platforms were great places to find and catch fish. Although popular articles about the "fabulous" fishing associated with offshore petroleum structures appeared in Gulf Coast newspapers and sport magazines in the 1960s and 1970s, it was not until the late 1970s that scientific research on the "rig" fishing phenomenon began to assess this special fishery. Ditton and Graefe (1978) and Dugas et al. (1979) described early work on the fishing directly associated with oil and gas production structures off the coasts of Louisiana and Texas. Figure 1 depicts oil and gas production areas within the Gulf of Mexico outer continental shelf (OCS) planning areas established by the U.S. Minerals Management Servicc (MMS) for leasing purposes.

In line with its responsibility to assess the environmental effects of offshore petroleum operations, the MMS began in the 1980s to investigate and document the effects of pervasive offshore oil and gas development on fisheries. Ditton and Auyong (1984) evaluated survey information collected for MMS from 1980 to 1981 on fishing patterns associated with several hundred production platforms scattered throughout the Louisiana

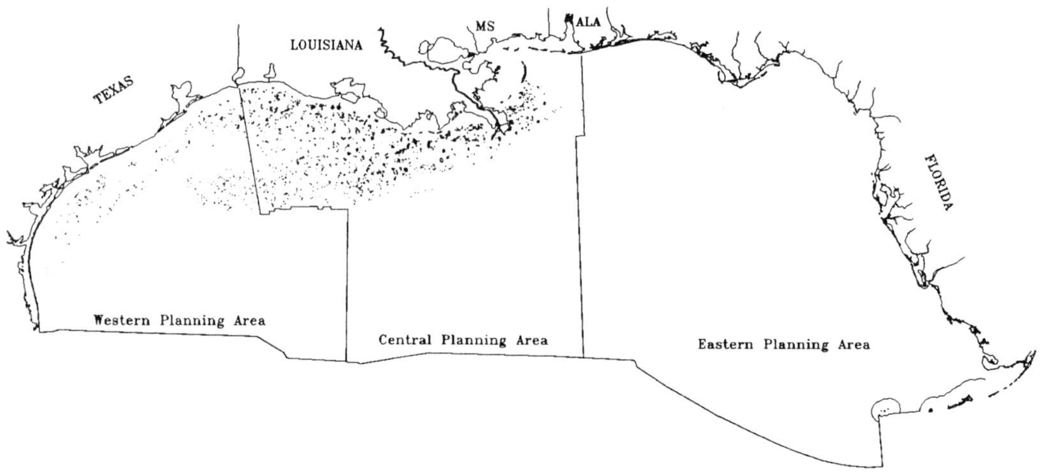

FIGURE 1.—Minerals Management Service Outer Continental Shelf Planning Areas and general locations of oil and gas production structures in the Gulf of Mexico.

offshore leasing area. Using 12,000 sightings of boat fishing activity near 164 major production platforms, they showed how access, shelf characteristics, transportation networks, day of the week, demographics, and the interrelationships among these factors influenced the amount, location, and timing of offshore platform fishing. The report also documented 45 separate incidents in which oil and gas industry personnel assisted more than 100 anglers and boaters. Assistance ranged from giving information and mooring assistance to providing first aid and life-saving rescues. Although conflict is often publicized, cooperation between fishermen and petroleum workers, common in the Gulf of Mexico, is seldom publicly acknowledged.

Gallaway and Lewbel (1982) used MMS-generated, site-specific data collected by Gallaway et al. (1981) to characterize the ecological communities associated with offshore platforms. Gallaway and Lewbel described and separated the offshore environment off the Louisiana coast into three separate biotic zones based on distinct, platform-associated biological communities and indicator fish species. Continental Shelf Associates also studied fishes associated with oil and gas structures off Louisiana. They described species diversity around production platforms and showed how environmental factors such as depth and current affected the location of specific fish in relation to the vertical and horizontal aspects of platform support legs (CSA 1982).

In 1985 the National Marine Fisheries Service (NMFS) reported the results of an agreement with MMS to collect data on recreational fishing associated with petroleum structures as part of the multiyear Atlantic and Gulf Coast Marine Recreational Fishery Statistics Survey. As part of this survey, Witzig (1986) found that over 70% of all recreational fishing trips that originated in Louisiana and extended more than 3 miles from shore targeted oil and gas structures in 1984. Furthermore, he determined that anglers who fished around oil and gas structures caught more, bigger, and more desirable fish (e.g., snapper, grouper, seatrout, mackerel) on average than marine recreational fishermen who fished in other areas (Witzig 1986).

Roberts and Thompson (1983) determined that sport divers in Louisiana valued the privilege of diving at petroleum production platforms at over \$150 a year per diver. Although no one has estimated the economic value of hook-and-line recreational and commercial fishing at petroleum platforms in the Gulf of Mexico, the economic value of artificial reefs in general is great. Milon (1988) estimated the present value of the offshore artificial reef system (33 reefs) of Dade County, Florida, at up to \$128 million.

Oil and Gas Structures as Artificial Reefs

Public forums organized and sponsored by the MMS in a series of annual information transfer meetings during the 1980s elicited additional information on the present and potential fishery uses of oil and gas structures (MMS 1982, 1984,

1985, 1986a, 1987). Federal and state conservation officials, private conservation groups, fishing organizations, oil and gas companies, university researchers, and others encouraged the conversion and use of oil and gas structures as artificial reefs.

The durability, stability, openness, and extensive surface area of oil and gas structures make them suitable for artificial reef development. Because active oil and gas platforms already function as artificial reefs, a structure's effect on fish and fishing can be assessed prior to permitting its later use as a reef on location. Phillips (1988a) demonstrated that artificial reefs can provoke conflict when competition between fishery user groups arises, despite the reefs' enhancement of fish and fishing opportunities; thus, reef development projects are best sponsored and sited by public agencies with fisheries development, management, and conservation responsibilities.

In 1983, the U.S. Secretary of the Interior created the "Recreation, Environmental Enhancement and Fishing in the Sea" (REEFS) Task Force to formulate a national policy on artificial reef development. The efforts of the Task Force culminated in the enactment of the National Fishing Enhancement Act of 1984 (NFEA; Title II of Public Law 98-623). This law established broad artificial reef development standards and encouraged the development of planned artificial reefs that would enhance fishery resources. The new law spawned the development of the National Artificial Reef Plan under the leadership of the U.S. Department of Commerce. Besides specifying general design, construction, siting, and maintenance criteria for artificial reefs, the plan identified oil and gas structures as potentially suitable materials for artificial reef development (USNMFS 1985).

In response to the national policy and plan, the MMS established an artificial reef policy "to encourage the conversion of selected obsolete oil and gas structures to artificial reefs on the OCS to enhance recreational and commercial fishing opportunities . . ." (Rigg 1985). In support of its policy, the MMS sought public comment on regulatory proposals designed to facilitate conversion of oil and gas structures to reefs and supported an evaluation in cooperation with the National Academy of Sciences of the social, economic, technological, and environmental effects of offshore oil and gas structure disposal (NRCMB 1985). The MMS Gulf of Mexico OCS Region published a report on rigs-to-reef development options (Reggio 1987); organized, supported, and cochaired a special session on rigs to reefs at the Fourth International Conference on Artificial Habitats for Fisheries (Reggio 1989); and advised Gulf Coast states on the development of artificial reef plans in which obsolete oil and gas structures were possible materials.

Artificial Structures and Reefs in the Gulf of Mexico

The Gulf of Mexico contains approximately two-thirds of the world's offshore oil and gas structures and 95% of the production platforms in waters off the coastal USA (NRCMB 1985). Production structures extend from the shoreline out to 140 miles. Over 30,000 wells have been drilled offshore, and more than 7×10^9 barrels of oil and 80×10^{12} ft^3 of gas have been produced from over 25 million federal acres under lease. Currently, about 5,000 of the 8,000 former offshore oil and gas leases are still active.

There are over 4,500 oil and gas structures in the Gulf of Mexico (Table 1). These structures supply a quarter of the annual U.S. production of natural gas and about one-eighth of its oil, and form the world's most extensive artificial reef system. Approximately 1,000 of the Gulf's 4,500 offshore petroleum structures are located in state waters off the coasts of Louisiana and Texas. Most of the Gulf's oil and gas structures (90%) lie in state and federal waters off the Louisiana coast, and only a few production platforms lie east of the Mississippi River in waters off Mississippi and Alabama. The rest are scattered off the Texas coast.

Other materials such as derelict ships and concrete rubble have been used extensively in the construction of over 150 permitted artificial reefs off Florida, Alabama, Mississippi, and Texas. Three-fourths of all the existing, dedicated, or permitted reef sites in the Gulf of Mexico are off the west coast of Florida and the Florida Keys. Petroleum structures and permitted artificial reefs have contributed to the Gulf's status as one of the nation's most popular offshore recreational fishing areas and have not adversely affected its consistent ranking as the nation's most productive commercial fishing zone (Schmied and Burgess 1987; USNMFS 1988). Louisiana and Texas, with over four decades of experience in offshore energy exploration, are responsible for most current OCS oil and gas production. Fishing off both these states has contributed significantly to the popular-

TABLE 1.—Numbers of permitted artificial reefs, rigs-to-reefs projects, and existing oil and gas structures, and weight and removal cost estimates for oil and gas structures by water depth in the Gulf of Mexico as of December 31, 1988.

Water depth (ft)	Artificial reefs[a]	Planned and constructed rigs-to-reefs projects	Oil and gas structures[b]	Average jacket[c] weight (tons)	Structure removal cost (millions $)[b]
0–20	7	0	1,175	<100	0.05–0.5
21–100	112	2	2,474	500–700	0.5–1.5
101–200	32	4	650	700–1,500	1–2.5
201–400[d]	6	4	306	1,000–5,000	5–15
401–2,000	0	0	24	5,000–50,000[e]	15–100
0–2,000	157	10	4,629		

[a]MMS (1986b, Visual 3); Pybas (1987); G. Serviss (Sarasota County Natural Resources Department, personal communication).
[b]National Research Council (NRCMB 1985), updated with information from MMS data base on OCS structure removals and installations 1984–1988.
[c]A platform jacket is the network of underwater steel membranes that support the platform deck areas above water.
[d]Lift capacity limitation of existing derrick barges requires that jackets be cut into pieces for onshore removal in this depth range and deeper.
[e]Shell's Bullwinkle, installed in 1,350 ft of water in 1988, was the heaviest platform jacket (50,000 tons) in the Gulf of Mexico as of December 31, 1988.

ity and productivity of the Gulf of Mexico as a fishing zone.

Proclamation of a national policy and plan stimulated interest in, and construction of, new and deeper artificial reefs along the Atlantic and Gulf coasts of the USA (Phillips 1988b). Artificial reef plans have been published or are under development for waters off all Gulf states (Jones et al. 1986; Wilson et al. 1987; H. Maddux, Texas Parks and Wildlife Department, personal communication). Reef-planning projects in the Gulf have addressed the potential for converting obsolete rigs to reefs and note limitations such as international agreements, water depth, liability, removal, transportation and maintenance costs, and defense concerns that inhibit more aggressive

TABLE 2.—Rigs-to-reefs projects in the Gulf of Mexico.[a]

Reef name (R)	Year constructed	Sponsor	Former owner	Distance to shore (miles)	Water depth (ft)	Materials
(1) S Tower	1979	Florida Department of Natural Resources	Exxon	35	105	Prototype subsea production template
(2) Pensacola	1982	Florida Department of Natural Resources	Tenneco	22	175	4-pile jacket Deck (detached)
(3) Marathon	1983	Alabama Department of Conservation and Natural Resources	Marathon	50	240	4-pile jacket Deck (detached)
Panama City						
(4a) Stage 1	1984	Bay County, Florida	U.S. Navy	12	100	16-pile jacket
(4b) Stage 2	1984	Bay County, Florida	U.S. Navy	3	60	8-pile jacket
(5) Shelter	1987	Louisiana Department of Wildlife and Fisheries	Oxy	90	238	8-pile jacket
(6) Snapper	1988	Louisiana Department of Wildlife and Fisheries	Chevron	28	104	8-pile jacket Deck (attached)
(7) Angel	1988	Louisiana Department of Wildlife and Fisheries	Exxon	120	310	8-pile jacket 8-pile jacket
			Mobil	120	310	8-pile jacket

[a]Three additional rigs-to-reefs projects have been permitted off Texas for drilling rigs sunk in accidents, and a large rigs-to-reef project has been constructed in the Atlantic Ocean off Miami and Fort Lauderdale from portions of three former Tenneco platforms from the Gulf of Mexico.

rigs-to-reefs developments. Experience gained through the creation of nine planned rigs-to-reefs projects (Table 2) and the amount of the mutual benefits available through recycling obsolete structures have provided incentives for steady progress toward more and better projects in or near offshore oil and gas fields.

In 1988 the Subcommittee on Safety of Navigation of the International Maritime Organization (IMO) endorsed a new set of "Guidelines and Standards for the Removal of Offshore Installations and Structures on the Continental Shelf and in the Exclusive Economic Zone." For the first time, these international guidelines explicitly provide for the creation of artificial reefs to enhance living marine resources when such reefs are in line with a country's established policies. Once this agreement is fully ratified by the IMO Assembly, the U.S. Department of Defense may relax its requirement for the complete upending of a structure before creating a reef at its former location. This increased flexibility would facilitate more creative, productive, and cost-effective methods of creating permanent artificial reefs out of obsolete oil and gas structures. Rigs-to-reefs projects completed by Marathon Oil (Dominguez 1983),

TABLE 2.—Extended

R	Dimensions (ft)	Weight (tons)	Donation to reef sponsor
(1)	125×96×60	2,200	Materials, transportation, construction
(2)	130×90×80 50×72×26	500 300	Materials, transportation, construction
(3)	250×140 Unknown	1,100 550	Materials, transportation, construction
(4a)	105×105	1,502	Materials, construction
(4b)	84×60	697	Materials, construction
(5)	260×120×100	3,400	Materials, construction, $250,000
(6)	160×100×63	1,150	Materials, construction, $50,000
(7)	210×170×86 250×114×66	3,100 2,100	Materials, transportation, construction, $250,000
	213×125×45	2,650	Materials, transportation, construction, $50,000

Cities Service/Oxy (Thornton and Quigel 1989) and Exxon (Anonymous 1989) have stimulated new removal techniques and cost-saving technologies that favor reef development as a disposal option. Louisiana and Texas, which have steadfastly supported OCS energy production, now have the greatest potential to work with the oil and gas industry in providing creative disposal opportunities that can perpetuate existing environmental benefits from petroleum structures. Both the Texas Senate Natural Resources Committee, which prepared a legislative report on rigs to reefs, and the Texas House Committee on Environmental Affairs (Agnich et al. 1988) conducted hearings calling for legislation to facilitate a rigs-to-reefs initiative in Texas. The proposed legislation was passed and signed into law in June 1989 with provisions calling for the establishment of an advisory council and artificial reef trust fund and the development of a state reef plan.

In the last 5 years of the 1980s, installations of oil and gas structures outpaced removals by three to one. MMS office records on offshore platform additions and removals between 1984 and 1989 indicate 834 structures were installed and 261 removed, a net addition of 573 structures. The National Research Council (NRCMB 1985) predicted that this trend will reverse itself before the turn of the century in the established production areas of the Gulf of Mexico, where over 2,500 structures are expected to become obsolete in the next 20 years. During 1988, 159 OCS structures were installed, 99 were removed, and 78 were designated for removal, pending approval or equipment scheduling for removal.

The Louisiana Program—A Working Model for Rigs to Reefs

In 1986, Louisiana became concerned about the removal of offshore oil and gas structures (Reggio et al. 1986) and proposed legislation to develop a comprehensive artificial reef program for the state. The program is responsible for siting, maintaining, and enhancing artificial reefs in both state and federal waters off the Louisiana coast, using, but not limiting itself to, retired oil and gas structures.

The Louisiana Artificial Reef Initiative (LARI) drafted legislation in response to the National Fishing Enhancement Act. The NFEA provided encouragement and guidelines to states for creating well-developed, well-organized artificial reef programs. The LARI, created in 1985 and composed of university, state, and federal officials;

members of recreational and commercial fishing interests; and representatives of Louisiana's oil and gas industry, modeled the state legislation after the NFEA. The Louisiana Fishing Enhancement Act (Act 100), which provided the authority to develop a comprehensive artificial reef program in Louisiana, was signed into law in June 1986.

Act 100 established Louisiana's Artificial Reef Development Program and named and outlined the duties of the three participating agencies. The program is administered by the Louisiana Department of Wildlife and Fisheries and conducted jointly with staff from the Center for Wetland Resources, Coastal Fisheries Institute, and the Louisiana Geological Survey at Louisiana State University. The act also created the Louisiana Artificial Reef Development Council composed of the Secretary of the Department of Wildlife and Fisheries, the Dean of the Center for Wetland Resources, and the Director of the Louisiana Geological Survey. The Council is responsible for overseeing development and implementation of the program.

The Act established the state of Louisiana as the permittee for the artificial reefs developed under the program's jurisdiction and appointed the Department of Wildlife and Fisheries as agent for the state. The state assumes responsibility for the reefs upon placement within the established reef permit area. The state, donors, and other participants constructing a reef under NFEA and Act 100 are absolved from liability, provided the terms and conditions of the authorizing federal artificial reef permit are met.

Act 100 mandated that a plan be drafted to establish the rationale and operational guidelines for the program, including the siting criteria for Louisiana's artificial reefs. The plan was accepted and endorsed by the 1987 Louisiana Legislature (Wilson et al. 1987).

With the plan in place for guidance, Louisiana began the lengthy process of identifying areas inappropriate for reef development. The process, known as exclusion mapping, excluded areas such as shipping lanes, traditional commercial fishing areas, pipeline corridors, restricted military zones, existing live bottoms, and other areas deemed unsuitable due to bottom type, hydrological conditions, and preference by other user groups (Christian 1984; D'Itri 1985; Myatt 1985).

Following exclusion of areas where reefs should not be developed, a series of public hearings was held across south Louisiana to outline the program and, with public input, to select areas where reefs should be located. As a result of those hearings, eight artificial reef planning areas (Figure 2) were chosen where specific artificial reef projects could be sited in Phase 1 of the program. These planning areas facilitate disposal planning by oil and gas companies and provide flexibility in specific site selection within the planning areas, thereby encouraging industry cooperation. At the suggestion of local fishermen, a ninth planning area (Ship Shoal) was added seaward of the cities of Houma and Morgan City. Phase 2 of the program will be implemented at a later date and will concentrate on siting reefs in Louisiana's inshore and nearshore waters.

Louisiana Act 100 did not authorize state general funds for the program, but established the Louisiana Artificial Reef Trust Fund. Oil and gas companies that donate structures to the program are asked to donate half the disposal savings realized through program participation for deposit into the Louisiana Artificial Reef Trust Fund. Based on average removal cost (NRCMB 1985) and the location of the state's artificial reef planning areas, Louisiana estimated that the oil and gas industry can save up to $1 million per structure, depending on water depth, by converting it into an artificial reef, compared to the cost of traditional onshore abandonment. The interest generated by the Artificial Reef Trust Fund is designated for program operation, research, and development, and the ultimate financial goal of the program is to be completely self-supporting.

Until such a time is reached, the program is funded by alternative sources. At present, Louisiana's main funding source is the Federal Aid in Sportfish Restoration Program. These funds are generated by a federal tax on recreational fishing gear that is returned to the states for recreational fishery enhancement projects, provided the states can supply a 25% match. Louisiana State University's Coastal Fisheries Institute provides the match. Other funding has been procured through industry grants for continued research into the feasibility of converting obsolete oil and gas structures into artificial reefs.

Since the Louisiana Artificial Reef Program was approved in 1987, three artificial reef sites made from the underwater support structures (platform jackets) from five obsolete petroleum platforms have been established offshore Louisiana (Table 2). The first reef was created in October 1987 when Oxy Corporation (formerly Cities Service) donated the jacket of a large eight-pile structure. The structure, located 90 miles off the coast of

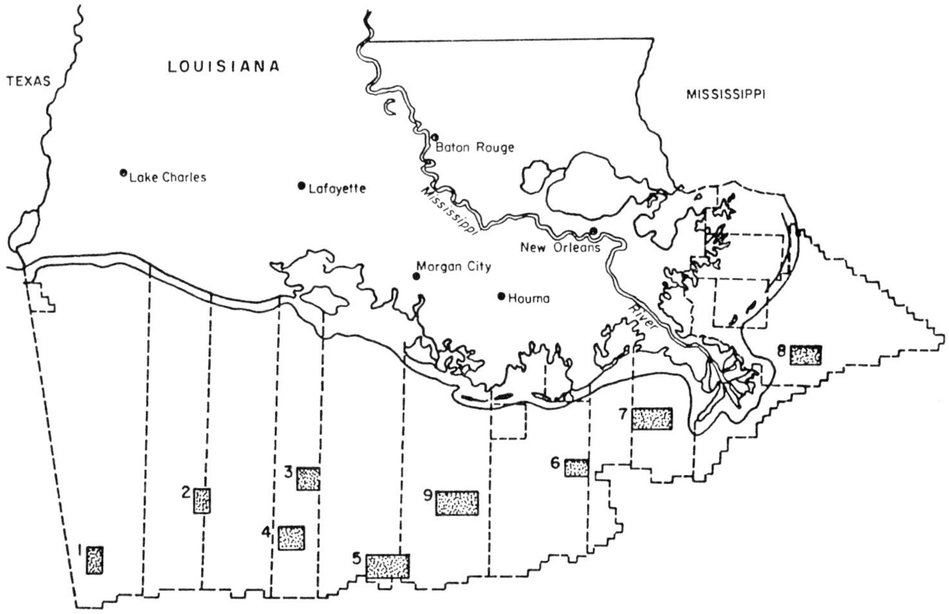

FIGURE 2.—Offshore Louisiana artificial reef planning areas: **(1)** West Cameron Planning Area, **(2)** East Cameron Planning Area, **(3)** South Marsh Island (76) Planning Area, **(4)** South Marsh Island (146) Planning Area, **(5)** Eugene Island Planning Area, **(6)** South Timbalier Planning Area, **(7)** West Delta Planning Area, **(8)** Main Pass Planning Area, **(9)** Ship Shoal Planning Area.

southwest Louisiana, was toppled in place in one of the nine planning areas endorsed by the Artificial Reef Council (Thornton and Quigel 1989).

Chevron USA followed suit 10 months later, toppling one of its former production platforms 28 miles off the southeast coast of Louisiana (Kasprzak 1989). Again, the structure was in a Council-approved planning area (South Timbalier, Figure 2) and was toppled in place in 102 ft of water.

Louisiana's third reef was a cooperative effort between Mobil Exploration and Producing and Exxon Corporation. Because the structures scheduled for removal were not in a planning area, the participating companies transported them from offshore Texas. Exxon towed two structures 25 miles to the West Cameron Artificial Reef Planning Area, where the company placed them within a permitted site on active Exxon and Mobil oil and gas leases. Mobil transported its structure over 200 miles from a location off Corpus Christi, Texas. Each company also has an actively producing structure scheduled to be abandoned within the next few years inside the state-permitted reef site. Placing the already abandoned structures near producing structures gives Mobil and Exxon the option of toppling their active structures in place once the structures cease production and are scheduled for removal.

Other offshore oil and gas operators have also expressed an interest in the program, and negotiations are in progress. Negotiations to obtain platforms and to determine the amount of donation are done on a case-by-case basis between the oil and gas operator and the state. The size, location, distance from shore, water depth, resale value, and proximity of the platform to the permitted reef site all affect the cost of converting a rig into a reef; thus, it is not always cost-effective for operators to participate (Pope 1988).

Federal and state governments, the oil and gas industry, as well as commercial and recreational fishermen, have been beneficiaries of Louisiana's artificial reef program. However, it will take the continued cooperation of the various state and federal agencies involved and the support of the Gulf user groups to ensure that Louisiana's program will enjoy continued success.

Conclusion

As we approach the 21st century, there are ever-increasing demands on marine fish and restrictions on fishing. Through state and industry leadership, we have an unprecedented opportunity in the Gulf of Mexico to create artificial reefs with oil and gas structures, which will concentrate marine life and enhance fishing. As noted in the widely endorsed National Recreational Fisheries Policy adopted in 1988 (USFWS 1988), "a concerted and diligent effort is required to maintain, restore, and increase the productivity of fish populations and habitats to provide for continuing public benefits. . . . The policy suggests this can best be accomplished through a process of government-to-private sector cooperation and coordination consistent with existing authorities and responsibilities." These words can and are being put into action in the Gulf of Mexico through state-recognized artificial reef programs.

References

Agnich, F., E. Kuempel, and R. Saunders. 1988. Rigs-to-reefs. Report of the Subcommittee on Rigs-to-Reefs of the Committee on Environmental Affairs. Texas House of Representatives, Interim Report to the 71st Texas Legislature, Section III, Austin.

Anonymous. 1989. Exxon donates obsolete oil platform jackets to Louisiana artificial reef site. Alligator 7(6). U.S. Department of the Interior, Minerals Management Service, New Orleans, Louisiana.

Christian, R. T. 1984. Transportation costs of artificial reef materials. Artificial Reef Development Center, Technical Report 4. (Sport Fishing Institute, Washington, D.C.)

CSA (Continental Shelf Associates, Inc.). 1982. Study of the effects of oil and gas activities on reef fish populations in the Gulf of Mexico OCS area. Executive summary and final report to Bureau of Land Management, New Orleans, Louisiana.

D'Itri, F. M., editor. 1985. Artificial reefs: marine and freshwater applications. Servis, Chelsea, Michigan.

Ditton, R. B., and J. Auyong. 1984. Fishing offshore platforms, central Gulf of Mexico: an analysis of recreational and commercial fishing use of 164 major offshore petroleum structures. U.S. Department of the Interior, Minerals Management Service, OCS Monograph/MMS 84-0006, Metairie, Louisiana.

Ditton, R. B., and A. R. Graefe. 1978. Recreational fishing use of artificial reefs on the Texas coast. Report of Texas A&M University to Texas Coastal and Marine Council, Austin.

Dominguez, R. 1983. Innovative techniques for economically removing and transporting oil and gas structures for reefs. Pages 143–145 *in* V. C. Reggio, Jr., and D. Larson, editors. Proceedings, fourth annual Gulf of Mexico information transfer meeting. U.S. Department of the Interior, Minerals Management Service, OCS Report/MMS 84-0026, Metairie, Louisiana.

Dugas, R., V. Guillory, and M. Fischer. 1979. Oil rigs and offshore sport fishing in Louisiana. Fisheries (Bethesda) 4(6):2–8, 10.

Gallaway, B. J., and G. S. Lewbel. 1982. The ecology of petroleum platforms in the northwestern Gulf of Mexico: a community profile. U.S. Fish and Wildlife Service Biological Services Program FWS-OBS 82/27. (Also: U.S. Bureau of Land Management, Gulf of Mexico Regional Office, Report OFR 82-03, New Orleans, Louisiana).

Gallaway, B. J., and six coauthors. 1981. Ecological investigations of petroleum production platforms in the Central Gulf of Mexico, volume 2. The artificial reef studies. National Technical Information Service, PB 82-167826, Springfield, Virginia.

Jones, J. I., and six coauthors. 1986. A plan for siting artificial reefs in the northern Gulf of Mexico. Mississippi–Alabama Sea Grant Consortium and Continental Shelf Associates, Inc., Sea Grant Publication MASGP-86-021, Ocean Springs, Mississippi.

Kasprzak, R. 1989. Turning over a new reef. Louisiana Conservationist (January–February):34–36.

Milon, W. J. 1988. The economic benefits of artificial reefs: an analysis of the Dade County, Fla., reef system. University of Florida, Sea Grant College Program, Gainesville.

MMS (Minerals Management Service). 1982. Offshore fishing workshop session. Pages 47–75 *in* V. C. Reggio, Jr., editor. Proceedings, 3rd annual Gulf of Mexico information transfer meeting. Texas A&M Research Foundation, TAMRF-MMS Contract AA851-CT1-55, College Station.

MMS (Minerals Management Service). 1984. Rigs-to-reefs session. Pages 102–163 *in* V. C. Reggio, Jr., and D. Larson, editors. Proceedings, 4th annual Gulf of Mexico information transfer meeting. U.S. Department of the Interior, Minerals Management Service, OCS Report/MMS 84-0026, New Orleans, Louisiana.

MMS (Minerals Management Service). 1985. Rigs-to-reefs session. Pages 295–341 *in* V. C. Reggio, Jr., and W. DuBose IV, editors. Proceedings, 5th annual Gulf of Mexico information transfer meeting. U.S. Department of the Interior, Minerals Management Service, OCS Study/MMS 85-0008, New Orleans, Louisiana.

MMS (Minerals Management Service). 1986a. Session IIF/rigs-to-reefs. Pages 99–119 *in* V. C. Reggio, Jr., and M. Fleetwood, editors. Proceedings, 6th annual Gulf of Mexico information transfer meeting. U.S. Department of the Interior, Minerals Management Service, OCS Study/MMS 86-0073, New Orleans, Louisiana.

MMS (Minerals Management Service). 1986b. Final environmental impact statement: proposed oil and gas lease sales 110 and 112. U.S. Department of the Interior, Minerals Management Service, OCS EIS 86-0087, New Orleans, Louisiana.

MMS (Minerals Management Service). 1987. Session 13—sea and shore: coastal recreation and petroleum development. Pages 323–348 *in* V. C. Reggio, Jr., and S. Gaudry, editors. Proceedings, 7th annual Gulf of Mexico information transfer meeting. U.S. Department of the Interior, Minerals Management

Service, OCS Study/MMS 87-0058, New Orleans, Louisiana.

Myatt, D. 1985. Artificial reef site selection to maximize recreational fishing benefits in the Gulf of Mexico. Pages 314–321 *in* V. C. Reggio, Jr., and W. DuBose IV, editors. Proceedings, 5th annual Gulf of Mexico information transfer meeting. U.S. Department of the Interior, Minerals Management Service, OCS Study/MMS 85-0008, Metairie, Louisiana.

NRCMB (National Research Council Marine Board). 1985. Disposal of offshore platforms. National Academy Press, Washington, D.C.

Phillips, S. H. 1988a. Artificial reefs and fishery conflicts: problems and opportunities for the sport fishing industry. Sport Fishing Institute, Artificial Reef Development Center, Washington, D.C.

Phillips, S. H. 1988b. Atlantic reef profile study. Reef briefs. Artificial Reef Development Center 15(Fall–Winter):1–3. (Sport Fishing Institute, Washington, D.C.)

Pope, D. L. 1988. The Louisiana artificial reef program. Louisiana Coastlines (October):1–2. (Louisiana Department of Natural Resources, Baton Rouge.)

Pybas, D. W. 1987. Atlas of artificial reefs in Florida. University of Florida, Florida Sea Grant Extension Bulletin SGEB-13, Gainesville.

Reggio, V. C., Jr. 1987. Rigs-to-reefs: the use of obsolete petroleum structures as artificial reefs. U.S. Department of the Interior, Minerals Management Service, OCS Report/MMS 87-0015, New Orleans, Louisiana.

Reggio, V. C., Jr., compiler. 1989. Petroleum structures as artificial reefs: a compendium. U.S. Department of the Interior, Minerals Management Service, OCS Study/MMS 89-0021, New Orleans, Louisiana.

Reggio, V. C., Jr., V. Van Sickle, and C. Wilson. 1986. Rigs-to-reefs. Louisiana Conservationist (January–February):4–7.

Rigg, J. B. 1985. Minerals Management Service Artificial Reefs Policy. Policy memorandum of the Associate Director for Offshore Minerals Management, January 3, 1985. U.S. Department of Interior, Minerals Management Service, Reston, Virginia.

Roberts, K., and M. Thompson. 1983. Petroleum production structures: economic resources for Louisiana sport divers. Louisiana State University, Center for Wetland Resources, Sea Grant Publication NOAA-84022701, Baton Rouge.

Schmied, R. L., and E. E. Burgess. 1987. Marine recreational fisheries in the southeastern United States: an overview. U.S. National Marine Fisheries Service Marine Fisheries Review 49(2):2–7.

Thornton, W. L., and J. C. Quigel. 1989. Rigs-to-reefs: a case history. Pages 77–87 *in* V. C. Reggio, Jr., compiler. Petroleum structures as artificial reefs: a compendium. U.S. Department of the Interior, Minerals Management Service, OCS Study/MMS 89-0021, New Orleans, Louisiana.

USFWS (U.S. Fish and Wildlife Service). 1988. National recreational fisheries policy. USFWS, Washington, D.C.

USNMFS (U.S. National Marine Fisheries Service). 1985. National artificial reef plan: NOAA (National Oceanic and Atmospheric Administration) Technical Memorandum, NMFS (National Marine Fisheries Service) OF-6:29–30, Washington, D.C.

USNMFS (U.S. National Marine Fisheries Service). 1988. Fisheries of the U.S., 1987. U.S. National Marine Fisheries Service Current Fisheries Statistics 8700.

Wilson, C. A., V. R. Van Sickle, and D. L. Pope. 1987. Louisiana artificial reef plan. Louisiana Department of Wildlife and Fisheries, Technical Bulletin 41, Baton Rouge.

Witzig, J. 1986. Rig fishing in the Gulf of Mexico—1984, marine recreational fishing survey results. Pages 103–105 *in* V. C. Reggio, Jr., and M. Fleetwood, editors. Proceedings, 6th annual Gulf of Mexico information transfer meeting. U.S. Department of the Interior, Minerals Management Service, OCS Study/MMS 86-0073, New Orleans, Louisiana.

American Fisheries Society Symposium 11:18–21, 1991

Improving Communication between the Oil and Fishing Industries

CRAIG FUSARO
Joint Oil/Fisheries Liaison Office
121 Gray Avenue, Suite 205, Santa Barbara, California 93101, USA

Abstract.—The Joint Oil/Fisheries Committee and Liaison Office were formed in 1983 by the offshore oil and fishing industries in south-central California in response to a growing number of space-use conflicts between the industries. In professionally mediated negotiating sessions, the Joint Committee developed the structure and functions of the Liaison Office, hired staff, and has overseen the operations of the office since 1983. The office facilitates interindustry communication, keeps records of conflicts and their resolution, and provides other services to the industries as directed by the Joint Committee. Work of the Committee and Liaison Office has included an Oil Service Vessel Traffic Corridor Program, A Manual for Geophysical Operations in Fishing Areas of South/Central California, recommendations to the California State Lands Commission regarding geophysical survey notification procedures, and a 3-year regional Fisheries Impact Program. The Liaison Office has also advised researchers in joint fact-finding efforts on the effects of geophysical survey acoustic energy pulses on fish dispersal and on the eggs and larvae of commercially valuable species.

Identification of Need

The nature and historical development of problems between the commercial fishing industry and offshore oil industry in south-central California have been described by Knaster (1986). Most of these problems were space-use conflicts between the oil industry in its various activities and facilities and fishermen using traditional fishing areas and techniques. For example, seismic survey vessels towed mile-long cables that tangled with buoys of lobster and crab traps and set nets, and platforms and pipelines were placed in traditional bottom trawl grounds. Seafloor debris from oil and gas construction or operations likewise was encountered in fishing grounds. Operators of oil industry and commercial fishing vessels and gear experienced right-of-way conflicts. The frequency and extent of such offshore space-use conflicts led both industries to seek improved knowledge of their counterpart's activities. In April 1983, the University of California Cooperative Extension Sea Grant Marine Advisor and the Mediation Institute organized a meeting between the two industries to discuss these problems. Participants identified several issues that could probably be resolved by further discussion, and each industry elected three representatives to continue the dialogue. This election produced what is now known as the Joint Oil/Fisheries Committee of South/Central California, more simply, the Joint Committee. It is now composed of five members from each industry.

One of the most immediate needs of both industries was improved communication. It was agreed that the lack of timely notice on when and where each industry planned to operate in the Santa Barbara Channel and Santa Maria Basin was in part responsible for many of the conflicts that had occurred by 1983. As oil and gas exploration and development expanded in the early 1980s, interactions between fishermen and oil industry operators increased. Although seismic survey notification procedures were developed for state and federal waters before 1983, the procedures for information exchange did not facilitate direct flow of information between the industries. Fishermen often had to expend much time and effort to get seismic survey information. Further, short-notice changes to seismic survey workplans were common. Thus, one of the first agenda items of the Joint Committee was the development of the Liaison Office to facilitate communication. After completing the formal agreement that outlined the structure and functions of the Joint Committee and Liaison Office, the Committee hired a marine biologist to staff the office in October 1983.

Structure and Function of the Liaison Office

The Joint Oil/Fisheries Liaison Office is a neutral entity in the conflict resolution process. The office does not attempt to advocate the position of either industry to the other, but is managed by both industries simultaneously, according to the guidelines and policies established by the Joint Committee. The Joint Committee believed that the Liaison Office should have two main goals: (1) to provide a network whereby members of one industry could readily reach the appropriate contact in the other industry, and (2) to provide a neutral meeting place at which the two industries could meet to discuss problems.

The agreement on the structure and functions of the Liaison Office described four principal activities. First, the Liaison Office should facilitate interindustry communication, specifically, official notices for activities such as seismic operations, public hearings, and other meetings of interest to both industries. The Office contacts fishermen who could be affected by a specific offshore oil and gas project, based on information that each fisherman may provide the Liaison Office about location and methods of fishing. The Liaison Office maintains confidentiality of such information and uses it only to determine which fishermen to contact. The Office also provides short-term notification about particular vessel operations, such as revisions to seismic operations and out-of-lane tanker activities. The Liaison Office notifies oil industry representatives and geophysical companies of specific fishing operations, areas, and peak fishing times to permit coordination between affected companies and fishermen (reverse notification). The Office also assists in setting up meetings between oil industry project proponents and potentially affected commercial fishermen and calls meetings to discuss specific issues.

Second, the Liaison Office should provide a clearinghouse for information for both industries. Information such as research studies relevant to oil industry and fishing industry interactions, proposed medium- to long-term projects related to oil exploration, development, or production, and information on existing facilities, projects, and operations of both industries are kept at the office.

Third, the Liaison Office should facilitate the filing of claims between fishermen and oil companies, geophysical survey operators, or federal, state, and local agencies. The Office assists fishermen in properly filing a claim, identifying the appropriate company or agency, using the correct forms, etc. The Office may only assist in the claims-filing process to the extent of assuring that appropriate procedure is followed; it is specifically directed by the Joint Committee not to participate in the substance of claims negotiations.

Finally, the Liaison Office is directed to keep records of issues discussed by the two industries and of reported conflicts at sea, and to maintain records on commercial fisheries activities for purposes of reverse notification. For a more complete discussion of Liaison Office functions, see *A Manual for Geophysical Operations in Fishing Areas of South/Central California* by the Joint Oil/Fisheries Liaison Office and Joint Oil/Fisheries Committee (1986).

Supplemental Activities of the Liaison Office

Other issues, activities, and functions have been added to the role of the Liaison Office on an as-needed basis since its original charter was developed. Many of these stem from further negotiations by the Joint Committee on the problems originally discussed in the April 1983 meeting.

Two such issues involved at-sea conflicts between vessels operating in the same space: vessel traffic and vessel right-of-way, two independent issues. Nearshore set gear fishermen using traps and nets complained of gear loss (buoys of traps or nets cut off) that occurred when crew and supply boat operators took line-of-sight routes between offshore platforms and drilling rigs and onshore crew and supply bases. The Liaison Office and Joint Committee developed and agreed upon a set of corridors ¼ nautical mile wide in which the oil industry would direct crew and supply boat traffic. The program is voluntary, and compliance is monitored by both industries. If vessels or fishing gear are found not in compliance, the observer is asked to report the condition to the Liaison Office. The Liaison Office contacts the identified party to discuss the conditions under which the event occurred and to remind the party about the development of the program through mutual agreement.

Another issue that occupied the Liaison Office in its early years was the conflicting use of set fishing gear, trawlers, and seismic survey boats and cables. Again by mutual agreement, the Liaison Office developed an information handbook for geophysical survey vessel staff and fishermen in the effort to increase communication and reduce gear loss and vessel right-of-way conflicts. *A Manual for Geophysical Operations in Fishing Areas of South/Central California* (Joint Oil/Fisheries Liaison Office and Joint Oil/Fisheries Committee 1986) followed a manual of similar name developed by the Oil/Fishing Group of Alaska for Alaska waters (Hanley 1984). The manual is a starting point for the reverse notification process by which oil and geophysical companies become informed about the particulars of various fishing fleets using the region.

Other committees were also established that dealt more with resource issues than with direct interindustry space-use conflicts. One of these committees became known as the Seismic Steering Committee, later, the Fish Dispersal Committee. This committee included representatives of the fishing, oil and gas, and geophysical industries, as well as various state and federal regulatory or permitting agencies involved in resource development or management offshore California. The Fish Dispersal Committee developed a joint fact-finding process to address the potential dispersal or decrease in catchability of fish when seismic surveys operate in fishing grounds. The Liaison Office served as a technical advisor to this committee, as well as to a later, similar committee, the Eggs and Larvae Committee, which investigated the potential effects of acoustic energy pulses from seismic exploration on the eggs and larvae of commercial fish and shellfish.

The Liaison Office also assisted the Joint Committee in developing a series of mitigation–compensation programs. As an outcome of years of difficult negotiations on the issues of compensation for or mitigation of perceived negative impacts of the oil industry on the fishing industry, a proposal was made jointly by the oil and fishing industries to the state of California. The Joint Committee believed that state and federal oil revenue-sharing moneys (funds distributed to coastal states pursuant to Outer Continental Shelf Lands Act [OCSLA] section 8[g], Public Law 83-212) were an appropriate source of funds to develop a series of innovative programs to mitigate some of the effects of multiple-use ocean development. After a long series of false starts and dead ends, the Joint Committee and Liaison Office successfully secured 8(g) funds for these programs. In 1989, the first year of a 3-year effort, program elements included:

- Area, Gear, and Technology Research and Development: The State Secretary of Environmental Affairs contracted for research and development on new fishing areas, gear, or technology to seek means by which the fishing industry might offset the areal loss of fishing grounds caused by oil industry exploration and development activities.
- Vessel and Gear Staging and Repair Space Program: The State Coastal Conservancy identified fishing ports in which space and facilities were needed and could be provided for the repair of gear damaged by seafloor debris or other oil industry equipment or operations.
- Vessel Safety Gear Grant Program: The Secretary of Environmental Affairs made grants of up to $5,000 per commercial fishing vessel over the 3-year program for vessel or personnel safety and survival equipment, to enable fishermen to more safely extend their range beyond areas of multiple-use conflicts.
- Seafloor Debris Cleanup Program: The State Lands Commission investigated areas that were known to contain snags or other seafloor obstructions, with the goals of producing a list of known snag areas for commercial fishermen, and, where possible, identifying the nature and ownership of such snags and removing them as advised.
- San Pedro Purse Seine Fleet Assistance: The Secretary of Environmental Affairs assisted the San Pedro purse seine fleet in a resource assessment and market evaluation for the sardine fishery.

The 1989–1990 program added a marketing assistance element to help in marketing fish and shellfish from the multiple-use regions in both domestic and international fish markets. A California Seafood Council may be formed as a result. Year three (1991) will add a fisheries development corporation, designed to provide a low-interest, revolving loan fund for fisheries infrastructure development in areas of oil and gas activity or perhaps to work in conjunction with the marketing assistance element. The specific format of the corporation has yet to be finalized.

The Liaison Office participated in the development of each of these program elements with the Joint Oil/Fisheries Committee and continues to be involved in their implementation and administration. The Liaison Office has also assisted in the development of complementary programs at the local level in Santa Barbara County, such as a Local Fisheries Enhancement Fund and a Local Fishermen's Contingency Fund. The Enhancement Fund was developed by Santa Barbara County to provide some measure of compensation for cumulative effects not included in the individual project permit process. The Contingency Fund was developed by the County to accomplish two goals: (1) to complement the Federal Fishermen's Contingency Fund, which does not cover gear damage or loss in state waters, and (2) to loan fishermen funds to keep fishing while they await the federal claim review process. The federal government historically has taken the better part of a year to reimburse fishermen for gear damage or loss. The county will loan fishermen funds within 15 d of a completed claim application to repair or replace gear and get them back in the water for the length of time that the federal government takes to determine a payment. Compensation for damage to gear in state waters is not a loan but a reimbursement for damages payment.

The Liaison Office began in 1983 as a 1-year experiment in the facilitation of interindustry communication and conflict resolution. This year, 1991, marks the eighth year of operations for the Liaison Office. This experience has shown that the functions outlined by the Joint Committee for the Liaison Office were indeed necessary and the Office has successfully implemented the tasks directed to it by the Joint Oil/Fisheries Committee.

References

Hanley, P., editor. 1984. A manual for geophysical operations in fishing areas of Alaska, 2nd edition. The Oil/Fisheries Group of Alaska, Anchorage.

Joint Oil/Fisheries Liaison Office and Joint Oil/Fisheries Committee. 1986. A manual for geophysical operations in fishing areas of south/central California. Joint Oil/Fisheries Liaison Office, Santa Barbara, California.

Knaster, A. 1986. Mediation and scientific issues. Environment (Washington, D.C.) 28(10):6–30.

American Fisheries Society Symposium 11:22–25, 1991

Effects of Nonexplosive Seismic Energy Releases on Fish

DILWORTH W. CHAMBERLAIN
Atlantic Richfield Company, 515 South Flower Street, Los Angeles, California 90071, USA

Abstract.—The effects of petroleum-exploration-related nonexplosive energy releases on fish and fisheries is of great concern to fishermen, regulatory agencies, and the general public. This concern has given rise to a number of scientific studies on these effects. Study results suggest a decrease in the ability of commercial fishermen to catch certain species of rockfish *Sebastes* spp. and some physical effects on individual fish near the energy source. It is concluded, however, that nonexplosive energy releases such as those produced by airguns do not cause significant harm to marine life.

There are three main sources of underwater ocean noise: water motion, which includes surf, breaking waves, rain, hail, and tides; anthropogenic noise, such as ships; and marine life (Wenz 1962). Other sources include earthquakes and natural explosions, thermal noise, and noise from sea ice. Of these sources, only noise made by humans is usually of environmental concern, because it is not a natural component of ocean ecosystems. The highest volumes of sound produced by anthropogenic sources, when measured at or near the sound source, approach the lower range of sound pressure levels of natural ambient ocean noises generated by earthquakes or explosions (Wenz 1962). These could be in the range of 60–80 dB or higher, and therefore could produce negative environmental effects.

One source of anthropogenic noise in the ocean is from petroleum exploration, development, and production, including noise from geophysical surveys. The objective of seismic geophysical exploration in the ocean, as on land, is to use sound waves to study the layers of rock beneath the earth's surface, often in attempts to locate oil and gas deposits. The principle of seismic exploration was derived from the science that deals with the recording of earthquakes and related phenomena, and it is done in aquatic environments by releasing energy pulses near the water surface. These energy pulses, emitted by acoustic pulse generators, pass through the water column, enter the earth underneath, are reflected back to the ocean surface by the underlying layers of rock, picked up by special microphones (hydrophones), and recorded. Through careful study of these data, geologists can better determine where petroleum deposits may be located.

Many people, including commercial fishermen, fishery scientists, regulatory agency personnel, and environmentalists, have expressed concern about the effects of seismic exploration on fish health and behavior and on fishing success. Specifically, these concerns are that geophysical seismic exploration injures or kills plankton and fish, scares fish away from fishing areas, or in some other way reduces fish vulnerability to capture.

Information on the effects of underwater acoustical energy waves on fish first dates from about 1944 (Gowanloch and McDougall 1945) and continues to the present. Early geophysical seismic techniques, which used explosives, injured fish and are no longer used in the USA for this reason. Injury from explosives was primarily caused by a very rapid increase in pressure, immediately followed by a negative pressure. Hill (1978) described this phenomenon. Injuries to fish included loss of scales, swim bladder damage, hemorrhaging, and rupture of internal organs (Pierce 1984). Lethal effects have been observed up to 61 m from the explosion of up to 9.1 kg of high explosive. Greater amounts of explosive, up to 544 kg, killed fish at distances up to 183 m. With very large amounts of explosives, up to 1.8 tonnes, fish mortalities occurred up to 366 m from the energy source (see review by Linton et al. 1985).

Nonexplosive energy-producing devices now in use have slower pressure rise times. Reviews of a number of studies of these nonexplosive devices include papers by Falk and Lawrence (1973), Schuijf and Hawkins (1976), Hill (1978), Wright (1982), Greene et al. (1985), and Linton et al. (1985). Nonexplosive energy devices produce seismic energy by various means including electrical spark (sparker), water implosion (water gun), hydraulically produced vibrations (vibroseis), gas exploder systems (ignited mixtures of propane or acetylene and oxygen), wire exploder (similar to sparker), pulsed echo sounding (hydrosonde), rapidly condensing steam vapor (vaporchoc), pulsing transducers (boomer), and release of compressed air (airgun). The airgun is the seismic energy source now used in about 90% of U.S. offshore geophysical surveys (L. G. Bowles, Halliburton Company, personal communication).

The acoustical signal from a seismic energy source is the structure (form) and strength (amplitude) of the underwater sound wave produced by a source such as the devices mentioned above, or an array of these (airguns or waterguns), in the form that is received by a receptor such as a fish, invertebrate, or hydrophone. The acoustical signal varies in frequency, amplitude, intensity, and velocity with the type of instrument that produces it and the amount of energy released.

Review of Nonexplosive Seismic Effects

Research into the effects of nonexplosive seismic energy sources began about 1970 (Kostyuchenko 1971). Eggs of European anchovy *Engraulis encrasicolus*, striped mullet *Mugil cephalus*, common carp *Cyprinus carpio*, and blue runner *Caranx crysos* were exposed to energy releases from TNT (trinitrotoluene) charges, an electric pulse generator, and an airgun (5-L capacity at 140 atmospheres working pressure). The airgun was suspended 0.5, 5, and 10 m in the water column above the test containers. Examination of the samples following exposure to the airguns showed that within a radius of 0.5 m, European anchovy eggs were damaged the most and common carp eggs the least. At a distance of 10 m, there was no injury. Survival of incubated eggs was 75.4% for those held at 0.5 m from the discharge, and those held at 10 m had the same survival rate as the unexposed eggs. Exposed larvae showed no pathological changes.

Smolts of coho salmon *Oncorhynchus kisutch* were subjected to the effects of a single airgun and an airgun array at distances of 1, 5, and 10 m. Exposure of coho salmon to one airgun of 328 cm^3 at 1 m distance and a depth of 3 m caused no observed mortality, injury, or distress. Exposure of coho salmon to the simultaneous discharge of nine airguns (3,933 cm^3) at 5- and 10-m distances gave similar results (R. J. Weinhold, Alaska Department of Fish and Game, and R. R. Weaver, Geophysical Services, Inc., personal communications).

In 1984, workers in Norway (J. Dalen and G. M. Knutsen, Havforskningsinstitutet, personal communication) exposed eggs, larvae, and juveniles of Atlantic cod *Gadus morhua* to energy releases from airguns and waterguns. Exposure of all three life stages to a small airgun (780 cm^3) resulted in no significant differences between test and control groups, except in older juveniles (110 d), which developed balance problems and recovered within a few minutes. Exposure of juvenile cod to a large airgun (8,603 cm^3) caused transient balance problems with recovery within a few minutes. Airgun exposures were at 1–10 m from the energy source and at 4-m depth. Juveniles exposed to the watergun at 2-m distance were killed immediately or died within 2–3 h. All had ruptured swim bladders and hemorrhaging. Thirty percent had ruptured abdominal walls. Fish exposed at 6-m distance developed only transient balance problems and recovered. Dalen and Knutsen (1986) also conducted behavioral studies using an array of 40 airguns towed at 6-m depth with a total chamber volume of 77,871 cm^3 operating at 2,000 pounds per square inch (psi). The airguns were fired every 10 s, which gave a firing interval of 25 m at 5 knots. Results showed a decrease in abundance of demersal and pelagic fish in the study area. The decrease in abundance ranged from 15 to 54%, depending on the species of fish.

Many species of rockfish exhibit shoaling (schooling) behavior and can be found at various depths around rocks or associated with underwater vegetation such as kelp. This shoaling behavior is often used to advantage by commercial and sports fishermen. In southern California, an airgun array was used to test the behavior of rockfish, *Sebastes* spp., shoals during and after the array passed over them (Greene 1985). The array used 28 airguns of various sizes with a combined volume of 65,548 cm^3. These were operated at 1,875 psi at a depth of 10 m. A firing interval of 9 s at a speed of 5.4 knots resulted in the array firing every 25 m. Results varied in the 10 tests that were conducted. In most of the tests, fish school heights diminished, and in at least 2 tests, the fishes essentially disappeared. Four schools diminished in height before airgun operation began. Some fish dispersed in the absence of seismic sounds, while new schools formed near the sound source.

Another California study exposed four life stages of the northern anchovy *Engraulis mordax* to airgun energy releases in the ocean (Holliday et al. 1987). Eggs, larvae, and adults were exposed to a series of airgun energy releases simulating a towed airgun array. Tests were conducted at distances of 2 m with 164- and 656-cm^3 airguns and at 3 m with 1,966- and 4,916-cm^3 airguns. All were operated at 2,000 psi. The most significant effect was a 35% reduction in the survival of 4-d-old larvae. This effect resulted from exposure levels three to four times those of a typical seismic array (40 airguns) passing over the fish at a distance of 3 m. Actual seismic surveys in the field operate at energy levels one-third to one-quarter of those that caused the largest effect, a 35% reduction in survival of 4-d-old larvae. The authors attributed this reduction in survival to interference or delay in the onset of feeding. Other observations included subtle effects on survival for the youngest larvae tested, and reduction in length of the larvae. No gross morphological damage was observed. Except for the effects noted above, extrapolation of the generally statistically nonsignificant data by Holliday et al. (1987) suggested that noticeable effects on eggs and larvae would only result from large numbers of multiple exposures to full seismic arrays.

Further studies were carried out in central California (Pearson et al. 1987) to determine the effects of airgun sounds on rockfish and the rockfish fishery. The study was in two parts. A behavioral experiment used a single airgun to measure response of different rockfish species. Several species gave alarm and startle responses upon sound exposure to a single airgun. Other behavioral responses included the collapse of rockfish schools to the ocean bottom. For other species, alarm reactions were more individual; some fish rose in the water column, while others eddied around or moved to the ocean bottom. There was evidence of habituation to sound emissions, because startle responses given at the beginning of the sound presentations were not maintained throughout the sound presentation. Also, alarm responses were not always maintained throughout the sound presentations. In the second part of the study, a study of catch success, the rockfish catch was reduced about 50% by sound emissions from a single airgun. However, because of the design of this experiment, the authors were unable to make any conclusions about the distance or length of time over which reduced catch might occur.

Discussion

The information available to date on the effects of nonexplosive energy releases from seismic geophysical exploration in the ocean has not demonstrated significant effects to fish. The major findings have been some behavioral changes and physical effects very near the source of the released energy, usually no more than 5–10 m away. The more important issue may be the economic effects on commercial fishing if fish schools are dispersed or fish do not take a hook when geophysical exploration activity is underway in the vicinity, rather than any significant physical effect on fish. With concern and adequate communication between the petroleum industry and commercial fishermen, potential economic effects can be substantially reduced.

A recent study (Pearson et al. 1988) on zoea larvae survival, development rate, and behavioral response of Dungeness crab *Cancer magister* showed no substantial effects after exposure to an array of seven airguns (13,765 cm^3) at 1-m distance. This and additional information on the effects of nonexplosive seismic energy releases on marine invertebrates have shown findings similar to those reviewed here for fish, except that dispersal behavior was not observed (Pierce 1984; Linton et al. 1985). This supports the conclusion that anthropogenic energy releases in the ocean from nonexplosive energy sources are not a significant source of harm to the marine environment. However, they may be a source of economic impact to commercial fishermen if reduced catches result from the exploration activity.

References

Dalen, J., and G. M. Knutsen. 1986. Scaring effects of fish and harmful effects of eggs, larvae and fry from offshore seismic explorations. ICA Associated Symposium on Underwater Acoustics (16–18 July 1986), Halifax, Nova Scotia.

Falk, M. R., and M. J. Lawrence. 1973. Seismic exploration: its nature and effect on fish. Canadian Fisheries and Marine Service Technical Report Series CEN T 73-9.

Gowanloch, J. N., and J. E. McDougall. 1945. Effects from the detonation of explosives on certain marine life. Louisiana Conservationist 3(1):3–6.

Greene, C. R., Jr. 1985. A pilot study of possible effects of marine seismic airgun array operation on rockfish plumes. Report to Joint Oil/Fisheries Seismic Steering Committee, Santa Barbara, California.

Greene, G. D., F. R. Engelhardt, and R. J. Paterson, editors. 1985. Proceedings of the workshop on effects of explosives use in the marine environment. Canada Oil and Gas Lands Administration, Environmental Protection Branch, Technical Report 5, Ottawa.

Hill, S. H. 1978. A guide to the effects of underwater shock waves on arctic marine mammals and fish. Canada Department of Fisheries and Oceans, Institute of Ocean Sciences, Pacific Marine Science Report 78-26, Sidney, British Columbia.

Holliday, D. V., R. E. Pieper, M. E. Clarke, and C. F. Greenlaw. 1987. The effects of airgun energy releases on the eggs, larvae, and adults of the northern anchovy (*Engraulis mordax*). Report of Tracor Applied Sciences to the American Petroleum Institute, API Publication 4453, Washington, D.C.

Kostyuchenko, L. P. 1971. Effects of elastic waves generated in marine seismic prospecting on fish eggs in the Black Sea. Hydrobiological Journal 9(5):45–48.

Linton, T. L., N. Hall, D. LaBomascus, and A. M. Landry. 1985. The effects of seismic sounds on marine organisms: an annotated bibloiography and literature review. Texas A&M University Sea Grant College TAMU-SG-86-604.

Pearson, W. H., J. R. Skalski, and C. I. Malme. 1987. Effects of sounds from a geophysical survey device on fishing success. Report of Battelle Memorial Institute and BBN Laboratories to U.S. Minerals Management Service, OCS Study MMS 87-0020 (Contract 14-12-0001-30273), Los Angeles.

Pearson, W. H., J. R. Skalski, S. D. Sulkin, and C. I. Malme. 1988. The effects of seismic energy releases on the zoeal larvae of the Dungeness crab (*Cancer magister*). Report of Battelle Memorial Institute to California Department of Fish and Game, Contract 6C-1943, Sacramento.

Pierce, C. 1984. The biological effects of marine seismic exploration activities: a literature survey. Master's thesis. California State University, Hayward.

Schuijf, A., and A. D. Hawkins, editors. 1976. Sound reception in fish. Elsevier, New York.

Wenz, G. M. 1962. Acoustic ambient noise in the ocean: spectra and sources. Journal of the Acoustical Society of America 34:1936–1956.

Wright, D. G. 1982. A discussion paper on the effects of explosives on fish and marine mammals in the waters of the Northwest Territories. Canadian Technical Report of Fisheries and Aquatic Sciences 1052.

American Fisheries Society Symposium 11:26–38, 1991

Effects of Oil Drilling on Deepwater Reefs Offshore California

GARY D. BREWER

U.S. Department of the Interior, Minerals Management Service, Pacific OCS Region
770 Paseo Camarillo, Camarillo, California 93010, USA

JEFFREY L. HYLAND

Marine Science Institute, University of California
Santa Barbara, California 93001, USA

DANE D. HARDIN

Kinnetic Laboratories, Inc.
307 Washington Street, Santa Cruz, California 95060, USA

Abstract.—Multiyear, multidisciplinary monitoring studies began in October 1986 off Point Conception and the adjacent Santa Maria Basin, California, to detect changes in deepwater benthic communities and determine if changes were associated with the discharge of drilling waste from offshore oil and gas activities. One year of seasonal studies on deepwater (105–213 m) reefs adjacent to Platform Hidalgo, conducted before drilling, suggested temporally stable benthic communities dominated by a turf of foraminifera and hydroids and large numbers of brittle stars, sea anemones, cup corals, polychaetes, and decapod crustaceans. Analysis of sediments and suspended particulate materials adjacent to Platform Hidalgo indicated that reef animals were exposed to contaminants associated with drilling wastes. Continued monitoring of potentially affected invertebrate reef communities and studies of the growth and life history of selected invertebrate taxa were designed to measure the nature and extent of oil and gas drilling impact. The monitoring program was supplemented during 1990 with studies on the species composition, distribution, movements, and biology of deepwater reef fishes. These parameters were related to the proximity of the platform and its waste discharges. The experimental design included the evaluation of sublethal factors associated with the potential accumulation and metabolism of metals and hydrocarbons by the fishes. Sampling techniques for commercially important fishes, including rockfishes (*Sebastes* spp.) and lingcod (*Ophiodon elongatus*), used remotely operated vehicles, hydroacoustic devices, fishing, and tagging. Results from the 1990 monitoring program are not reported here.

As part of a commitment to study the long-term effects of outer continental shelf (OCS) oil and gas activities on the marine environment, the U.S. Minerals Management Service (MMS), Pacific OCS Region is conducting multiyear environmental studies to assess the impacts of oil and gas development and production platforms. These efforts have been stimulated by concerns that routine U.S. Environmental Protection Agency (EPA)-permitted discharges, while not acutely toxic, may alter the structure of benthic communities or cause sublethal stress to marine organisms (Capuzzo 1987; Neff 1987). The effects of exploratory drilling operations on soft-bottom communities, which have been well documented, are short-term and minimal, while the potential long-term effects of chronic discharges from multiwell development and production platforms are still uncertain (National Research Council 1983, 1985; Boesch and Rabalais 1987).

After soliciting recommendations for study designs at a public workshop (MMS 1982), then surveying potential sites for long-term studies (SAIC 1985; Piltz 1986), the MMS began a benthic monitoring program in 1986. The goals were to detect spatial and temporal changes in soft (unconsolidated) and hard substrate (reef) invertebrate communities and evaluate whether any observed changes were related to drilling activities (Brewer et al. 1987). Experimental and control stations were established around two sites of proposed oil and gas drilling (Platforms Hidalgo and Julius) to monitor long-term, potentially subtle changes in sediment chemistry and texture and any concomitant changes to benthic communities. Results of initial sampling during the predrilling phase of the 5-year California Monitoring Program (CAMP) were described by Hyland and Neff (1988).

The CAMP study sites off Point Conception and in the Santa Maria Basin were chosen, in part, because the study region was free of the potentially confounding influences of large rivers and major sewage outfalls. There was no oil and gas development drilling in the region until April 1986 when drilling from Platform Irene (located north of Point Arguello) began.

Biological communities associated with soft substrates are monitored at stations 400 m to 10 km from the proposed site of Platform Julius and at a series of three cross-shelf transects, positioned to monitor the broader Santa Maria Basin region (Figure 1). Originally scheduled for 1987, the installation of Julius has been stalled by delays in the permitting of an onshore, crude oil processing facility. Seasonal, pre-impact data were collected from October 1986 to May 1989; additional sampling has been suspended until a definitive date is established for platform installation and drilling.

The effects of drilling wastes on deepwater, invertebrate reef communities are examined at Platform Hidalgo. Hidalgo is the most northern of a three-platform group (Hermosa–Harvest–Hidalgo) off Point Conception (Figure 2). Hidalgo was positioned in early 1986, and drilling began in November 1987. The first 7 of 53 potential wells had been drilled by February 1989, and by that

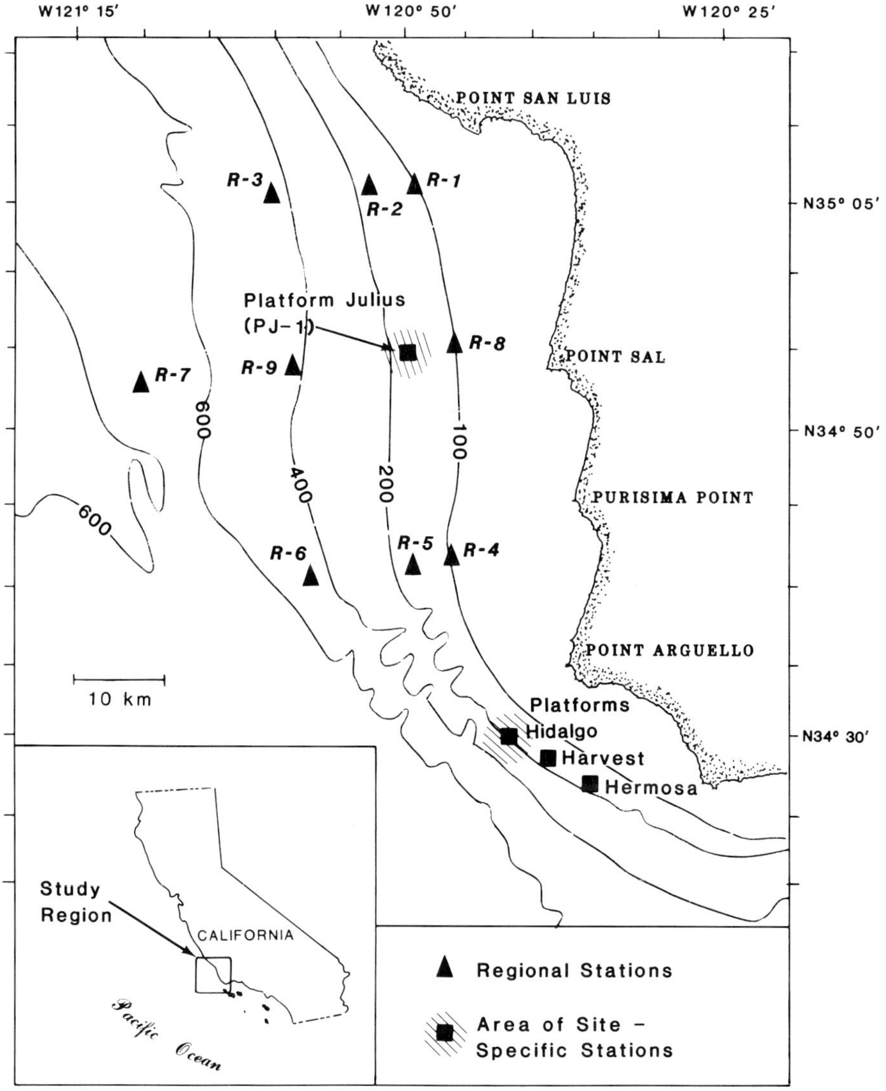

FIGURE 1.—Location of the CAMP study region including Platforms Hidalgo, Harvest, and Hermosa, and proposed Platform Julius. Locations of regional soft substrate stations are represented by R-1 to R-9. Station PJ-1 is the proposed site of Platform Julius.

date, 800 tonnes of drilling wastes had been discharged. Drilling at Harvest (November 1986) and Hermosa (January 1987) preceded drilling at Hidalgo. In addition to drilling fluids and drill cuttings, small amounts of treated sewage and other materials may have been released from the platform (Table 1). These substances also have the potential to influence nearby reef communities. No produced water, which is saline fossil water associated with oil-bearing reservoirs, is discharged from existing platforms in the region. Sampling of reef communities began at Hidalgo in October 1986 and has continued through October 1990.

In this paper, we describe preliminary results of selected invertebrate and chemical monitoring studies at Hidalgo from October 1986 to May 1988 and discuss the rationale for supplementing benthic work with studies that began in 1990 of the potential effects of platform discharges on fishes and fisheries.

Methods

Sampling and analytical techniques for the CAMP study were described by Hyland and Neff (1988). The experimental methodology followed Green's (1979) criteria for an "optimum impact study design," including rigorous spatial and temporal controls (see Carney 1987 for a critique on experimental design associated with OCS impact analyses). Parameters characterized and monitored included benthic invertebrate assemblages; the chemistry of platform discharges, sediments, sediment pore waters, and animal tissues; sediment grain size and mineralogy; sediment trans-

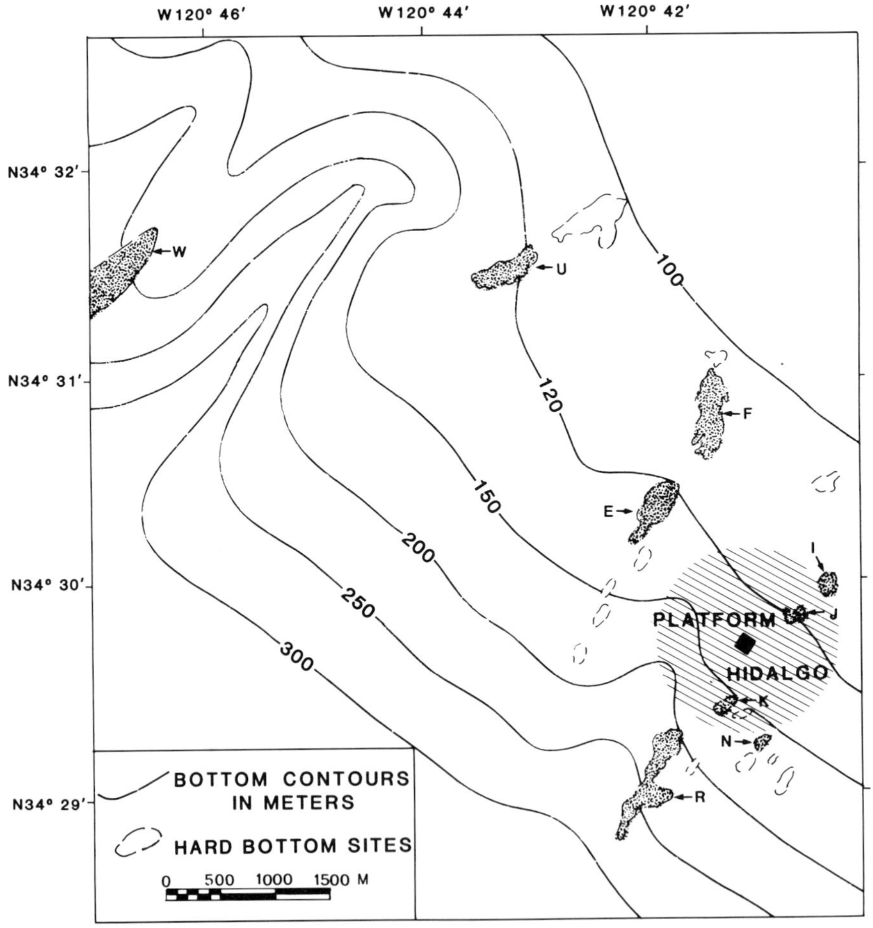

FIGURE 2.—Location of hard substrate (reef) stations around Platform Hidalgo, showing exposed rocky features in black. Reefs selected for long-term monitoring are designated as sites W, U, F, E, I, J, K, N, and R.

TABLE 1.—Major permitted discharges and potential disruptive agents associated with offshore oil and gas drilling and production (after Neff et al. 1987).

Substance	Description or treatment
Drill cuttings	Clay to gravel-sized particles of sedimentary rock that are cut by the rotating drill bit as it penetrates the earth. Trace metals and hydrocarbons may be present. Several hundred tonnes may be discharged per well, depending upon the depth of the well.
Drilling fluids (muds)	Mixtures of barium sulfate, natural clays, lignosulfonates, sodium hydroxide, specialty additives, and some metals, including Cr, Cd, Cu, Fe, Pb, Hg, Ni, and Zn. Muds are used to cool and lubricate the drill bit, maintain formation pressures to prevent blowout, and force drill cuttings to the surface. Several hundred tonnes may be discharged per well, depending upon the depth of the well.
Cooling water, deck drainage, ballast water	Treated as necessary in an oil–water separator before discharge.
Domestic sewage	Given activated sludge treatment before discharge.
Sacrificial anodes, corrosion, anti-fouling paints	Small amounts of Al, Cu, Hg, In, Sn, and Zn may leach from the platform structure and drill string.
Produced water	Saline "fossil water" associated with oil-bearing geological formations. Hydrocarbons and metals may be present in varying concentrations. Produced waters are treated, as necessary, in an oil–water separator before discharge. Volumes range from 0 to 1.5 million L/well daily.

port; water currents and other oceanographic features.

The reefs under study were isolated boulder fields and rocky outcroppings surrounding Hidalgo. Two types of rocky habitat were sampled. Low-relief habitat consisted of rocks that ranged from approximately 0.2 to 0.5 m in vertical relief, and high-relief habitat consisted of rocks greater than 1.0 m in vertical relief. We used sediment grab samples and sediment trap samples, collected before and after drilling began at Hidalgo, as indicators of hydrocarbon and metal exposure to the reefs. Sediments adjacent to each of nine reefs surrounding Hidalgo were collected for chemical analysis with a 0.1-m^2 Van Veen grab during October 1986, May 1987, and October 1987 (Table 2; Figure 2). For comparison, we also present data for sediment samples that were collected concurrently from CAMP regional stations with a 0.25-m^2 Hessler–Sandia box corer. Sediment traps were deployed and retrieved at selected Hidalgo reefs to examine reef exposure to particulate drilling waste contaminants. Baited animal traps also were deployed on soft-bottom areas adjacent to Hidalgo and at distances of 1 and 4 km from the platform to capture mobile reef invertebrates for tissue analysis. Beginning in October 1988, additional organisms were collected for tissue analysis with a remotely operated vehicle (ROV) at stations J and U (Figure 2). Analytical and quality control procedures for analysis of hydrocarbons and metals in these various types of samples were detailed in Boehm et al. (1988) and Crecelius (1988).

Concurrent with the sediment sampling, reef invertebrate macrofauna were sampled from 60 photoquadrats taken randomly at each site. Photographs were taken from the ROV, Recon IV, while a pilot guided the vehicle over each reef via a remote television camera. Photoquadrats consisted of 70-mm color slides taken about 1.4 m from the rock surface so that each photoquadrat covered an area of about 0.7 m^2. The photoquadrats were analyzed by a random point–contact method, whereby 50 points were randomly placed on the photoquadrat; the proportion of points that touched a taxon was used to estimate percent cover of that taxon. Counts of noncolonial organisms were also made and normalized to the amount of rock surface (Hardin et al. 1988). Although the photographic techniques were not designed to sample nekton, fishes sometimes appeared in the photoquadrats, and we have included this information.

Results and Discussion

Metals and Hydrocarbons

We summarize preliminary data that address the potential input of drilling-related materials to the reefs adjacent to Hidalgo in this section. A thorough description of metals, hydrocarbon components, and hydrocarbon diagnostic ratios in sediments and sediment traps before and after drilling was begun is beyond the scope of this report and is reported elsewhere (Steinhauer and Imamura 1990).

Concentrations of hydrocarbons as total alkanes (TALKs), total polycyclic aromatics (TPAHs), and 11 metals in sediments adjacent to

TABLE 2.—Locations of hard-bottom sites near Platform Hidalgo (PH: stations E to W), regional soft-bottom sites (stations R-1 to R-9), and Platform Julius (station PJ-1) in the Santa Maria Basin.

Station	Latitude Longitude	Depth (m)	Relief type	Distance from PH (km)
		Hard-bottom sites		
PH	34°29.70′N 120°42.14′W	132		
E	34°30.26′N 120°42.76′W	119	Low	1.4
F	34°30.81′N 120°42.36′W	105	Low	2.1
I	34°29.99′N 120°41.68′W	107	Low	0.9
J	34°29.82′N 120°41.82′W	117	Low	0.5
K	34°29.37′N 120°42.26′W	160	High	0.7
N	34°29.21′N 120°42.05′W	166	Low	0.9
R	34°29.11′N 120°42.67′W	213	Low–high	1.4
U	34°31.48′N 120°43.51′W	113	Low	3.9
W	34°31.52′N 120°45.86′W	195	Low–high	6.6
		Soft-bottom sites		
R-1	35°05.83′N 120°49.16′W	91		67.7
R-2	35°05.50′N 120°53.40′W	161		68.5
R-3	35°05.30′N 121°00.90′W	409		71.8
R-4	34°43.01′N 120°47.39′W	92		25.9
R-5	34°42.69′N 120°50.83′W	154		27.5
R-6	34°41.40′N 120°57.90′W	410		32.4
R-7	34°52.90′N 121°10.30′W	565		60.7
R-8	34°55.30′N 120°45.87′W	90		47.8
R-9	34°53.68′N 120°59.12′W	410		51.4
PJ-1	34°55.79′N 120°49.91′W	145		49.7

each of the study reefs before drilling at Hidalgo are shown in Table 3. Concentrations of TALKs reflect hydrocarbon inputs from biogenic, petrogenic, and anthropogenic sources; PAHs are of interest because they contain key compounds, some highly toxic, that are unique to fossil fuels. Because of its high concentration in drilling fluids, barium (Ba), in the form of barium sulfate, is used as a convenient tracer of drilling discharges, but is virtually nontoxic (Neff 1987).

With the exception of high levels of TPAHs at stations J, K, and R, concentrations of TPAHs, TALKs, and metals at the Hidalgo stations fell within the background range of hydrocarbons and metals at CAMP regional stations throughout the Santa Maria Basin. Overall, the metal concentrations were typical of relatively uncontaminated fine-grained sediments offshore southern California (SAIC 1985). The cause of the relatively high levels of TPAHs at stations J, K, and R is not clear; these stations are close to Hidalgo, and activities associated with platform installation and predrilling may have been responsible. As indicated below, TPAHs at these same stations fell to background concentrations during subsequent sampling periods.

Sediment concentrations of Ba, TALKs, and TPAHs during October 1986, May 1987, and October 1987 suggested some temporal variability (Figure 3). While Ba concentrations at regional stations remained relatively constant, there was a significant increase of mean Ba at Hidalgo stations from October 1986 (752 μg/g, SE = 36) to October 1987 (859 μg/g, SE = 25) (based on result of a t-test for paired observations, $P \leq 0.05$). Measurements of the 10 other metals showed little temporal change during the same predischarge period. Levels of TALKs varied, apparently seasonally, at both regional and Hidalgo stations. Data for TPAHs were less clear. Relatively high and variable mean concentrations of TPAHs were found at Hidalgo stations in October 1986 (0.42 μg/g, SE = 0.20) and at regional stations in May 1987 (0.30 μg/g, SE = 0.08). Mean TPAHs were much lower at Hidalgo stations during May 1987 (0.07 μg/g, SE = 0.01) and October 1987 (0.11 μg/g, SE = 0.01) sampling periods. At regional stations, TPAHs were relatively low during October 1986 (0.08 μg/g, SE = 0.01) and October 1987 (0.11 μg/g, SE = 0.02).

Data on sediment trap composition were confined to two sampling periods, before drilling at Hidalgo (October 1986–May 1987), and after drilling began (January 1988–May 1988). Interpretation of the data was also limited by a disappointing return of sediment trap samples ($N = 2$) for the predrilling period. Nevertheless, the mean concentration of Ba increased from 658 μg/g (SE = 6.00) for the Hidalgo predrilling period to 2,190 μg/g (SE = 221, $N = 7$) after drilling began. Unlike Ba, mean concentrations of TALKs in

TABLE 3.—Total alkane hydrocarbons (TALKs), total polycyclic aromatic hydrocarbons (TPAHs), and metals in sediment samples collected adjacent to reef stations at Platform Hidalgo (stations E through W) compared with regional soft substrate stations (R-1 through R-7) and Platform Julius (PJ-1) during October 1986. Each value (μg/g) is the mean of triplicate analyses. No data are available from regional stations R-8 and R-9 for October 1986.

Station or statistic	TPAHs	TALKs	As	Ba	Cd	Cr	Cu	Pb	Hg	Ni	Ag	V	Zn
					Platform Hidalgo								
E	0.111	1.3	6	908	0.62	93	16	14	0.084	44	0.127	65	64
F	0.100	1.6	6	754	0.55	106	18	16	0.084	49	0.140	99	72
I	0.089	1.5	7	782	0.54	115	18	14	0.063	49	0.170	74	70
J	1.485	2.1	6	798	0.55	112	19	16	0.056	48	0.147	48	65
K	0.373	0.9	14	610	0.41	147	13	12	0.065	30	0.107	67	85
N	0.066	0.9	16	553	0.37	166	12	11	0.038	30	0.101	49	87
R	1.415	2.3	6	779	0.74	119	13	14	0.058	33	0.098	62	59
U	0.096	1.9	6	761	0.63	104	19	15	0.126	48	0.137	85	68
W	0.061	0.9	7	819	0.61	187	12	20	0.063	38	0.110	61	59
Mean	0.422	1.5	8	752	0.56	128	16	15	0.071	41	0.126	68	70
SE	0.197	0.2	1	36	0.04	11	1	1	0.008	3	0.008	5	3
					Regional Stations R-1 to R-7 and PJ-1								
Mean	0.080	1.7	6	688	0.72	129	17	13	0.063	41	0.117	78	67
SE	0.012	0.2	1	33	0.14	23	3	1	0.006	3	0.021	6	6

sediment traps decreased from 23.7 μg/g (SE = 2.78) to 8.96 μg/g (SE = 1.49) during the same periods. Similarly, mean concentrations of TPAHs declined in sediment traps from 0.98 μg/g (SE = 0.09, October 1986–May 1987) to 0.06 μg/g (SE = 0.01 January 1988–May 1988).

In summary, the data on Ba indicate strongly that reef stations adjacent to Platform Hidalgo were influenced by drilling wastes because (1) we found increasing concentrations of sediment Ba at Hidalgo stations from 1986 to 1987, relative to background levels and relative to concentrations of Ba at regional stations, and (2) we found an increase of Ba in sediment traps between predrilling samples and samples collected after drilling began at Hidalgo. The increase in sediment Ba that occurred before drilling began at Hidalgo suggests another source of drilling wastes. The most likely origin of the Ba was Platform Harvest, located 3.0 km southeast of Hidalgo (Figure 2); 19 wells were drilled at Harvest from October 1986 to May 1988, after which drilling was temporarily suspended. Another potential source of the Ba was Platform Hermosa, 7.3 km southeast of Hidalgo, where 10 wells were drilled from January 1987 to May 1988, after which drilling was temporarily suspended. Currents in the Point Conception region were highly variable, but the mean flow at Hidalgo was parallel to the isobaths, shifting from a predominately northward flow to a southward flow on a seasonal basis (Savoie et al. 1988). Hence, the potential exists for the transport of fine drilling particulates to Hidalgo from the more southern platforms. Other studies have shown that drilling muds may be transported long distances, as indicated by increased sediment Ba. For example, Jenkins et al. (1989) found Ba concentrations in excess of 34,000 μg/g adjacent to an exploratory well in the Santa Barbara Channel; Ba levels diminished to background concentrations (700–900 μg/g) at distances of 500 to 2,500 m from the well.

Interpretation of the variable hydrocarbon data must await future analyses. The TALKs and the TPAHs in sediments and sediment traps are potentially influenced by both drilling related and nondrilling related sources. SAIC (1985) reported trace to moderate levels of petroleum contamination for the Santa Maria Basin before development drilling in the region had begun. Natural seeps, atmospheric deposition, and tanker operations may be significant sources of hydrocarbons, while industrial and municipal wastewaters and river discharges are probably minor contaminant sources to the area. Levels of TALKs may be influenced, in part, by seasonal inputs of plant material or plankton. Additional data on the chemistry of platform discharges, sediments, and sediment trap samples (including component PAHs and ALKs and various hydrocarbon diagnostic ratios) have recently been reported by Steinhauer and Imamura (1990).

Benthic Invertebrates

Hard substrate epifauna was dominated by a Komokoiacea (foraminifera)–Hydroida turf (Table 4), followed by unidentified brittle stars (oph-

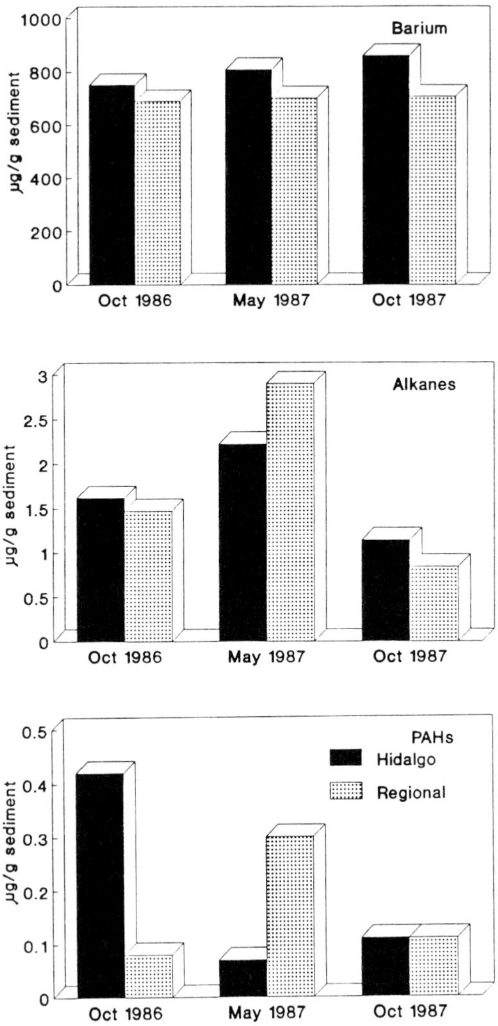

FIGURE 3.—Sediment concentrations of barium, alkanes (TALKs), and TPAHs at Platform Hidalgo reef sites and regional soft substrate stations (including R-1 to R-9 and PJ-1).

iuroids), feather star (*Florometra serratissima*), pink anemone (*Amphianthus californicus*), plumose anemone (*Metridium senile*), white coral (*Lophelia californica*), cup corals (*Paracyathus stearnsii* and *Desmophyllum crista-galli*), sea fan (*Lophogorgia chilensis*), unidentified feather duster worm (Sabellidae), tunicate (*Pyura haustor*), unidentified crab (Galatheidae), and lamp shell (*Terebratulina* sp[p]). Mean densities of the dominant reef organisms varied little during the three predrilling sampling periods (Figure 4). Data on percent cover suggested similar stability.

The Bray–Curtis (unweighted pair-group) index (Bray and Curtis 1957; Swartz 1978), which compares species abundances among sampling sites, was used as a measure of similarity among reef stations. Two station-groups were identified for the October 1987 sampling period (Figure 5). One cluster included shallow to intermediate depth, low-relief stations (F, J, E, N, I, and U), and the second cluster included intermediate depth to deep, low- and high-relief stations (W, R, and K). The Bray–Curtis index for the October 1986 and May 1987 sampling periods generally showed similar station clusters, except that station N (intermediate depth, low relief) was grouped with the intermediate depth to deep stations (W, R, and K).

Organisms that showed a significant positive correlation (based on direct counts with $P \leq 0.05$) with reef depth included pink anemone, white coral, and cup coral (*Desmophyllum crista-galli*), feather duster worm, mottled anemone (*Stomphia didemon*), crab, and (based on percent cover) an unidentified shelf sponge. Similarly, two cup corals (*Caryophyllia* sp[p]. and *Paracyathus stearnsii*) were negatively correlated with depth.

Overall, benthic community data collected during the 1-year predrilling sampling period indicated seasonally stable invertebrate reef communities around Hidalgo; depth was a major factor influencing the distribution of the dominant taxa. The height of the rocky relief may also govern distribution, but the separate effects of reef depth versus reef height were difficult to evaluate because reefs with greater than 1 m relief were found only at the intermediate depth and deep stations.

The apparent increase of Ba in sediments and sediment traps indicates that the reefs adjacent to Hidalgo were exposed to an influx of drilling materials. Continued monitoring of the communities through 1990 should provide more definitive information on how the reef communities respond to additional low-level, periodic doses of drilling wastes. In addition to community parameters, CAMP investigators are evaluating potential changes in the growth, reproductive potential, and life history of several of the dominant taxa that are expected to be most sensitive to the drilling wastes. Lissner (1988) summarized information on the feeding modes, growth rates, longevity, reproduction, dispersal, recruitment, and general biology of deepwater reef communities off California. Lissner (1988) suggested that deepwater corals may be particularly sensitive to fine sediments such as those from drilling fluids. Filter and suspension feeders may ingest suspended

TABLE 4.—Mean percent cover and density (normalized counts) of the most abundant reef taxa in high- and low-relief habitat during October 1987 (unid. = unidentified).

High-relief habitat			Low-relief habitat		
Taxon	Mean	SE	Taxon	Mean	SE
Percent cover					
Komokoacea–Hydroida mat	64.3	1.2	Komokoacea–Hydroida mat	79.3	0.6
Sea anemone (*Amphianthus californicus*)	5.6	0.4	Brittle stars, unid.	5.5	0.3
Brittle stars, unid.	2.7	0.2	Feather star	2.8	0.3
Cup coral (*Desmophyllum crista-galli*)	2.4	0.3	Cup coral (*Paracyathus stearnsii*)	1.6	0.1
Feather duster worm, unid.	2.0	0.2	Feather duster worm, unid.	1.0	0.1
Coral (*Lophelia californica*)	2.0	0.5	Cup coral (*Caryophyllia* sp[p].)	0.9	<0.1
Cat shark egg cases	1.4	0.3	Brittle star (*Ophicantha diplasia*)	0.7	<0.1
Sponge, white encrusting	1.4	0.1	Plumose anemone	0.6	0.1
Sponge, shelf	1.0	0.2	Sponge, white encrusting	0.6	<0.1
Feather star	0.5	0.2	Sea squirt	0.4	<0.1
Density					
Sea anemone (*Amphianthus californicus*)	50.8	2.8	Brittle stars, unid.	56.8	1.9
Brittle stars, unid.	44.4	2.1	Cup coral (*Paracyathus stearnsii*)	52.6	2.8
Crab (Galatheidae)	20.9	0.9	Cup coral (*Caryophyllia* sp[p].)	17.4	0.7
Feather duster worm	14.7	0.8	Cup coral (*Balanophyllia elegans*)	9.5	1.0
Cup coral (*Desmophyllum crista-galli*)	14.3	1.3	Feather duster worm, unid.	7.9	0.7
Cat shark egg cases	11.6	3.0	Lamp shell	6.3	0.7
Sea fan	5.4	0.5	Crab (Galatheidae)	4.4	0.3
Tunicate	2.9	0.4	Tunicate	2.4	0.2
Cup coral (*Carophyllia* sp[p].)	2.2	0.2	Cat shark egg cases	1.6	0.3
Rockfishes	0.7	0.1	Rockfishes	0.6	0.1

particulates with adsorbed metals and hydrocarbons. Hence, the cup corals (*Paracyathus stearnsii* and *Caryophyllia* spp.) that were characteristic of low-relief reefs (Table 4), and white coral and cup coral (*Desmophyllum crista-galli*) that were relatively abundant on high-relief reefs, may be appropriate taxa for the monitoring of long-term effects.

Preliminary data on tissue samples from the sea hare (*Pleurobranchaea californica*), captured in animal traps around Hidalgo, indicated that levels of Ba in its soft tissues increased from 3–4 μg/g in October 1986 to 6–16 μg/g in October 1987 (E. Crecelius, Battelle Ocean Sciences, personal communication). The two- to fivefold increase in tissue Ba correlated with a 14% increase in sediment Ba during the same period. Jenkins et al. (1989) found significant increases in Ba in the tissues of stout cyclocardia (a clam, *Cyclocardia ventricosa*) and ice cream cone polychaete (*Pectinaria californiensis*) downcurrent of an exploratory well in the Santa Barbara Channel. Jenkins et al. (1989) examined the tissue and subcellular distribution of the accumulated Ba and concluded that most of the Ba was compartmentalized by the animals as insoluble barium sulfate in granular pellets, and less than 0.1% of the Ba accumulated in the subcellular cytosol. Neff et al. (1989b) found that some marine animals exposed to drilling fluids accumulated metals, but that most of the metal actually remained in the animal's gut or gills as unassimilated barite particles. The study by Jenkins et al. (1989) and others (reviewed by Neff 1987) indicated that Ba and other metals in drilling fluids were generally nontoxic because of their limited bioavailability.

To date, CAMP has found consistently low levels of TPAHs (average < 0.005 μg/g) in the tissues of sea hare. Studies will continue of the bioaccumulation of chemical contaminants in sea hare and other species such as crabs (*Cancer* spp.) captured by the animal traps. In addition, the manipulator arm and suction pump on the ROV will be used to collect the cup coral *Paracyathus stearnsii* and California sea cucumber (*Parastichopus californicus*), both abundant in the study area. The latter species is of particular interest because it can consume large amounts of contaminated surficial sediment during normal feeding.

Fishes

Fishes occurred in low numbers and in only 9% of the photoquadrats during the October 1986 sampling period. Taxa included several rockfishes (*Sebastes* spp.), hagfish (*Eptatretus stoutii*), flounders (Pleuronectidae), combfish (*Zaniolepis* spp.), lingcod (*Ophiodon elongatus*), sanddabs (*Citharichthys* spp.), skates (*Raja* spp.), plainfin midshipman (*Porichthys notatus*), cusk eels

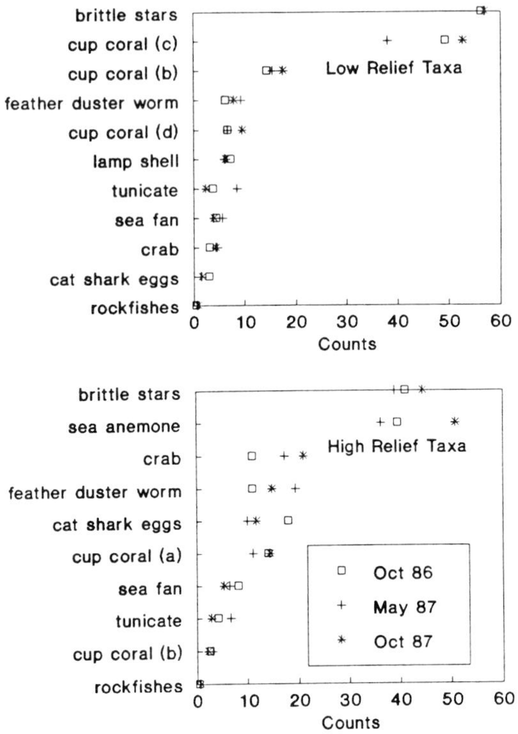

FIGURE 4.—Relative abundance of hard substrate taxa (normalized by area) identified from photoquadrats on low-relief (0.2–0.5 m height) and high-relief (>1 m height) features at Platform Hidalgo reef stations. Cup corals a, b, c, and d refer to *Desmophyllum crista-galli*, *Caryophyllia* sp(p)., *Paracyathus stearnsii*, and *Balanophyllia elegans*, respectively.

(Ophidiidae), and gobies (Gobiidae). Large numbers of cat shark (Scylorhinidae) egg cases were observed on high- and low-relief substrates (Table 4). There was no consistent trend in total fish abundance by reef depth or relief height, although certain taxa such as lingcod were observed only at deeper stations (Figure 6).

As a supplement to the limited photoquadrat data on fishes occurring on the reefs near Hidalgo, the relative abundance of reef-associated rockfishes as sampled by (1) direct observations and simultaneous video and still photographic surveys of deepwater reefs (55–241 m) from a manned submersible in the Santa Maria Basin and western Santa Barbara Channel (SAIC 1985), and (2) catches made by commercial hook-and-line (setline) gear over deepwater reefs (82–183 m) on the central California coast just north of the Santa Maria Basin (Pearson et al. 1987) are summarized in Table 5. Clearly, these data are not directly

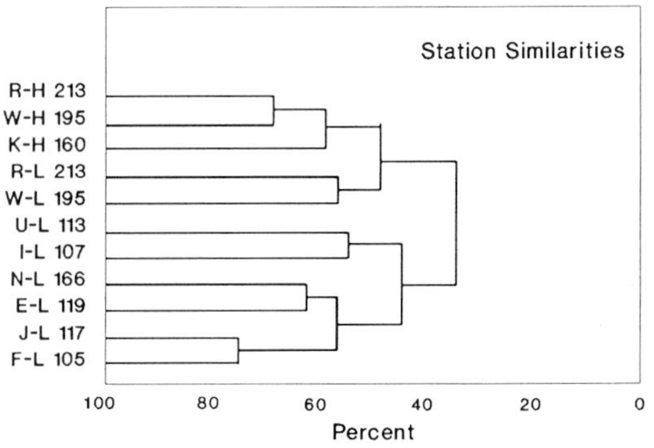

FIGURE 5.—Bray-Curtis index of similarities in species abundances among Platform Hidalgo reef stations for samples collected during October 1987. Stations (R, W, K, U, L, I, N, E, J, and F) include designations for low (L) or high (H) relief and for station depth (m).

comparable because sampling was conducted in different areas, at different times, and with different sampling goals. Nevertheless, the variability in the rankings of the rockfish taxa is, in part, a reflection of sampling bias for each of the methods. Such sampling bias is one of the problems to be overcome in the design of meaningful studies on the effects of oil and gas platforms to fishes and fisheries.

The original rationale for CAMP studies on benthic invertebrate communities, exclusive of the nekton, was based on the following: (1) The benthos is a sink for the accumulation of particulate drilling wastes. Although pelagic waters may receive periodic pulses of contaminants, such disturbances are short-lived. Concentrations of metals and hydrocarbons are diluted to back-

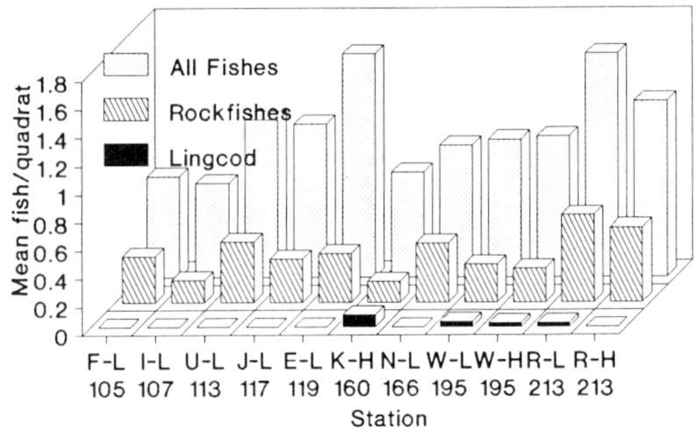

FIGURE 6.—Relative abundance of total fishes, rockfishes, and lingcod (normalized by area) identified from photoquadrats at Platform Hidalgo reef stations. Stations (F, I, U, J, E, K, N, W, and R) include designations for low (L) or high (H) relief and for station depth (m).

TABLE 5.—Most frequently occurring rockfishes associated with hard substrate features on the central California coast as sampled via CAMP ROV photoquadrats, direct observations and video from a manned submersible (SAIC 1985), and commercial hook-and-line gear (Pearson et al. 1987). Taxa are listed by relative abundance under each category.

Common name	*Sebastes* species
ROV photoquadrats	
Greenstriped rockfish	*elongatus*
Calico rockfish	*dalli*
Honeycomb rockfish	*umbrosus*
Greenspotted rockfish	*chlorostictus*
Greenblotched rockfish	*rosenblatti*
Rosy rockfish	*rosaceus*
Submersible	
Olive or yellowtail rockfish	*serranoides* or *flavidus*
Flag rockfish	*rubrivinctus*
Rosy rockfish	*rosaceus*
Greenstriped rockfish	*elongatus*
Starry rockfish	*constellatus*
Blue rockfish	*mystinus*
Copper rockfish	*caurinus*
Chilipepper	*goodei*
Squarespot rockfish	*hopkinsi*
Commercial hook-and-line gear	
Chilipepper	*goodei*
Vermilion rockfish	*miniatus*
Boccacio	*paucispinis*
Yellowtail rockfish	*flavidus*
Greenspotted rockfish	*chlorostictus*
Canary rockfish	*pinniger*
Yelloweye rockfish	*ruberrimus*
Rosy rockfish	*rosaceus*
Starry rockfish	*constellatus*
Bank rockfish	*rufus*
Greenstriped rockfish	*elongatus*
Widow rockfish	*entomelas*
Copper rockfish	*caurinus*
Cowcod rockfish	*levis*
Speckled rockfish	*ovalis*
Flag rockfish	*rubrivinctus*
Rosethorn rockfish	*helvomaculatus*
Greenblotched rockfish	*rosenblatti*
Olive rockfish	*serranoides*

ground levels quickly, often within a few tens of meters of the discharge pipe (Neff 1987). (2) The relative immobility of most benthic invertebrates makes them more susceptible to long-term exposure to contaminated sediments compared with free-ranging animals or animals suspended in the water column. (3) Collecting quantitative data on free-ranging animals is problematic compared with more precise techniques for collecting data on stationary invertebrates associated with soft substrates (with box corers) and hard substrates (using photoquadrats). Even if an unbiased sampler was available, the mobility of fishes and even some reef invertebrates such as the feather star and sea anemones makes the experimental approach to impact studies more difficult. For example, conclusions regarding the uptake (or lack of uptake) of contaminants from a point–source discharge are tenuous if the distributional history of the species under study is unknown.

Nevertheless, the concern persists that commercially valuable finfish may be at risk from platform discharges. To help resolve this controversy, beginning in 1990, MMS supplemented CAMP with a study at Hidalgo entitled "Effects of an OCS Oil and Gas Production Platform on Rocky Reef Fishes and Fisheries." The planned 3-year project encompasses the major themes of (1) spatial variation of deepwater reef fishes in the area around Hidalgo, (2) sublethal effects of platform discharges on selected reef fishes, and (3) ecological relationships of fish guilds associated with the platform and adjacent deep water reefs. These themes, as well as potential, state-of-the-art sampling and analytical techniques, were addressed at a workshop in December 1988. Recommendations from the workshop (MMS 1988) were used to help plan the study.

Although drilling at Hidalgo has begun and effects on fishes may have already occurred, the area was selected for studies on fishes because of its physiographic setting, with isolated reefs at varying distances from the platform; past and continuing commercial fishing activity on the reefs; the availability of pre- and postspud (first drilling) environmental data from the area; and the lack of alternative sites where development drilling was likely in the foreseeable future.

The study of platform effects on reef-associated fishes was planned, in part, because fishes are known to aggregate on or over rocky outcroppings and topographic features (Gunderson and Sample 1980). Commercial hook-and-line and setline and stationary gill-net fisheries take advantage of such behavior at the Hidalgo reefs and elsewhere. Such behavior, that is, a limited home range, provides the opportunity for tenable hypothesis testing of platform effects on reef-associated species (at least relative to fishes that predominate over topographically featureless soft substrate). For example, yellowtail rockfish (*Sebastes flavidus*) returned to a specific topographic feature after being displaced as much as 22 km (Carlson and Haight 1972). Tagged olive rockfish (*S. serranoides*) exhibited relatively little movement among adjacent reefs off the coast of southern California (Love 1980).

To overcome the bias of any one sampling technique, studies of fish distribution among the reefs at Hidalgo use a combination of (1) high-quality video from an ROV, (2) fixed and mobile hydroacoustic equipment, and (3) commercial hook-and-line and setline gear. Movements of fishes among the reefs are assessed using "breakaway" hook tags. These methods test the hypothesis that the distribution and abundance of selected fish taxa among the reefs are not related to the proximity of the platform or to platform discharges, or both.

Studies of sublethal impacts to Hidalgo reef fishes involve the uptake, accumulation, and metabolism of metals and hydrocarbons (and hydrocarbon metabolites) that are associated with platform discharges. Because studies have shown that metals are generally not bioavailable and do not pose a threat to fishes (Neff et al. 1989a), the investigations concentrate on hydrocarbons. Seasonal collections of fishes at experimental reefs (i.e., near the platform or influenced by platform discharges) and control reefs are made to evaluate potential effects to biochemical pathways (e.g., elevated levels of micronuclei in the blood) and cells (e.g., genetic lesions). Life history parameters, including recruitment variability among the reefs, are also considered.

The general ecology of fishes at the platform structure (i.e., an artificial reef) and the natural reefs is being assessed. The structure and functional relationships of fish guilds at the platform are compared and contrasted with natural reef-associated assemblages. Foods–feeding and predator–prey relationships are examined as are movements of fishes between the platform and the reefs.

In conclusion, deepwater rocky habitats off California are of special interest to resource agencies. The reefs are colonized by unusual assemblages of animals that are unable to grow on soft substrates. Many of the invertebrate taxa inhabiting the reefs are slow-growing, long-lived, and thought to be intolerant of environmental change. Few species that inhabit the reefs have been studied (Lissner 1988). Aggregations of commercially important fishes are a common feature of offshore rocky outcroppings.

Decisions about permitting drilling operations adjacent to deepwater reef communities are hampered by a poor understanding of the responses of the reef-associated animals to drilling-related discharges. There is little evidence of acute toxicity from the discharge of water-based drilling muds from single, exploratory wells. Poorly understood, however, is the potential for sublethal effects from cumulative, low-level doses of toxicants, especially PAHs, from multi-well development and production platforms such as Hidalgo (Boesch and Rabalais 1987). Questions persist about the possible burial or smothering of sessile animals by cuttings discharges and the clogging of sensitive respiratory or feeding structures by fine-grained particulates in drilling fluids. Invertebrate larvae that normally settle and grow on hard substrate might be affected by increased sedimentation rates and changes in substrate texture or chemistry. Changes in bottom substrate may not affect adult animals directly, but benthic communities may change through time due to the settlement of a different assemblage of planktonic young, better adapted to colonize the altered bottom conditions. Although past studies have detected evidence of drilling-fluid contamination (by increased Ba concentrations) in sediments several kilometers downcurrent of drilling discharges, concomitant changes in benthic communities have been limited to distances of several hundred meters (Neff 1987). Effects on benthic communities may be primarily due to changes in the physical nature of the bottom, resulting from the accumulation of both drill cuttings and epifauna dislodged from the platform's legs and supports, rather than to any chemical or toxicological effect. To date, our studies of invertebrate reef communities near Hidalgo show no evidence of community change or sublethal effects attributable to platform activities. We note, however, that the detection of nearfield impacts at Hidalgo is limited by the location of the closest reef, which is 500 m from the platform.

Decision makers are also confronted with issues relating to the physical presence of offshore platforms, both as artificial reefs and in space-use conflicts with fishermen. Although the aggregation of fishes around offshore platforms is well known, the presumed benefits are not without controversy (Buckley 1989; Polovina 1989). If a platform simply aggregates fishes without increasing overall fish production, then the redistribution of the fishes may result in over-harvesting if fishing is enhanced at the platform and reduced availability of fishes at nearby natural reefs. Fishermen continue to target the rocky outcrops near Platform Hidalgo. Indeed, fishermen objected emphatically to the placement of in situ oceanographic instruments by CAMP investigators near the Hidalgo reefs because the samplers threatened to interfere with fishing lines and nets. Apparently, the reefs still yield profitable catches despite past and ongoing oil and gas activities in the area.

The installation of Platform Hidalgo resulted in a temporary fishing exclusion zone that extended about 2 km around the platform site. In addition, pipeline barges and support vessels were used to lay pipelines from Hidalgo to shore-based processing facilities and resulted in the temporary exclusion of fishermen from a 2 km-wide corridor. Construction lasted several months.

The displacement of fishermen from their normal fishing grounds is yet another factor (see Fusaro 1991, this volume) in the complex, environmental policy equation that must weigh the benefits and risks of offshore oil and gas development. We believe that our studies (planned to continue through at least 1993) of the long-term dynamics of benthic communities in the Santa Maria Basin, including studies of the role of oil and gas platforms as sources of environmental contaminants and as habitats for fishes, will provide definitive information that will be applied to these important decisions.

References

Boehm, P., M. Steinhauer, and A. Uhler. 1988. Chemical analysis of hydrocarbons and related materials in sediment, pore water, and tissues. Pages 6.1–6.120 *in* J. Hyland and J. Neff (1988).

Boesch, D. F., and N. N. Rabalais, editors. 1987. The long-term effects of offshore oil and gas development: an assessment and a research strategy. Elsevier, New York.

Bray, J. R., and J. T. Curtis. 1957. An ordination of the upland forest communities of southern Wisconsin. Ecological Monographs 27:325–349.

Brewer, G. D., F. Piltz, and J. Hyland. 1987. Monitoring changes in benthic communities adjacent to OCS oil production platforms off California. Pages 1593–1597 *in* Oceans 87, proceedings, volume 5. Coastal and estuarine pollution. Marine Technology Society, Washington, D.C.

Buckley, R. M. 1989. Habitat alterations as a basis for enhancing marine fisheries. California Cooperative Oceanic Fisheries Investigations Reports 30:40–45.

Capuzzo, J. D. 1987. The biological effects of petroleum hydrocarbons in the sea: assessments from experimental microcosms. Pages 343–410 *in* Boesch and Rabalais (1987).

Carlson, H. R., and R. E. Haight. 1972. Evidence for a home site and homing of adult yellowtail rockfish, *Sebastes flavidus*. Journal of the Fisheries Research Board of Canada 29:1011–1014.

Carney, R. S. 1987. A review of study designs for the detection of long-term environmental effects of offshore petroleum activities. Pages 651–696 *in* Boesch and Rabalais (1987).

Crecelius, E. 1988. Chemical analysis of trace metals in sediments, pore waters, and tissues. Pages 7.1–7.37 *in* J. Hyland and J. Neff (1988).

Fusaro, C. 1991. Improving communication between the oil and fishing industries. American Fisheries Society Symposium 11:18–21.

Green, R. H. 1979. Sampling design and statistical methods for environmental biologists. Wiley, New York.

Gunderson, D. R., and T. M. Sample. 1980. Distribution and abundance of rockfish off Washington, Oregon and California during 1977. U.S. National Marine Fisheries Service Marine Fisheries Review 42(3–4): 2–16.

Hardin, D., R. Spies, and T. Parr. 1988. Hard bottom epifaunal assemblages. Pages 5.1–5.26 *in* J. Hyland and J. Neff (1988).

Hyland, J., and J. Neff, editors. 1988. California OCS Phase II Monitoring Program: year one annual report. U.S. Department of the Interior, Minerals Management Service, Pacific OCS Region, OCS Study MMS 87-0115, Los Angeles.

Jenkins, K. D., S. Howe, B. M. Sanders, and C. Norwood. 1989. Sediment deposition, biological accumulation, and subcellular distribution of barium following the drilling of an exploratory well. Pages 587–608 *in* F. R. Engelhardt, J. P. Ray, and A. H. Gillam, editors. Drilling wastes. Elsevier, New York.

Lissner, A., editor. 1988. Review of recovery and recolonization of hard substrate communities of the Outer Continental Shelf. U.S. Department of the Interior, Minerals Management Service, Pacific OCS Region, OCS Study MMS 88-0034, Los Angeles.

Love, M. S. 1980. Isolation of olive rockfish, *Sebastes serranoides* populations off southern California. U.S. National Marine Fishery Service Fishery Bulletin 77:975–983.

MMS (Minerals Management Service). 1982. Recommendations for a Pacific Outer Continental Shelf monitoring program for OCS oil and gas development activities. Summary and recommendations of a conference, March 30 and March 31, 1982. U.S. Department of the Interior, MMS, Pacific OCS Region, Administrative Report, Los Angeles.

MMS (Minerals Management Service). 1988. Effects of an OCS oil and gas production platform on rocky reef fishes and fisheries. Proceedings report of a public meeting–workshop, December 14, 1988, to U.S. Department of the Interior, MMS, Pacific OCS Region, Contract 10217, Los Angeles.

National Research Council. 1983. Drilling discharges in the marine environment. National Academy Press, Washington, D.C.

National Research Council. 1985. Oil in the sea. Inputs, fates and effects. National Academy Press, Washington, D.C.

Neff, J. M. 1987. Biological effects of drilling fluids, drill cuttings and produced waters. Pages 469–538 *in* Boesch and Rabalais (1987).

Neff, J. M., R. G. Breteler, and R. S. Carr. 1989a. Bioaccumulation, food chain transfer, and biological effects of barium and chromium from drilling muds by flounder, *Pseudopleuronectes americanus*, and lobster, *Homarus americanus*. Pages 461–479 *in* F. R. Engelhardt, J. P. Ray, and A. H. Gillam, editors. Drilling wastes. Elsevier, New York.

Neff, J. M., R. E. Hillman, and C. D. Hunt. 1989b. Bioaccumulation of trace metals from drilling mud barite by benthic marine animals. Pages 461–479 *in* F. R. Engelhardt, J. P. Ray, and A. H. Gillam, editors. Drilling wastes. Elsevier, New York.

Neff, J. M., N. N. Rabalais, and D. F. Boesch. 1987. Offshore oil and gas activities potentially causing long-term environmental effects. Pages 149–174 *in* Boesch and Rabalais (1987).

Polovina, J. L. 1989. Artificial reefs: nothing more than benthic fish aggregators. California Cooperative Oceanic Fisheries Investigations, Reports 30:37–39.

Pearson, W. H., J. R. Skalski, and C. I. Malme. 1987. Effects of sounds from a geophysical survey device on fishing success. U.S. Department of the Interior, Minerals Management Service, Pacific OCS Region, OCS Study 87-0020, Los Angeles.

Piltz, F. 1986. Monitoring long-term changes in biological communities near oil and gas production platforms. Pages 856–861 *in* Oceans 86, Proceedings, volume 3. Monitoring strategies symposium. Institute of Electrical and Electronics Engineers, New York, and Marine Technology Society, Washington, D.C.

SAIC (Science Applications International Corporation). 1985. Assessment of long-term changes in biological communities in the Santa Maria Basin and western Santa Barbara Channel—Phase I. Volume II, synthesis of findings. Report to U.S. Department of the Interior, Minerals Management Service, Pacific OCS Region, Contract 30332, Los Angeles.

Savoie, M. A., P. J. Kinney, and R. L. Bernstein. 1988. Physical oceanography. Pages 8.1–8.69 *in* J. Hyland and J. Neff (1988).

Steinhauer, M., and E. Imamura, editors. 1990. California OCS Phase II Monitoring Program: year three annual report. U.S. Department of the Interior, Minerals Management Service, Pacific OCS Region, OCS Study MMS 90-0055, Los Angeles.

Swartz, R. C. 1978. Techniques for sampling and analyzing the marine macrobenthos. National Technical Information Service, PB-281 631, Springfield, Virginia.

American Fisheries Society Symposium 11:39–41, 1991

Introduction to Papers on the Beaufort Sea

The construction of causeways in the nearshore Beaufort Sea and their potential effects on amphidromous fish is an issue that has been discussed extensively during the past 3 years. Concerns have been expressed that causeways have the potential for affecting fish populations by (1) physically blocking or impeding migrations or (2) altering the physical environment and affecting migrations, feeding, and rearing. At the present time, three causeways, West Dock, East Dock, and Endicott, have been built in the Prudhoe Bay–Sagavanirktok River area. Additional causeways for development of nearshore oil reservoirs have been proposed, and the potential development of small causeways to support offshore oil and gas operations has been analyzed by Minerals Management Service (MMS) in each environmental impact statement (EIS) for the Beaufort Sea.

To best understand the potential effects of offshore oil and gas development on other uses of marine and estuarine areas—such as subsistence fishing—many aspects of the marine environment need to be investigated and coalesced into EIS analyses. Developing a multiple-use principle for making our nation's marine waters available to all segments of society requires using the best available information to assess the effects of one activity on other activities. Therefore, scientific information on many facets of the physical and biological environments of the Beaufort Sea is used to evaluate and balance the progression of human activities on the coastal North Slope of Alaska.

To determine the potential effects of fixed structures within the nearshore zone of the Beaufort Sea, many aspects of the biological environment need to be examined. Temperature and salinity tolerances, growth rates, energy assimilation, feeding and migration patterns, and reproductive capacities are all factors that need to be considered when assessing the potential risks to North Slope fish populations from nearshore structures.

Environment

The nearshore Beaufort Sea is a shallow-water coastal system with a rugged, ice-scoured coastline. The land and nearshore sediments have an underlying permafrost that provides some stability to the overlying sediments. The coast is incised by numerous small streams and rivers. Two of the larger rivers along the Alaskan Beaufort Sea are the Sagavanirktok and Colville rivers. The largest river in the region is the Mackenzie River in Canada (Figure 1).

The North Slope Beaufort Sea area is within the Arctic Circle and icebound for most of the year. Extended ice cover can be found in the nearshore Beaufort Sea between October and June. However, during July, August, and September, the sea is ice free near shore, and fish can be found in the nearshore coastal zone. The summer winds are typically from the easterly direction, but the winds can be variable and quick changing.

Because of the shallow water near shore and low tidal amplitude (<30 cm) of the area, the movement of water in the coastal Beaufort Sea is predominantly wind driven. Ice breakup typically occurs in late June or early July, and freshwater input to the coastal system from snow and ice melt in the river drainages is usually highest in late June. Easterly winds during the ice-free times set up a wind-driven movement of nearshore coastal water in an east-to-west direction. The input of fresh water from the coastal river systems decreases the salinity of this nearshore coastal band, and, because of thermal input to the shallow-water system, temperatures are elevated. In effect, under the right wind and freshwater input conditions, this nearshore coastal band may be compared to a river that separates the coastline and the colder, higher-salinity water off shore.

Biology

The main species of concern are the amphidromous fishes along the North Slope coast that are the basis for a small, but important, subsistence fishery. Most of the species, such as the least cisco *Coregonus sardinella*, broad whitefish *Coregonus nasus*, and Arctic char *Salvelinus alpinus* have reproducing populations along the Alaskan coast. However, Arctic ciscoes *Coregonus autumnalis* in the Alaskan Beaufort Sea are thought to originate from the Mackenzie River in Canada's Northwest Territories. No reproducing population of Arctic ciscoes has been identified from the Alaskan North Slope, and the population of Arctic ciscoes along the Alaskan Beaufort Sea

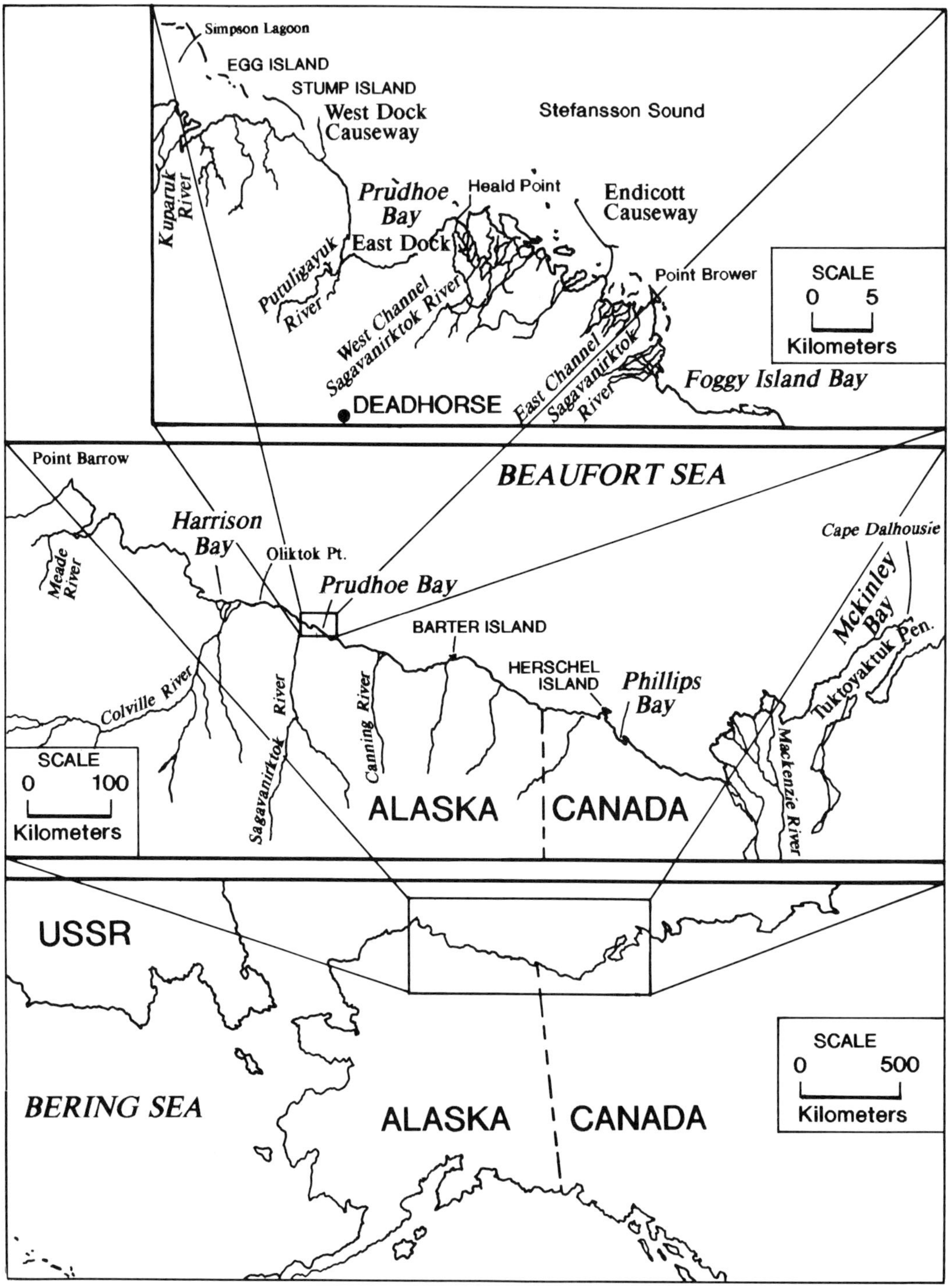

FIGURE 1.—Causeways and geographical features of the nearshore Beaufort Sea. Figure provided by B. J. Gallaway, LGL Ecological Research Associates.

coast is apparently replenished annually by immigration of young-of-the-year from the Canadian Mackenzie River. The proportion of the Mackenzie River young-of-the-year that migrates into Alaskan waters, however, is unknown.

The nearshore coastal band of low-salinity, higher-temperature water is an important habitat for all amphidromous fish along the Alaskan Beaufort Sea coast. This brackish band provides the major feeding habitat during the ice-free months and is an important migration path for young Arctic ciscoes from the Mackenzie River. The popular hypothesis is that the usually east-to-west movement of the nearshore coastal band provides a wind-aided migration path for young Arctic ciscoes to Alaskan rivers. Interannual variability of young-of-the-year Arctic ciscoes that undergo the wind-aided migration to the Alaskan coastal waters is very high. Therefore, the representative size of each year-class found along the Alaskan coast (approximately ages 0–8) also varies greatly from year to year.

It is thought that mature Arctic ciscoes use this nearshore band for returning to the Mackenzie River to spawn. During the first year in which they spawn, adults from the Alaska coast may migrate eastward to the Mackenzie River from as far away as the Colville River, beginning their movements early in the open-water season. However, most studies along the Beaufort Sea coast have found few mature Arctic ciscoes during the summer.

Concerns

Alteration of the physical oceanography regime of the nearshore band could be detrimental to Arctic cisco and other amphidromous species. Solid-fill structures could deflect or create discontinuities in the band of relatively warm, brackish water that occurs in the open-water season along the Alaskan Beaufort Sea coast. The deflection or discontinuity could affect migrations, the availability of feeding habitat, overwintering success, and fecundity of Arctic cisco and other amphidromous fish species. Although individual nearshore structures may not have an appreciable effect on fish populations, multiple structures—or structures not effectively designed for their location—may have a cumulative or proportionally greater effect on local fish populations. The migrations that Arctic ciscoes are thought to undergo probably make them more susceptible to the effects of causeway construction than fish that have local spawning populations.

To best address the concerns that have been raised, many of the ecological factors that determine amphidromous fish population dynamics have been investigated. The breadth of scientific research that has been developed ensures that the best available scientific information on these species is used in balancing multiple uses of the nearshore Beaufort Sea waters. To this end, the following research topics have contributed valuable insight into the effects of human activities on the nearshore Beaufort Sea environment.

American Fisheries Society Symposium 11:42–80, 1991

The Endicott Development Project—Preliminary Assessment of Impacts from the First Major Offshore Oil Development in the Alaskan Arctic

BENNY J. GALLAWAY

LGL Ecological Research Associates, Inc.
1410 Cavitt Street, Bryan, Texas 77801, USA

WILLIAM J. GAZEY

W. J. Gazey Research
1214 Camas Court, Victoria, British Columbia V8X 4R1, Canada

JOSEPH M. COLONELL

Woodward-Clyde Consultants
701 Sesame Street, Anchorage, Alaska 99503, USA

ALAN W. NIEDORODA

Hunter Services, Inc.
Post Office Box 1703, Gainesville, Florida 32602, USA

CHRISTOPHER J. HERLUGSON

BP Exploration (Alaska), Inc.
900 East Benson Boulevard, Anchorage, Alaska 99519, USA

Abstract.—The Endicott Causeway was constructed in 1985 to develop an offshore oil field in the nearshore Beaufort Sea off the mouth of the Sagavanirktok River near Prudhoe Bay, Alaska. Based upon experiences at the West Dock Causeway, located across Prudhoe Bay, 25 km west of the Sagavanirktok River, there was concern that the oceanographic changes induced by the new causeway would result in significant adverse environmental effects on anadromous fish and their habitat. The major concerns were that (1) the causeway would cause cold (<5°C), marine (≥26‰) bottom water to be advected to the shoreline along the west side of the causeway on east winds (east winds occur 60% of the time in summer), and (2) if these sharp hydrographic changes occurred, they could lead to increased mortality of fish. These changes could increase mortality by (1) blocking the migration of young-of-the-year Arctic ciscoes *Coregonus autumnalis* from Canada to the Colville River, and (2) by adversely affecting the growth and general condition of the fish. An environmental impact statement (EIS) was prepared for the project and concluded that such effects were unlikely. The EIS conclusions were by no means universally accepted, leading to the largest monitoring program for a single development in North America. Oceanographic and biological samples have been taken daily at up to 30 net sites and weekly at up to 28 oceanographic sampling transects around the causeway each summer, beginning in 1985. This paper reports the results of our reevaluation and analyses of the 1985–1987 data, originally collected by the Envirosphere Company, pertinent to the major issues. The 1985–1987 data show that cold, marine bottom waters have not been advected along the west side of the causeway to the shore. There is, however, a wake effect on the west side of the tip of the causeway that causes vertical transport of bottom water to the surface, affecting an area of about 500 hectares. In 1985–1987, marine bottom water was present in the affected area about 2–6% of the time. The effects of the causeway on the movements of young-of-the-year Arctic ciscoes were estimated from the temporal and spatial distribution of catches and related to synoptic meteorological conditions. The migrations appear to be a result of wind (current)-governed transport, and the fish have moved through and around the Endicott and West Dock causeways with little impedance. Analysis of growth rates and condition levels show no significant reductions; if anything, an increase occurred during 1985–1987. The increases appear to be a response to the progressively longer periods of high temperature that occurred in each successive year. Our conclusions represent preliminary judgements based upon only 3 years of data for fish that have a minimum of 7 years between generations. More data are needed to evaluate these and other potential effects, and such data are being collected. However, there is no evidence that significant adverse effects are occurring; thus, the possibility that the fish are suffering severe negative effects from the Endicott Causeway seems low.

In the early 1980s, explorers for the oil and gas industry located a large reservoir of oil and gas in the Endicott Formation in the Alaskan Beaufort Sea just east of Prudhoe Bay. The discovery of the formation, 3.46 km below the sea offshore the Sagavanirktok River delta, followed several years of exploration and geophysical studies, which began about 1976. With BP Exploration (Alaska) Inc. (BPX; formerly Standard Alaska Production Company) as the operator, a joint venture of eight companies proposed to develop this reservoir, estimated to contain 280–423 million barrels of oil (45–67 billion L) and 1.9–2.1 trillion ft^3 (54–59 trillion L) of natural gas. On 3 September 1982, BPX submitted preliminary draft permit applications to the U.S. Army Corps of Engineers (USACE) to begin development, using offshore drilling islands connected to the shore by a solid-fill causeway. After reviewing the applications, the USACE determined that an environmental impact statement (EIS) would be required for the development project, considering past experience at a similar project, West Dock.

West Dock, the first major causeway constructed in the region, is a 4-km-long causeway on the northwestern side of Prudhoe Bay (Figure 1). It was constructed perpendicular to shore at the end of a barrier island–lagoon system and extends to water depths of 3–4 m. The presence of cold, high-salinity waters on the west side of the causeway had been periodically documented since the mid-1970s; these cold, marine waters also occasionally advected to the shore and into the lagoon system. Near the shore, salinity and temperature levels on the west side of West Dock were sometimes as much as 19.8‰ more saline and 8.1°C cooler than levels on the east side of the dock (Woodward-Clyde Consultants 1982, 1983; Biosonics, Inc. 1984). These differences were only slightly lower than those reported across the seaward segment of the causeway (23.2‰ and 9.0°C). On occasions when marine waters were advected close to shore, movement of fish eastward from the Colville River to Prudhoe Bay was blocked (for a summary see Fechhelm et al. 1989), and an apparent habitat discontinuity was established (Envirosphere Company 1987). Some background information will help to explain the significance of these observations.

Background

A major concern about causeways is their effect on circulation and hydrography, which in turn can affect anadromous fish. Evaluation of these effects requires an understanding of some important natural phenomena, especially the coastal upwelling and marine intrusion processes that govern the availability of anadromous fish feeding habitat during summer, as well as the overall extent and quality of anadromous fish habitat available along the Beaufort Sea coast of western North America.

Oceanographic Processes

The extreme arctic climate is modified along the Alaskan coast of the Beaufort Sea by the presence of the sea (Walker et al. 1980). The climate along the coast during the open-water season is characterized by strong, persistent surface winds, low precipitation, and extensive cloud cover, primarily in the form of stratus and fog (Hummer 1987). Numerous studies have shown that wind is a major factor influencing the physical and biological environment of Alaskan Beaufort Sea water masses. During the open-water period, winds typically drive the nearshore surface currents at about 3% of the wind speed (Aagaard et al. 1981; Kozo 1984). Although annual and within-summer variability is great, the dominant winds during the open-water season are from the east-northeast, especially during early summer in June and July. As the season progresses, there is an increased frequency of westerly winds, which shift toward the west-southwest late in summer.

The prevailing east–west direction of the winds causes alongshore flows and the creation of a nearshore band of low-salinity or brackish water that is, on average, markedly warmer than water further off shore. River discharge into shallow nearshore waters provides fresh water to dilute the seawater and a source of heat that contributes to the warmer temperatures. However, the major factor affecting water temperature in the shallow, nearshore zone is air temperature.

The integrity of this nearshore band is not only governed by factors such as proximity to rivers, depth, air temperature, and alongshore flows, but also by the processes that result in exchange of offshore marine and nearshore fresh–brackish waters in an across-shore fashion. These events are also governed by the winds.

On east winds, surface waters along the Beaufort Sea coast are transported offshore as a result of Ekman transport. This process causes a surface divergence along the coast, and thus a nearshore depression of sea level. The offshore transport of surface waters is balanced by a return flow of

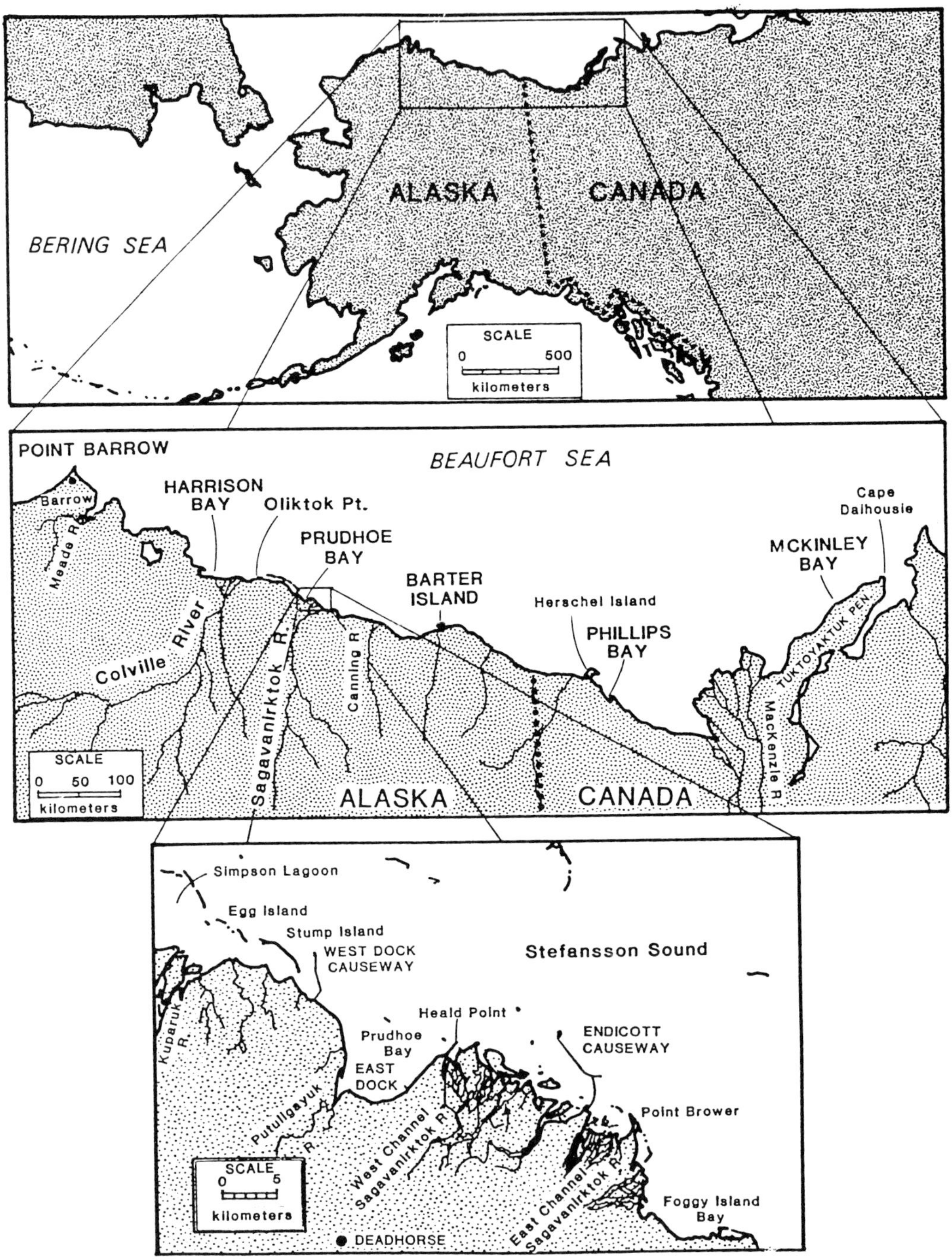

FIGURE 1.—Map of the Prudhoe Bay region of Alaska showing the location of area causeways.

marine water at depth, which sometimes penetrates to the surface in areas with depths of 2–3 m or less. In this paper, this process is referred to as marine upwelling, a regional-scale event. Causeways and natural coastal features perpendicular to local bathymetry also deflect surface flows off shore and produce local return flows of marine bottom water, but the spatial extent of this effect is limited to the immediate vicinity of the structure (Colonell and Niedoroda 1990).

The location of marine bottom water under warmer nearshore waters varies in proximity to shore and depth, primarily as a function of the wind and, to a lesser extent, as a function of the degree of stratification (determined by river discharge) and proximity of ice cover. Upwelling of cold, saline bottom waters to or near the surface regularly occurs to about 2-m depths on east and northeast winds and, under some conditions, even shallower (Britch et al. 1983). Wind-controlled upwellings occur naturally across the entire front edge of the Sagavanirktok River delta, across Prudhoe Bay, along the seaward side of Stump Island, and through the many passes between barrier islands into the Simpson Lagoon system, from Stump Island to Oliktok Point (Matthews 1979; Colonell and Weingartner 1982; Mangarella et al. 1982; Savoie and Wilson 1983, 1984, 1986). Once at the surface, these high-salinity waters are further mixed and advected to other areas by the wind.

Causeways do not appreciably enhance these upwellings, but rather provide an alternative mechanism by which high-salinity, cold bottom water can be carried to the surface, mixed, and advected away from the structure by wind. The causeway acts as a barrier to water movement, resulting in the formation of an eddy in the wake at the tip of the structure. The eddy causes a vertical mixing and transport of bottom water to the surface. This is called the wake effect. When Ekman transport has caused marine bottom waters to flow into the causeway area, these higher salinity waters are thus brought to the surface. From there they continue to mix and are advected to other areas and further mixed by winds, thereby increasing salinity levels in the nearshore zone and downwind of the causeway, at least to some extent.

High river discharge into coastal waters also contributes to offshore surface flows on light east winds. The sharp density discontinuities between the fresh surface and underlying brackish or marine waters permit the fresh surface water to decouple from underlying water of greater salinity and flow to the north (to the right of the wind, via Ekman transport) until mixing at the interface between the two water masses increases the friction such that the water flow is once more aligned with the wind. This effect is little if at all affected by causeways, because it is primarily a function of density stratification and wind direction and strength.

Salinity of surface water generally increases with both distance offshore and depth; the differences between nearshore and offshore surface salinities decrease as the season progresses due to reduced levels of river discharges, decrease in ice cover, and increased mixing. Under west winds, offshore surface waters surge shoreward in a phenomenon that has been termed a marine intrusion. This flow is balanced by marine bottom waters that are "downwelled" to the offshore. These surge events typically occur by late July, and they cause marked increases in salinity levels in nearshore habitats.

As described above, the volume of high-salinity water near shore that may be transported shoreward is mostly a function of the strength of (1) the onshore–offshore salinity gradient, (2) the presence of upwelled water at or near the surface nearshore (both of which are regional-scale effects, little influenced by causeways), and (3) the presence of high-salinity water due to the wake effects of causeways and other promontories. As elaborated by Colonell and Gallaway (1990), the causeway contribution is considered to be small by some and large by others. The latter belief stems from studies around West Dock. After reaching the surface and mixing, high-salinity water in the lee of the West Dock Causeway is advected by east winds to the west into Simpson Lagoon (Figure 1). This effect is reduced by freshwater discharge from the Kuparuk River, 11 km west of West Dock. Additional highly saline water flows into the lagoon from marine upwellings, entering eastern Simpson Lagoon between West Dock and the east end of Stump Island, through the Egg Island channel between the barrier islands facing the Kuparuk River delta, and, in fact, through inlets along the entire length of the island chain forming Simpson Lagoon (Colonell and Gallaway 1990). About 85% of the water exchange in the eastern lagoon system occurs through the Egg Island inlet (Savoie and Wilson 1986).

The high-salinity water in Stump Island Lagoon has been hypothesized to be the main source of the marine water that intrudes into Prudhoe Bay a few hours after a shift to westerly winds. However, this water would require 20 h to reach the bay; because marine water enters Prudhoe Bay within a few hours of a wind shift, the water must originate east of West Dock, and north of the shallow sill across the bay (Colonell and Gallaway

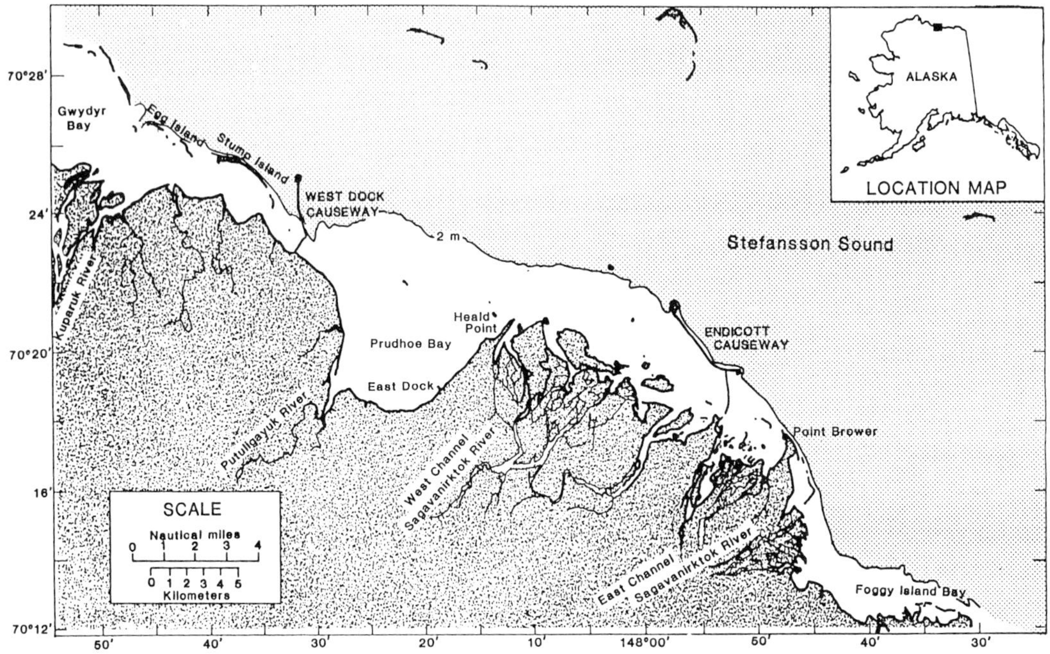

FIGURE 2.—Approximate distribution of alongshore habitat used by anadromous fish (white area from shore to 2-m depth contour) and marine environment (dotted area seaward of the 2-m depth contour).

1990). Considering the volume of water required to raise salinity and the short time over which intrusions can occur in Prudhoe Bay, the main source of the marine water can only be from off shore. While it is true that high-salinity water from the lagoon system west of West Dock ultimately does move around the causeway and enter Prudhoe Bay, the volume of high-salinity water attributable to wake effects of the causeway is not large in comparison to the contribution from offshore sources (Colonell and Gallaway 1990).

Anadromous Fish Summer Coastal Habitat

Craig (1989) provides the following description of anadromous fish coastal habitat in the nearshore zone of the Beaufort Sea. Scientific names in brackets have been added to the original account.

> The nearshore zone is marked by a series of bays, lagoons, deltaic mudflats, and narrow barrier islands. A biologically important feature of the nearshore zone is the occurrence of relatively warm and brackish water (5–10°C, 10–25‰) that frequently lies adjacent to the shoreline in summer. . . . This estuarine zone extends over much of the length of the coast and is often distinctly different from adjacent marine waters (−1 to 3°C, 27–32‰). This nearshore zone provides a transportation corridor for fishes not fully adapted to the marine environment as well as an important feeding habitat for anadromous and marine fishes such as Arctic cisco [*Coregonus autumnalis*], least cisco [*C. sardinella*], humpback whitefish [*C. pidschian*], broad whitefish [*C. nasus*], Arctic char [*Salvelinus alpinus*], fourhorn sculpin [*Myoxocephalus quadricornis*], and Arctic cod [*Boreogadus saida*]. In winter, the estuarine band is absent, and nearshore waters freeze solid to a depth of about 2 m.

In the Prudhoe Bay region of Alaska, the estuarine zone typically extends to depths of 2–3 m (Figure 2), but it can be larger or smaller, depending upon specific environmental conditions.

The three key determinants of anadromous fish habitat quality during summer are temperature, salinity, and prey. The effects of oceanographic and nearshore circulation processes on the temperature and salinity characteristics of the habitat were described above. A fact not mentioned above is that the prey upon which most of the species of anadromous fish depend are of marine origin (Truett 1981). The transport of prey from off shore into the nearshore zone is in large part dependent upon the marine upwelling–intrusion process, and the dispersal of prey throughout the nearshore zone is greatly influenced by alongshore flows. Griffiths and Dillinger (1981) showed that if the exchange of offshore–inshore water through the passes between the barrier islands of the Simpson Lagoon system was stopped, prey densities in the lagoon system would drop to levels that would not support predators within 7–10 d.

Although arctic anadromous fish are known to be limited in distribution by high salinity (absolute levels vary among species and size-groups), water temperature is probably more important to their well-being (Brett 1971). Temperatures above 5°C are considered to be critical for fish to achieve good growth or even to swim well. Bernatchez and Dodson (1985) found that coregonid fishes (cisco *Coregonus artedii* and lake whitefish *C. clupeaformis*) were not good swimmers compared with most salmonids, and their swimming performance was much lower, and energy requirements higher, at 5°C than at 12°C. Fechhelm et al. (1983) experimentally determined that small Arctic ciscoes (125 mm) preferred temperatures between 12 and 14°C, and their feeding capability was reduced and schooling behavior altered at temperatures of 5°C and lower.

Because of their relative intolerance to low temperature and high salinity, arctic anadromous fish must overwinter in the fresh or brackish waters of rivers and river deltas. They also spawn in these habitats. During summer, the estuarine delta areas at the mouths of rivers also provide refugia into which the fish using more brackish and food-rich areas can retreat in the face of strong marine intrusions. River delta areas are thus very important habitats in summer as well as in winter. This is especially true for small broad whitefish, which appear to be limited in their coastal distribution to the immediate vicinity of their source river (Colonell and Gallaway 1990).

Summer coastal habitat for anadromous fish in the western North American Arctic is more likely subdivided into discrete geographic areas rather than occurring as a continuous coastal band, as might be inferred from the above account from Craig (1989). Two lines of evidence support this contention. The first is the contrast in the distributional patterns of anadromous fishes, especially least cisco, broad whitefish, and humpback whitefish, versus those of Arctic char and Arctic cisco along the Beaufort Sea coast. The first group occurs in association with the Meade, Colville, and Mackenzie rivers, with a broad disjunction in distribution between about Herschel Island and the Canning River (Figures 1, 3). This region is occupied by Arctic cisco and Arctic char, which range across most of the Beaufort Sea coast

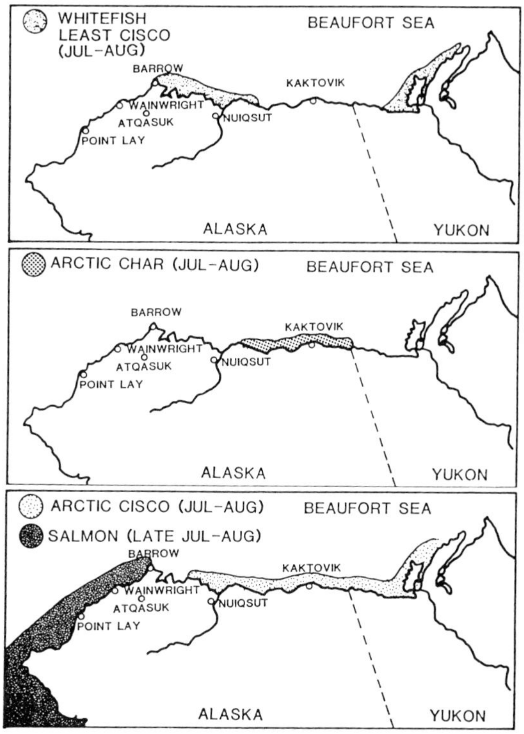

FIGURE 3.—Zones of highest abundance for broad whitefish, humpback whitefish, and least cisco, for Arctic char, and for Arctic cisco and salmon (*Oncorhynchus* spp.) along the Beaufort Sea coast of western North America (diagrammatic, adapted from Craig 1989).

(Figure 3). The rivers emptying into the sea between Herschel Island and the Canning River contain overwintering and spawning populations of Arctic char in upstream, warmwater spring areas, but the deltas and river channels elsewhere are shallow and mostly freeze to the bottom during winter (Craig 1989). Spawning populations of Arctic char are also found in some of the drainages associated with the Colville and Mackenzie River basins.

The only known spawning populations of Arctic cisco are associated with the Mackenzie River basin (Dryden et al. 1973; Nelson et al. 1987; Dillinger 1989). Although Arctic ciscoes do overwinter in other rivers such as the Colville and Sagavanirktok, catches of this species between Herschel Island and Canning River during the open-water season are typically dominated by either young-of-the-year (late summer) or adult and subadult specimens near sexual maturity (early to midsummer: Griffiths et al. 1975; Griffiths

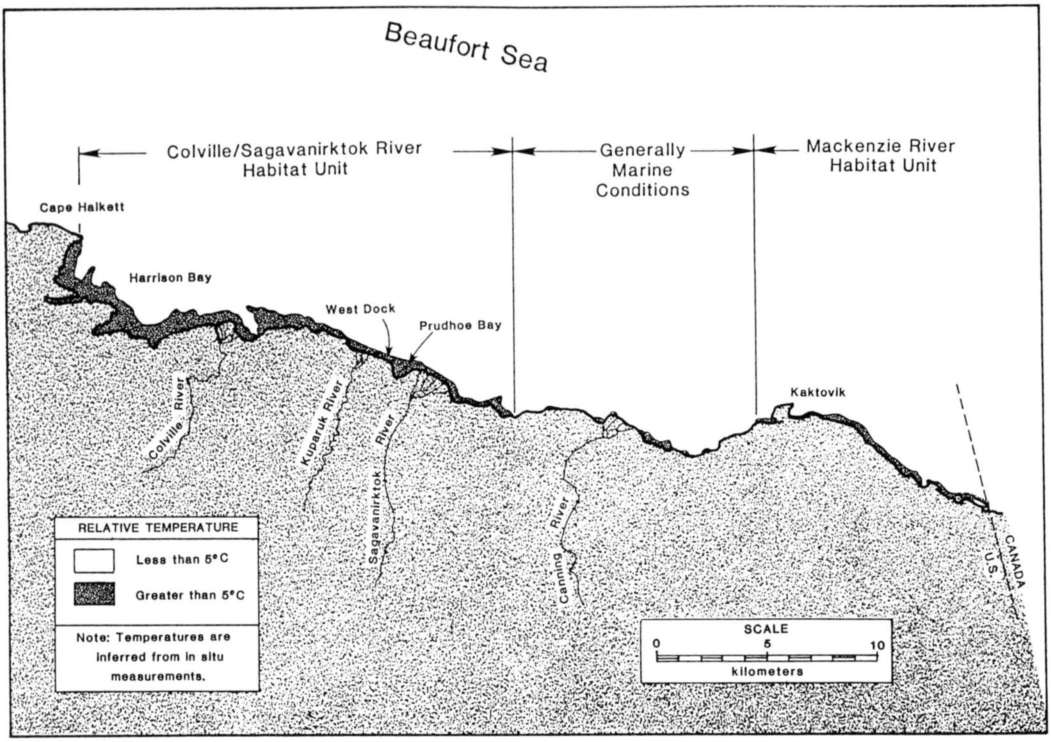

FIGURE 4.—Discontinuity of the coastal estuarine band delineating the Colville–Sagavanirktok River Habitat Unit. Temperature data are from NOAA-9 polar-orbiting satellite advance very high resolution radiometer (AVHRR) 2 August 1985.

1983). This size–age pattern is consistent with the hypothesis that the catches reflect westward transport or migration of young-of-the-year Arctic ciscoes into Alaskan waters and their early season return migration 7 years later to the Mackenzie River for spawning. This, combined with the general absence of Arctic cisco spawners in the Colville River delta and their abundance in the Mackenzie River, is one of the important foundations of the Mackenzie-origin hypothesis for Arctic cisco (Gallaway et al. 1983).

The disjunction in the distribution pattern of least cisco, broad whitefish, and humpback whitefish corresponds to an area where overwintering habitat is scarce, where water depths increase markedly within a short distance of shore, where coastline is exposed (not protected by barrier island chains), and where river discharge levels are relatively small. Satellite photography of this unused area during the open-water season often shows cold, marine waters extending to the shore (Figure 4). This condition also occurs regularly along the exposed coastline west of Harrison Bay to Barrow, Alaska (Figure 1). Such exposed areas are characterized by habitat conditions unfavorable to the anadromous fish species in question, which prefer warmer, low-salinity waters (Fechhelm et al. 1983; Griffiths and Gallaway 1983; Neill et al. 1983; Craig et al. 1985).

These two marine areas of coastline bound the Colville–Sagavanirktok River Habitat Unit (Figure 4). The eastern and western ends of the habitat are under the direct influence of the discharges from the two largest rivers on the Alaskan Beaufort Sea coast, the Colville and Sagavanirktok rivers. The area between these two rivers is protected by barrier islands and also receives the discharge of the Kuparuk River. These factors contribute to good anadromous fish summer habitat (warm, low-salinity waters, rich in food, and near low-salinity refugia). About 3,400 km^2 are included in the habitat unit.

This habitat is small, however, compared with the area under the direct influence of the Mackenzie River, one of the world's largest rivers (Figure 5). This area is defined as the Mackenzie

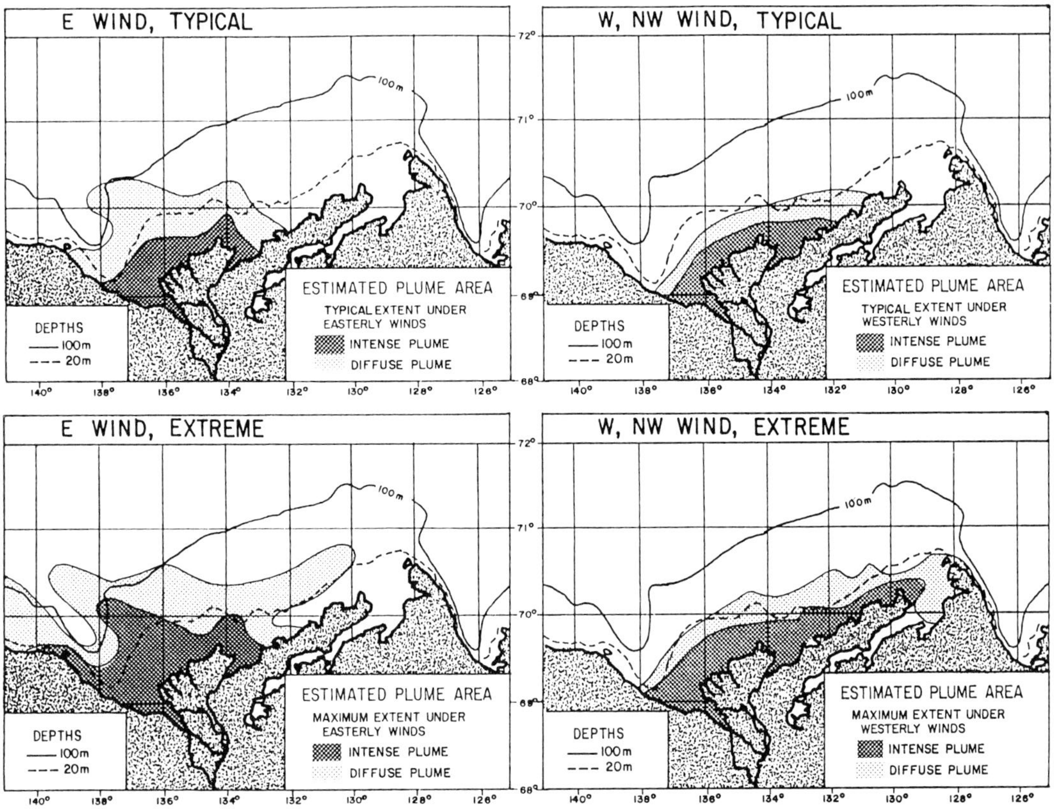

FIGURE 5.—Distribution of the Mackenzie River plume under typical and extreme winds (from Thomson et al. 1986).

River Habitat Unit and extends from near Herschel Island eastward to near Cape Dalhousie (Figure 1, Thomson et al. 1986). Under extreme east winds, the plume from the river can extend well into Alaskan waters, providing a migration corridor across the North Slope (Figure 5). The plume can also extend well offshore; the intense plume shown in Figure 5 is found to depths of 3–5 m, and the diffuse plume to depths of 2–10 m (Thomson et al. 1986).

From the above discussion, three geographically discrete habitats for anadromous fish may be described, the largest associated with the Mackenzie River plume, which typically affects an area extending from Herschel Island eastward to about Cape Dalhousie (Thomson et al. 1986; see also Figures 1, 5). The next largest habitat occurs between Harrison Bay and the Canning River in Alaska (see Figures 1, 4), and a third, much smaller area probably occurs in the vicinity of Barrow, Alaska, in association with the Meade River and nearby lagoons (Figure 1).

Even within the habitats described above, conditions should not be considered continuous, uninterrupted zones of suitable habitat over time and space. Within these areas, upwellings of cold, marine bottom waters can be carried nearshore on east winds, contributing to periodic marine intrusions along the coastline, especially in the vicinity of natural (e.g., Oliktok Point, Barnes et al. 1977) and artificial promontories (e.g., causeways, Colonell and Gallaway 1990), at the lagoon inlets between barrier islands (e.g., Matthews 1979; Colonell and Weingartner 1982; Mangarella et al. 1982; Savoie and Wilson 1983, 1984, 1986), and probably around other topographic features (e.g., natural and artificial islands, although no data currently exists to document this). The ebb and flow of suitable temperature and salinity conditions in nearshore waters greatly influence movements and habitat use patterns of anadromous fish (see review in Fechhelm et al. 1989). However, the upwellings and marine intrusions that affect anadromous fish movements and habitat use pat-

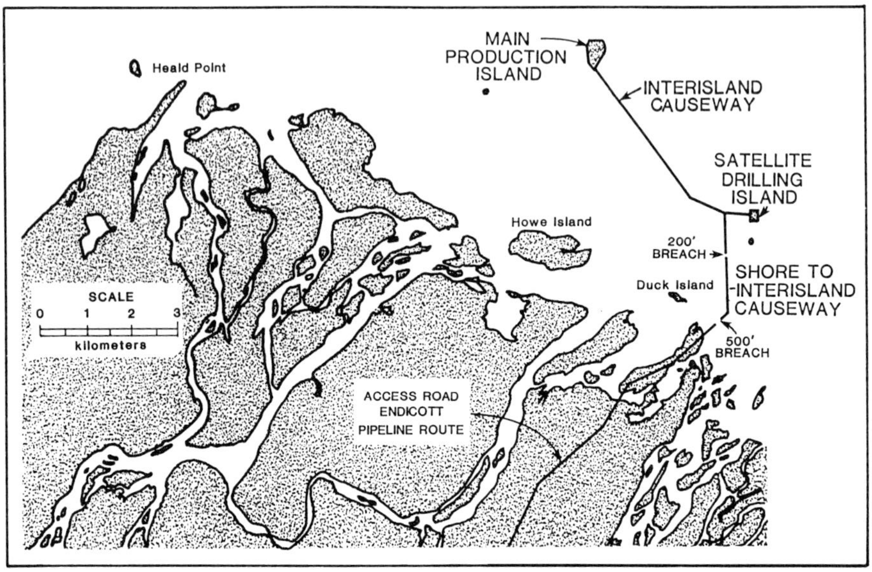

FIGURE 6.—Final design of the Endicott Causeway Development.

terns are critical because they supply the marine prey species in the nearshore zone that are dispersed by alongshore flows into habitats suitable to the fish.

Project Design

The final Endicott Project design included two gravel drilling islands (main production island, MPI, and a satellite drilling island, SDI) located offshore the Sagavanirktok River delta in waters just outside the 2-m isobath (Figure 6). These islands are connected by a solid-fill causeway (interisland causeway); a breached causeway (mainland causeway) extends from the shoreline to the interisland causeway to provide access between the drilling islands and the shore. The entire gravel causeway network is about 7.7 km long.

Two breaches were incorporated into the final design of the mainland causeway: a 152-m (500 ft) breach near shore and a 61-m (200 ft) breach near the intersection of the interisland and the mainland causeways (Figure 6). Most of the causeway construction was done during winter 1985; the 213 m of breaching was completed during winter 1986. During the interim period, temporary bypass causeways provided passage around the permanent breach locations (Figure 6). The temporary causeways were also breached. At the nearshore site (location of the 152-m breach), total breaching of the temporary bypass causeway was 30 m (100 ft) plus 20 culverts 1.8 m (6 ft) in diameter to allow fish passage. At the offshore site (61-m breach location), the temporary causeway included a 15-m (50 ft) breach and 8 culverts 1.8 m in diameter.

Controversial EIS Predictions

The environmental impact statement (EIS) for the Endicott Development was based upon site-specific baseline studies conducted in 1981 and 1982 (Dew 1982; LGL 1982; Woodward-Clyde Consultants 1982; Gallaway and Britch 1983). Additional data were available from the 1977–1980 Outer Continental Shelf Environmental Assessment Program (OCSEAP) studies in the nearby Simpson Lagoon, from the 1981–1984 Waterflood Project Monitoring Studies at West Dock, from the 1983–1984 Lisburne Studies within Prudhoe Bay, and from an array of other studies in the region (see Slaybaugh et al. 1989 for a complete listing of historical studies conducted in the region).

The foremost issues concerning anadromous fish included

(1) the degree to which causeway-induced water quality degradation (increased salinity, decreased temperature) would occur as a result of the marine upwellings on the lee side of the causeway, considering the frequency and duration of such events;

(2) the degree to which either the physical

presence of the causeway or causeway-induced water quality changes would impede the migration of young-of-the year Arctic ciscoes from Canada to the Colville River, and, if blockage occurred at the Endicott Causeway, the potential for the carrying capacity of the Sagavanirktok River delta to be exceeded, causing high mortality; and

(3) the degree to which the causeway-induced changes in water quality would degrade the foraging habitat in the area and lead to decreased growth and fitness and higher mortality of anadromous fish using the habitat.

In response to the first concern, the EIS (USACE 1984) concluded that high-salinity water that might occur in the lee of the MPI on the west side of the causeway would not be advected to the shoreline under east winds as had been observed at West Dock. This conclusion was based on the location of the causeway in shallow depths (≤2 m) at the mouth of the Sagavanirktok River, the second largest river on the North Slope.

The second potential effect of the causeway, on young-of-the-year Arctic cisco migration, was also addressed in the EIS. The modeling assessment (Fechhelm and Gallaway 1983) indicated that, even though passive drift of the small fish on strong currents would allow many fish to pass the barrier, some fish would become thermally trapped or blocked on the east side of the Endicott Causeway between the shore and the interisland causeway segment. This conclusion was responsible, in large part, for the inclusion of breaches in the mainland segment of the causeway to ensure fish passage. With this modification in project design, no significant effects on Arctic cisco migrations were expected (USACE 1984).

Given the results of the oceanographic modeling in the EIS, the effects of causeway-induced changes in water quality were not expected to be of sufficient magnitude to impair growth or condition of anadromous fish. It was also believed that fish could readily avoid habitats with unfavorable temperatures and salinities, the low-salinity refuge at the mouth of the river would not be significantly compromised, and adequate levels of prey would be present in areas of favorable water quality.

These EIS predictions were by no means universally accepted. The ensuing controversy over the permitting and construction of the Endicott Development brought about one of the largest environmental monitoring programs in North America, beginning in 1985 and continuing to the present. The purpose of these studies has been to assess the EIS predictions, document the actual effects of the development, and evaluate the effectiveness of the 213 m of breaching.

The controversy associated with the development reached a new high in March 1988, when the USACE formally discontinued the anadromous fish studies after only 3 years of a planned 7-year study. The rationale for this action was that significant effects to anadromous fish habitat had occurred and that mitigation was required. Once again, there was no consensus concerning the USACE's interpretation of the results. BPX elected to continue the anadromous fish studies, partially because the North Slope Borough (NSB) permit for the project contained a fish-monitoring requirement that would not have been met in the absence of the USACE program.

Purpose and Objectives

During 1988–1989, the environmental and biological data from the Endicott Project monitoring program for the period 1985–1987, as well as data available from the 1988 studies, were analyzed. The 1985–1987 field-sampling programs were conducted by the Envirosphere Company, and final data tapes were submitted each year to the USACE and BPX. However, final reports describing the results of these programs have not, at the time of this writing (September 1990), been prepared for any year other than 1985 (Envirosphere Company 1987). LGL Alaska Research Associates, Inc., has conducted the fish-monitoring program since 1988; and, as a necessary step, obtained, analyzed, and reported the data for 1985–1987 (LGL et al. 1990).

The purpose of this paper is to evaluate the three controversial issues addressed in the EIS based upon the results of our analyses of the 1985–1987 data, as well as the available results from the 1988 study. The recorded frequency of cold, high-salinity bottom water caused by the causeway in the nearshore zone and the distribution of this water in 1985–1987 are described first. Next, the apparent effects of the Endicott Causeway on the migration patterns of young-of-the-year Arctic ciscoes through the area in 1985–1987 are presented. Finally, the effects of causeway-induced habitat changes on key anadromous fish species are evaluated by examining growth and condition data (1985–1987) and population levels (1982 versus 1988) observed during the monitoring studies.

The specific objectives of the oceanographic and biological analyses were to

(1) identify the number of times that cold, marine bottom water upwelled near shore was transported to the surface due to the causeway's effect on circulation;

(2) determine if such water was advected into nearshore habitats in a manner that produced a salinity discontinuity in the alongshore zone of the delta that might compromise the refugia and present a physiological barrier to fish movement across the delta;

(3) determine the movement patterns of young-of-the-year Arctic ciscoes through the area during 1985–1987 and relate the observed patterns to meteorological events and the Endicott Causeway;

(4) define the growth achieved each summer by the small cohorts of Arctic cisco and broad whitefish (the present distributional patterns of these species and size-groups suggest that they are locally resident during the summer);

(5) define the condition levels of Arctic cisco, broad whitefish, Arctic char, and least cisco populations present in the project area during 1985–1987; and

(6) estimate the population levels for small Arctic ciscoes and broad whitefish in 1988, comparing these to the levels observed in the 1982 baseline study (Griffiths et al. 1983).

For objective (3), the EIS predicted that young-of-the-year Arctic ciscoes would successfully circumvent the causeway, there would be no thermal-trapping effect on the east side of the causeway in the nearshore area, and there would be no appreciable increase in the overwintering population in the lower Sagavanirktok River delta due to causeway blockage.

In this paper, we used the early- and late-season catch data for young-of-the-year Arctic ciscoes for each year to infer fish movement through the area and the likelihood that appreciable numbers of the fish overwintered in the Sagavanirktok River delta. The catch patterns were also compared to meteorological conditions to test the concept that the migrations were, in large part, wind-governed dispersals.

For objectives (4) and (5), the premise was that if habitat degradation or reduction attributable to the causeway was occurring (whether directly detected or not), it should be reflected in the growth rates of small fish and the condition of the populations overall. Clear growth patterns were evident from the length data for small Arctic ciscoes and broad whitefish. Growth of large anadromous fish is difficult to determine from size measurements, because of the overlapping size ranges of older year-classes. For these fish, condition probably provides the best response variable for detecting possible habitat effects.

Abundance trends will be ultimately important in the evaluation of project effects, at least over the long term. Passive sampling with fyke nets provided an index to fish movement, although the index may not have been related to absolute abundance in all instances. Mark–recapture studies of resident fish provided a better means of determining abundance trends over time. In this paper, we compare abundance levels of small Arctic ciscoes and broad whitefish in 1988 with levels observed in the baseline studies (Griffiths et al. 1983).

Study Area and Field-Sampling Methods

During each year of the 1985–1987 Endicott Monitoring Program, several aspects of the physical and biological environment were studied. These aspects included oceanographic and other physical processes (e.g., meteorology, river discharge, sedimentation, and ice breakup–freeze-up); anadromous fish distribution patterns, growth, and condition; and fish prey distribution and abundance. The data we used to meet the objectives of this paper were collected as part of the anadromous fish and oceanographic monitoring studies conducted by the Envirosphere Company during 1985–1987.

Anadromous fish studies.—In the anadromous fish studies of 1985–1987, 39 sites in the project area were sampled with fyke nets (Figure 7). Of these sites, 28 were sampled in 1985, 31 in 1986, and 20 in 1987 (Table 1). Samples were collected outside the Endicott Project area at Oliktok Point near Harrison Bay in 1985 and 1987 (Figure 1, Table 1).

Two general types of fyke nets were used, one along shorelines and the other in offshore habitats. Offshore nets were used to sample 3 (stations 5, 7, and 13) of the 28 locations sampled in the Prudhoe Bay area in 1985; 11 (stations 5, 7, 13, 60, 61, 62, 63, 64, 65, 66, and 69) of 31 sites in 1986; and 6 (stations 5, 62, 63, 65, 66, and 71) of 20 sites in 1987. The balance of the sampling effort each year used shoreline nets. Detailed descriptions of the exact types of net at each station in each of the study years 1985–1987 can be found in Cannon et al. (1987) and LGL et al. (1990).

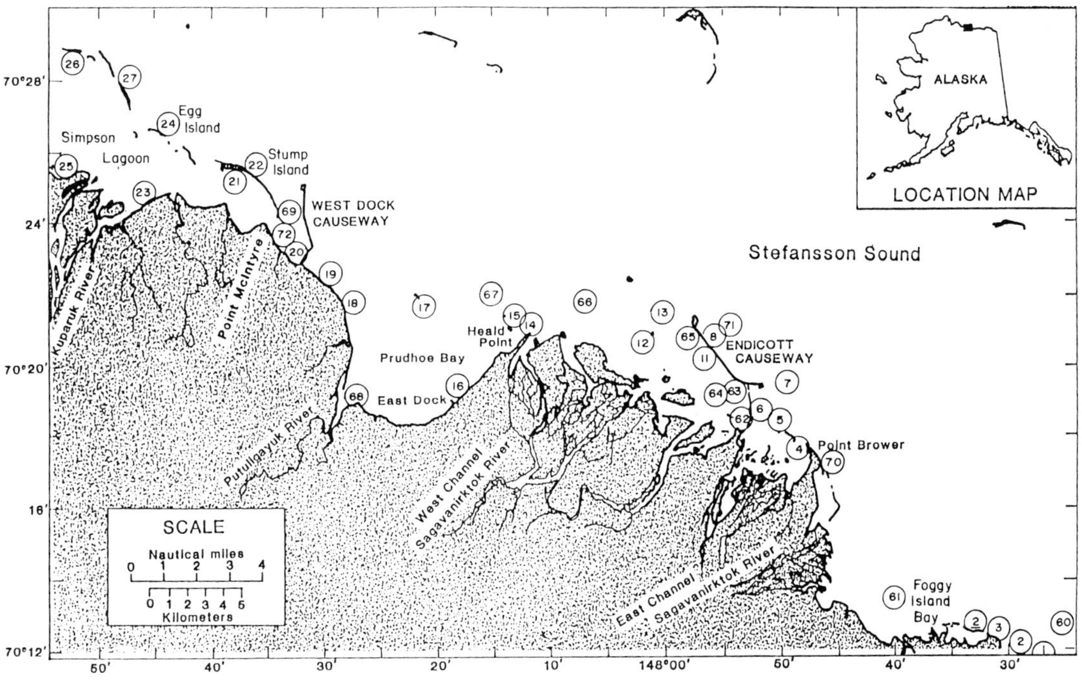

FIGURE 7.—Locations of fyke net sampling stations in the Endicott Project area, 1985–1987. See Table 1 for list of stations sampled each year.

The offshore net consisted of a lead terminating at a pair of double-cod-end traps, whereas the shoreline net lead terminated in a single double-cod-end trap (Figure 8). In both cases, the direction of fish movement could be estimated from the side of the net in which the fish were trapped and from obvious directional shifts in abundance across the area. Comparisons of abundance based upon fyke net catches can be difficult. Fyke nets are passive sampling devices, and catch depends on fish moving through the area. Accordingly, catches may not always be representative of actual fish abundance. Catch efficiency among nets or even between sides of the same net may differ depending upon many factors such as topography, circulation patterns, hydrographic conditions, the type of activity pattern exhibited by the fish, and the behavior of the fish upon encountering the net.

Fyke net sampling was conducted from 12 June to 15 September 1985, from 19 June to 15 September 1986, and from 18 June to 15 September 1987. Nets were checked daily, weather and ice conditions permitting. Ice and weather conditions in late June through mid-July seldom allowed sampling throughout the entire study area. During late July to mid-September, more complete sampling

TABLE 1.—Locations of fyke net sampling stations in the Endicott project area and years of station use. See Figure 7 for locations of individual stations.

	Year		
Region	1985	1986	1987
Foggy Island Bay	1, 2, 3	1, 2, 60, 61	70
East Endicott	4, 5, 6	4, 5, 6	4, 5, 6
North Endicott	7, 8, 13	7, 8, 13, 66	8, 66, 71
West Endicott	11, 12	11, 12, 62, 63, 64, 65	11, 62, 63, 65
Niakuk	14, 15	14, 15, 67	15, 67
Prudhoe Bay	16, 17, 18, 19	16, 17, 18, 19, 68	16, 18, 19
West of West Dock	20, 23, 25	20, 23, 25, 69	72
West of Barrier Islands	21, 22, 24, 26, 27	22, 26	22
Oliktok Point	28, 29, 30		28, 29

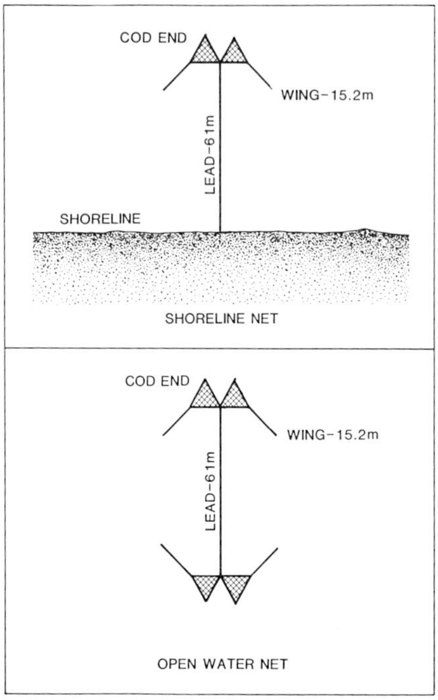

FIGURE 8.—Diagrammatic designs of shoreline and open-water fyke nets for fish sampling, 1985–1987.

was usually accomplished across the study area. The incomplete sampling early in the season made assessments of fish movement and habitat use patterns even more difficult during this period than in other periods, when more comparable data were obtained across the entire study area.

Biological data were collected at each net and included catch by species and size within species in each cod end of the net over the sampling interval and measurement of anadromous fish fork lengths. In addition, specimens of anadromous species were collected at regular intervals to provide data on stomach content, age, and length–weight relationships. Water temperature and salinity were measured each time the net was checked, along with other hydrographic data (turbidity, wave height, wave direction, water level, etc.).

In 1985, surface water temperatures at the fyke net stations were measured to the nearest 0.5°C with hand-held thermometers calibrated to a standard thermometer in the laboratory. Salinity levels in 1985 were determined in the laboratory with a YSI Model 32 Conductivity Meter. In 1986 and 1987, surface and bottom temperature and salinity levels were measured in the field with a YSI Model 33 Salinity Meter calibrated weekly to certified standards.

Water temperature and salinity data from the fyke net sampling programs gave the best temporal resolution of hydrographic conditions during 1985–1987. The stations sampled were representative of the variety of habitats available to anadromous fish in the study area. The fyke net measurements did not, however, provide the resolution necessary to estimate the spatial extent and depth distribution of the observed hydrographic properties. We obtained this information from the conductivity–temperature–depth (CTD) surveys conducted as part of each year's oceanographic programs (Hachmeister et al. 1987; LGL et al. 1990).

Oceanographic studies.—The CTD sampling design included 17, 27, and 19 transects for 1985, 1986, and 1987, respectively (Figure 9). Surveys were conducted from 23 July to 10 September in 1985, from 30 June to 14 September in 1986, and from 12 July to 14 September in 1987. The interval between surveys was intended to be 4 d, weather and ice conditions permitting. The actual interval ranged from 1 to 7 d. Complete or partial surveys were conducted on 23, 22, and 16 occasions in 1985, 1986, and 1987.

Surface-to-bottom water temperature and salinity data were collected each year with portable Applied Microsystems STD-12 instruments. The instruments internally recorded data at 10-cm depth intervals. The STD-12 instruments were calibrated at least once every 2 weeks against precision reversing thermometers and against water samples for which salinity levels were determined in the laboratory. Hachmeister et al. (1987) and LGL et al. (1990) provide additional details concerning CTD sampling.

Analysis Methods

During 1988–1989, we obtained tapes of original data from each year of the Endicott Monitoring Program, verified them (by comparing tape data to data listings in appendices of the program reports and to copies of original data forms in the case of discrepancies), and analyzed them as described below.

Oceanographic analyses.—The nearshore system abruptly changes from an almost freshwater to a more estuarine system in late July. Mixing from upwelling–intrusion cycles results in a general salinity increase over time, and stratification rapidly develops in the nearshore zone when winds are low, especially around the mouths of

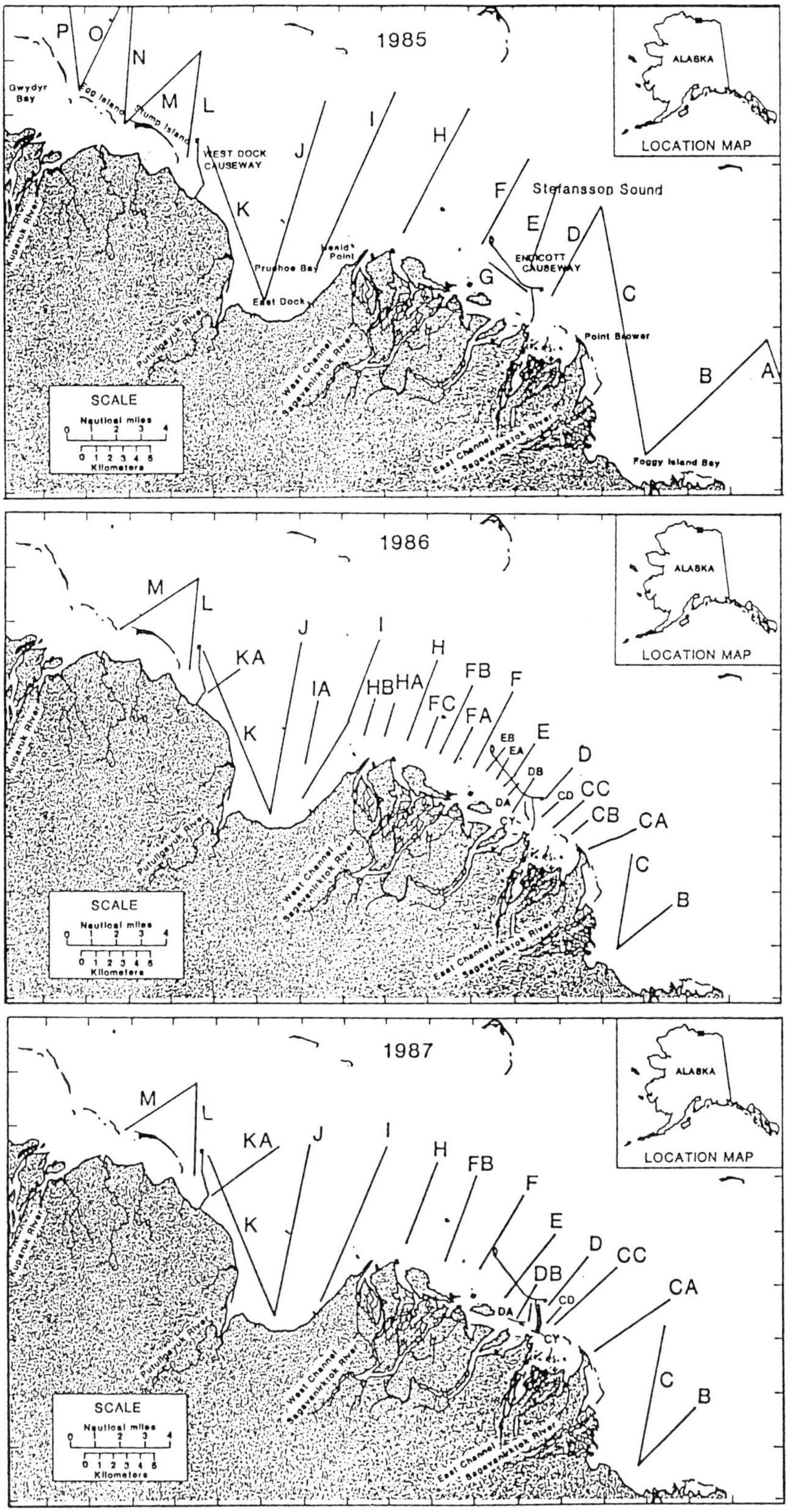

FIGURE 9.—Conductivity–temperature–depth sampling transects in the Endicott Project area, 1985–1987.

rivers. However, the salinity differences between the upper and lower water masses are not as great, nor the pycnocline (the plane separating water layers of different densities) as well defined as when high-salinity, offshore bottom waters are directly upwelled into the nearshore zone of 2-m depths or less. Under these conditions, upwelled bottom water can be identified by salinity levels of 26‰ or greater, typically about 30‰.

The ambient background salinity varies considerably, depending mainly upon the frequency, severity, and duration of previous upwelling–intrusion cycles and river discharge patterns. Because these events are governed by winds and rainfall, causeways should have little effect on background salinity levels unless (1) the wake effect significantly increases the introduction and mixing of high-salinity bottom water into the nearshore system that would not otherwise occur due to density differences between the water masses, or (2) localized circulation changes around the causeway result in the transport of high-salinity bottom water considerably more shoreward than would otherwise occur. Both situations could cause increased salinity levels in the Sagavanirktok River delta, which might degrade the quality of the feeding habitat of broad whitefish and the refuge habitat of some of the other species, and which might narrow the width of the low-salinity movement corridor across the Sagavanirktok River delta.

The basic hypothesis underlying the oceanographic analyses was that the location of the leading edge of high-salinity bottom waters upwelled from offshore would be governed by factors other than the causeway. It was believed that this hypothesis could be tested by comparing the locations of the leading edge of the high-salinity bottom water along the bathymetry "up- and downstream" of the causeway during upwelling events. If the causeway was having an effect, the leading edge should have occurred in shallower depths downstream of the causeway compared with upstream.

Next, even if no marked causeway effect was evident through analysis of the nearshore location of the high-salinity bottom water, the wake effect was considered as a source for introducing this water to the surface where it could be advected into nearshore habitat by winds. It was believed that the frequency and duration of the wake effect would provide an index to the magnitude of any causeway effect.

Because of the wake effect, a surface expression of higher-than-ambient surface salinity is always present in the lee of the MPI under east winds if the water column is stratified—a condition that occurs most of the time around river deltas when winds are low. Usually, the differences in salinity between bottom and surface water layers are small, less than 5‰ (Fechhelm et al. 1991, this volume). The potential for causeway effect is greatest when marine bottom water (salinity ≥ 26‰) has upwelled into the shallow nearshore zone near the tip of the causeway. In this paper, we determine the frequency of cases involving surface expression of 26‰ bottom water at the tip of the causeway. Under these conditions, the high density differences between surface and bottom waters would also enhance decoupling of water masses, and promote greater shoreward flows if such flows are occurring.

The daily temperature and salinity level measurements from each fyke net sampling station were used to detect the surface expression of upwelled bottom waters. These data were plotted as histogram bars on a map of the study area at the location of the nets. The salinity histograms were shaded to depict inclusion in one of four ranges (<15‰, 16–20‰, 21–25‰, and ≥26‰). These salinity ranges were defined based upon the data base describing anadromous fish habitat use patterns. This data base has grown exponentially since the Endicott EIS was prepared in 1982–1983. The results of these habitat studies (e.g., Dames and Moore 1988) have consistently shown that waters with salinities above 26‰ are little used by any species or size-group of the anadromous fish in question; some species and size-groups decrease in abundance in waters with salinity in excess of 20‰ (e.g., large broad whitefish, least cisco, small Arctic char); and small broad whitefish seem to avoid waters more saline than 15‰. Results of 1988 laboratory studies of small Arctic ciscoes, mostly age 1, have shown salinity changes of 30‰ and above to be lethal, but no apparent stress of any kind was observed when the fish were subjected to instantaneous salinity changes below 26‰ (Fechhelm et al. 1991).

For the oceanographic maps, hourly wind speed and direction were also plotted on each day's map to show the relationship of hydrographic changes to meteorological conditions. The oceanographic maps were produced with Hypercard® on an Apple Macintosh computer. Each map displayed the daily hydrographic and meteorological conditions in a pictorial format that could be viewed sequentially, yielding a dynamic view of changes over time and space. These maps were used to determine the frequency of time that high-salinity water was evident at the surface in anadromous fish habitat, particularly at stations 11, 65, and 13 in the lee west of the MPI where high-salinity water from the wake effect would be evident.

For each date on which high-salinity bottom water (≥26‰) was evident at the surface from the fyke net data, the CTD data for that date or the closest date (CTD profiles were taken about every 4 d) were reviewed to determine the shoreward extent of high-salinity bottom water across the region and if this water west (downstream) of the Endicott Causeway appeared more shoreward than east (upstream) of the causeway. Collectively, these physical data provided the basis for defining the frequency, duration, and spatial extent of anadromous fish habitat affected by the causeway. Anadromous fish habitat was defined as the nearshore zone, that is, the zone from the shore to the 2-m depth contour.

Biological analyses.—Because size-groups within species respond differently to environmental factors, the first step in the anadromous fish analyses was to identify size-cohorts. These were obtained from examination of the yearly length-frequency plots from all fyke nets; the data were organized by date of capture and length in millimeter increments. For smaller size-groups with distinct modes but overlapping size ranges, the analytical methods of MacDonald and Green (1985) for fitting mixtures of distributions were applied. Four size-groups or cohorts were defined for Arctic ciscoes and broad whitefish, while Arctic char and least ciscoes were classified only as small (<250 mm fork length) or large (>250 mm fork length).

The next step was to calculate catch per unit effort (CPUE) for each cohort. Values for CPUE were calculated directly from the catch and effort data files for each cod end of the net. Unmeasured fish were prorated into the identified size-cohorts before the standardization by effort. Effort was defined as the number of hours fished for each cod end of the net divided by 24 h. Catch was defined as the number of fish within each cohort or species counted over the sampling interval.

Annual differences in abundance and distribution of young-of-the-year Arctic cisco were estimated by comparing total CPUE among the three years 1985–1987 and by comparing the relative abundance patterns exhibited among stations by side of net for each year. Differences in daily catch patterns by side of net within each year were then correlated to differences in meteorological conditions. Mean daily CPUEs for each net side, averaged across all stations, were plotted in histograms to visually define periods of high and low catches. For each of these periods, a progressive vector plot of the wind speed and direction was prepared, and the relationships between wind vectors and catch defined by inspection. The premise was that periods of high catches on the east side of the nets should be associated with easterly winds, and vice versa. Lastly, unusual catch patterns for individual stations were examined for relationships to the wind patterns and other environmental factors, namely temperature and salinity.

More quantitative analyses were not conducted because wind speed and direction are only gross generalizations of overall current patterns, and catch levels in fyke nets depend on more factors than abundance alone. The historically defined relationship between wind and currents in the nearshore Beaufort Sea is that currents typically flow at about 3% of the wind speed, are aligned with the winds in shallow water but have a tendency to move to the right of the wind as depth increases, and usually follow the bathymetry (e.g., Aagaard et al. 1981). However, local topographic features can cause the generalized flow patterns to vary, especially along the irregular shorelines where the fyke nets were placed.

Fyke net catch data are also problematic. Catch rates depend upon fish movement—to be caught, fish must be both present and moving. It is also likely that catch efficiency of fyke nets changes as nets fill; that is, there may be a saturation effect. Lastly, there was usually a marked difference in the time scale between the fyke net sampling interval (24 h) and the accompanying physical measurements (e.g., hourly for wind speed and direction). At the moment, however, passive sampling devices represent the only successful techniques for capturing arctic anadromous fish in the shallow, nearshore areas where they appear to concentrate. As noted by Neill and Gallaway (1989), more emphasis needs to be placed upon developing valid mechanistic models that can be used to interpret catches made under these conditions.

Growth data (daily increases in length) for each cohort in each year were fitted to a von Bertalanffy growth curve with a length scale parameter instead of the usual time scale parameter, that is,

$$L = L_\infty - (L_\infty - L_0)e^{-kt};$$

L = mean length at time t;
L_∞ = the hypothetical ultimate mean length (asymptotic length);
L_0 = mean length when $t = 0$ (June 15)—the length scale parameter;
t = time in days (June 15 = 0); and
k = the growth coefficient.

The growth curves were regressed on the data with the Marquardt (1963) algorithm to provide a least-squares estimation of the required nonlinear parameters of the above equation. Significant differences between the parameters were determined at the 95% confidence level with t-tests following Gallucci and Quinn (1979). Even though parameter differences can be tested, there is no reliable significance test for a nonlinear growth curve (Draper and Smith 1981). However, linear analysis of covariance was also conducted on the log-transformed growth data to evaluate the results suggested by the comparisons of the nonlinear parameter estimates.

Length–weight data were subjected to covariance analysis for each of the anadromous species to distinguish if there was a change in condition (i.e., were fish heavier or lighter at a given length on average) among years (1985–1987). The regression model was

$$\log_e(W) = \log_e(a) + b \log_e(L) + \varepsilon;$$

a and b are constants; L is length; W is weight; and ε is the error term.

Population levels for small broad whitefish and Arctic ciscoes using the study area were estimated from mark–recapture data available from the 1988 study with the sequential Bayes algorithm described by Gazey and Staley (1986). This method is particularly useful for low recapture levels and can be used to determine if the assumptions upon which mark–recapture estimates are based are likely being met, as well as the nature of any violations (Gazey and Staley 1986).

Results and Discussion

Environmental conditions varied among the three years 1985–1987. Meteorological conditions during July–15 August were highly similar each year; progressive vector plots show protracted periods of northeasterly wind vectors (Figure 10). However, after August 15, the progressive vector plots were not highly similar among years (Figure 11). In 1985, northeasterly winds predominated until near the end of the season. In 1986, strong west and southwest winds prevailed from about 16 to 23 August, reversing to strong northeast winds between 23 August and 2 September (Figure 11). During 3–8 September 1986, the winds were light. The wind vector from 3 to 6 September 1986 was to the northwest, then variably easterly, northerly, and northeasterly during 6–8 September, until winds switched to the west-southwest on 8 September (Figure 11). Between 10 and 11 September 1986, winds were relatively strong out of the south, then switched to the west. From 13 to 15 September 1986, they were relatively strong from the east (Figure 11).

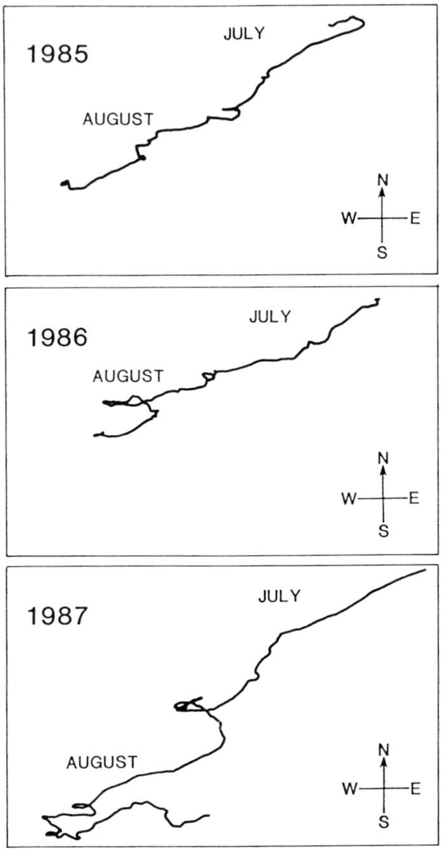

FIGURE 10.—Progressive vector plots of winds in the Endicott Project area, July–15 August, 1985–1987.

Northeasterly winds generally prevailed from July to 18 August 1987, followed by an extended period of strong westerly and southwesterly winds that lasted until 31 August (Figure 11). Between 1–4 and 6–8 September, winds were from the northeast and east respectively, inter-

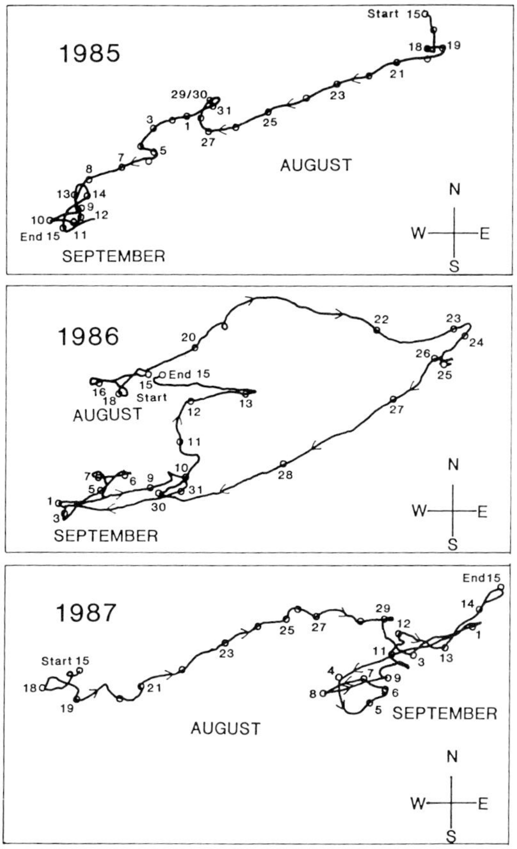

FIGURE 11.—Progressive vector plots of winds in the Endicott Project area, 15 August–15 September, 1985–1987.

rupted by a brief period of first north and northwest then southwest winds. From 8 to 15 September, the progressive vector plot generally extends to the northeast or north, indicating southwest and south winds (Figure 11). As will be shown later, the differences in the late summer wind patterns greatly affected the movements of young-of-the-year Arctic ciscoes.

River discharge increased during each year of the study (Figure 12). The most pronounced difference among the years was in 1987, when the highest discharge during the open-water season occurred late in the season. During most of 1987, the discharge was also more evenly split between the east and west channels than during 1985 and 1986.

The differences in meteorological and river discharge conditions among years appeared to greatly influence the mean temperature and salinity in the coastal zone of the project area. The periods of time that mean surface water temperatures at the study area fyke nets were above 5°C increased each year, 1985–1987 (Figure 13). In 1985, the best growth conditions (highest mean temperatures) occurred from about the first week in July through about the first week in August, after which a marked decline was evident. In 1986, mean surface water temperatures were in the 6–10°C range from early July through about 20 August (Figure 13). Warm (6–10°C) temperatures prevailed to 27 August in 1987.

As described above, upwelling of marine bottom water into the nearshore zone occurs on east and northeast winds. When east winds abate or switch to the west, high-salinity water surges shoreward, increasing nearshore salinities and greatly compressing the nearshore estuarine band. Typically, the first significant marine intrusion of the year occurs near the end of July, resulting in a steplike pattern of salinity increase. This steplike increase is reflected in the time series of smoothed daily average salinity levels from measurements at the study area fyke net stations (Figure 14). Prior to the third week of July in each year, mean salinity levels were less than 10‰, but after that date mean salinity levels were generally 10–20‰. The late-season decreases in study area salinity, particularly evident in 1987 (Figure 14), were mainly associated with high river discharge (Figure 12).

Surface Expression of High-Salinity Water

The nearshore locations of the 2- to 3-m isobaths in the western lee of the main production island (MPI) at Endicott and in the area west of West Dock are such that marine bottom water naturally comes very close to the mainland (see Figure 2). Surface expression of these waters is enhanced by the wake effect at the tip of the causeways, which promotes vertical transport throughout the water column. In 1985, 1986, and 1987, surface waters west of the MPI (stations 11, 65, and 13; see Figure 7) with a salinity of 26‰ or greater were documented by the fyke net sampling program to occur on only 3–5 dates each year. The 1985 dates were 10–11, 14 August and 9, 12 September; CTD data were taken on 8–9 and 12–13 August and 8–10 September. In 1986, the three occasions were 21 July and 15–16 August; the CTD data of 22 July and 15 August can be used to delineate the spatial extent of the upwellings. In 1987, surface upwellings west of the MPI

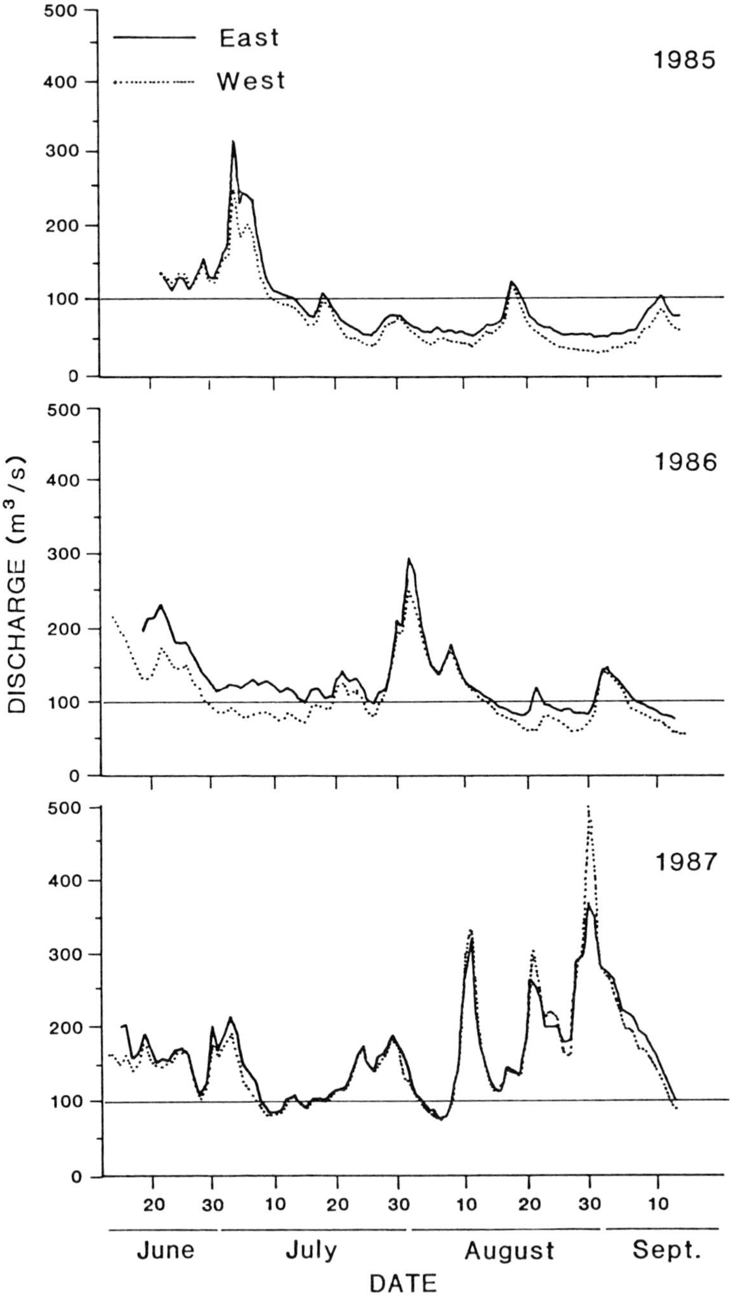

FIGURE 12.—River discharge levels through the east and west channels of the Sagavanirktok River in summer, 1985–1987.

occurred on 19, 20 July and 6 August; CTD data were available for 18, 22 July and for 7 August.

The frequency that high-salinity (>26‰) waters were recorded at the surface in the western lee of the MPI of the Endicott Causeway was 6% of the time in 1985, 2% in 1986, and 4% in 1987. The low frequency of such events provides some index of the expected magnitude of the effects from the Endicott Causeway. Over most of the three summers, high-salinity bottom waters were not

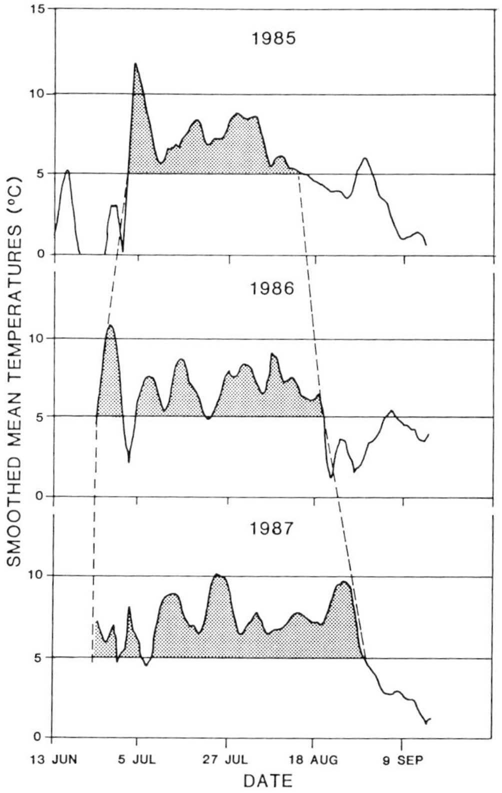

FIGURE 13.—Mean daily surface water temperature in the Endicott Project area in summer, 1985–1987. Stippled area shows periods when water temperature was 5°C or higher.

present in the lee of the Endicott Causeway for transport to the surface from the wake effect of the structure.

Spatial Extent of High-Salinity Water on the Sagavanirktok River Delta

The spatial extents of high-salinity water around the Endicott Causeway documented for 21–22 July and 15–16 August 1986 appear to be well defined by the CTD surveys made on 22 July and 15 August. These data provide good examples of the extent and nature of the causeway effect under extreme conditions. The upwelling event in mid-August 1986 was the largest and most severe documented by the monitoring program during 1986 and 1987, but most of the effect occurred seaward of the 2-m depth contour. The effects of the causeway in the August event were overwhelmed by the regional-scale effect. The spatial extent of high-salinity water observed for each of these 1986 events is described below.

21 July 1986.—The upwelling of marine bottom waters to the 2–3-m depth contour followed 18 d of predominantly moderate-to-light east and northeast winds, although strong westerly winds occurred for about 12 h on 18 July. Surface salinities around the Endicott Causeway and across the Sagavanirktok River delta were less than 10‰ on 19 July, and increased to near 20‰ in the lee of the MPI (but not elsewhere on the delta) on 20 July. By 21 July, the data at two fyke nets in the lee of the MPI showed surface salinities in excess of 26‰, whereas salinity at the other stations sampled across the delta remained below 10‰. On 22 July, only one station in the lee of the MPI had a salinity in the range of 20–25‰; surface salinity at the other station was less than 10‰. By 23 July, the salinity in the entire area was less than 10‰. Strong northeast winds coincided with the observed reduction in surface salinity over 22–23 July. This reduction in salinity levels was likely caused by mixing of the water column and demonstrates that the proportion of high-salinity water in the area was small.

Transect (CTD) data for 22 July 1986 were taken for onshore–offshore transect series D–I. These transects were located from the interisland-to-shore causeway west to the eastern margin of Prudhoe Bay (see Figure 9). The surface expression of upwelling on 22 July was restricted to the lee of the interisland causeway, especially in the area adjacent to the middle third of the causeway (Figure 15). The 15, 20, and 25‰ surface contours converged in this same area, sharply delineating the eastward extent of high-salinity surface waters along the causeway. The high-salinity (>25‰) pool of surface water occupied a tear-shaped area of about 414 hectares, about 8.5% of the total available habitat less than 2 m deep west from the mainland causeway to Heald Point. This entire surface effect is attributable to the wake effect of the causeway. Inshore of the high-salinity plume, an approximately 3.65-km-wide corridor of surface waters less than 15‰ extended across the delta. Seaward of the interisland causeway on 22 July, the low-salinity surface plume extended along the causeway for its entire length, flowing north and northwest beyond the MPI before mixing to 20‰ (Figure 15).

The 25‰ bottom salinity contour inshore of the interisland causeway impinged against the causeway just east of the point of the tear-shaped area and extended to the northwest, moving seaward

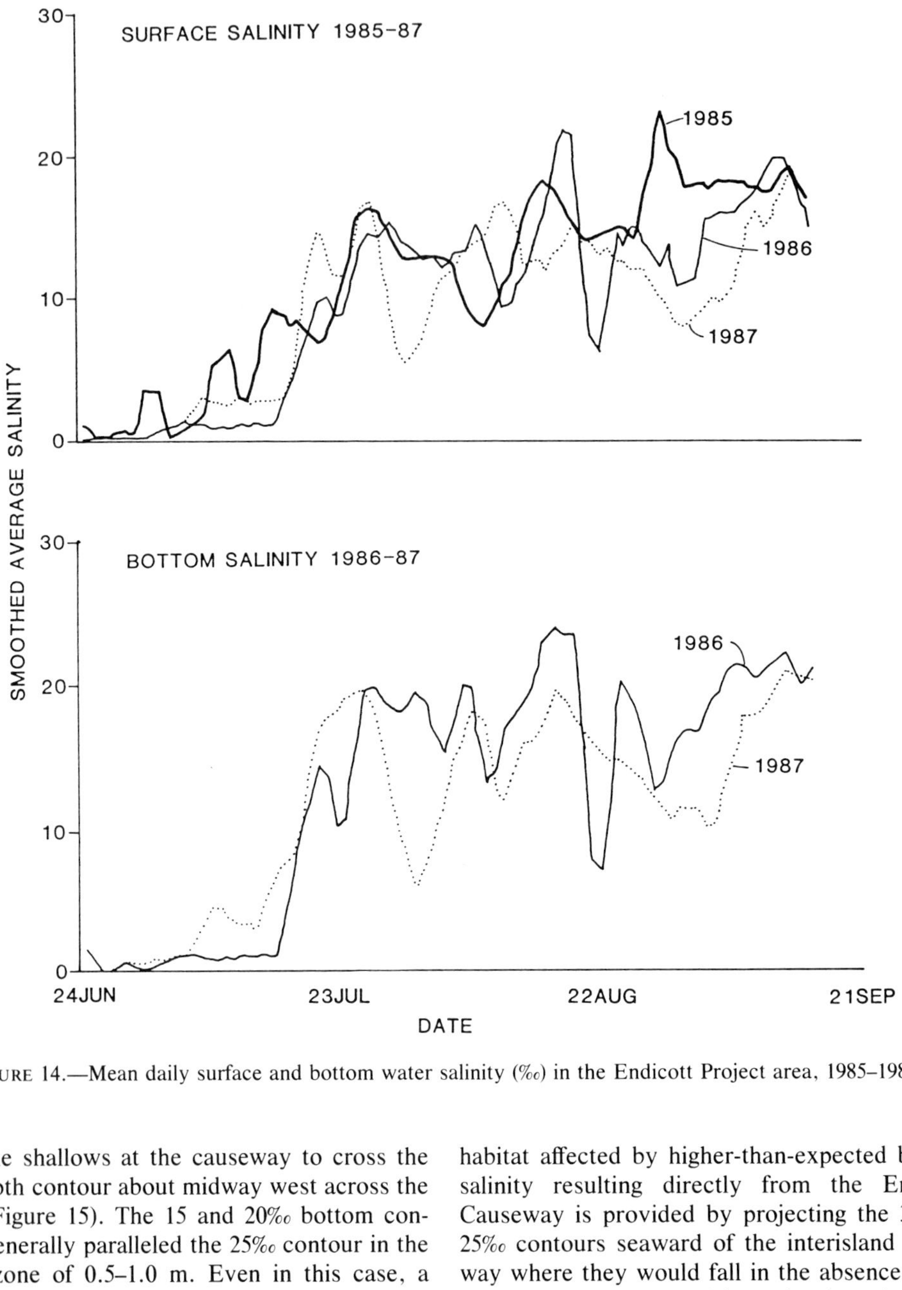

FIGURE 14.—Mean daily surface and bottom water salinity (‰) in the Endicott Project area, 1985–1987.

from the shallows at the causeway to cross the 1-m depth contour about midway west across the delta (Figure 15). The 15 and 20‰ bottom contours generally paralleled the 25‰ contour in the depth zone of 0.5–1.0 m. Even in this case, a 1.82–3.65-km-wide corridor of low-salinity water extended on the bottom across the delta. The 25 and 30‰ bottom contours on the seaward side of the causeway intersected the structure somewhere near or east of the MPI.

The total area shoreward of the interisland causeway affected by bottom salinities in excess of 25‰ on 22 July 1986 was about 1,267 hectares, or 27% of the nearshore (<2 m) habitat from the causeway west to Heald Point. An estimate of the habitat affected by higher-than-expected bottom salinity resulting directly from the Endicott Causeway is provided by projecting the 30 and 25‰ contours seaward of the interisland causeway where they would fall in the absence of the eastward and shoreward incursion into the lee of the causeway (Figure 15). Of the 27% nearshore habitat affected overall, about 8%, or 378 hectares, of the bottom effect appears directly attributable to the causeway.

The observed effect on surface and bottom waters lasted only 1–2 d. Fyke net and CTD data from 23 July showed that an abatement of high-salinity conditions occurred as winds continued from the northeast but increased in strength and

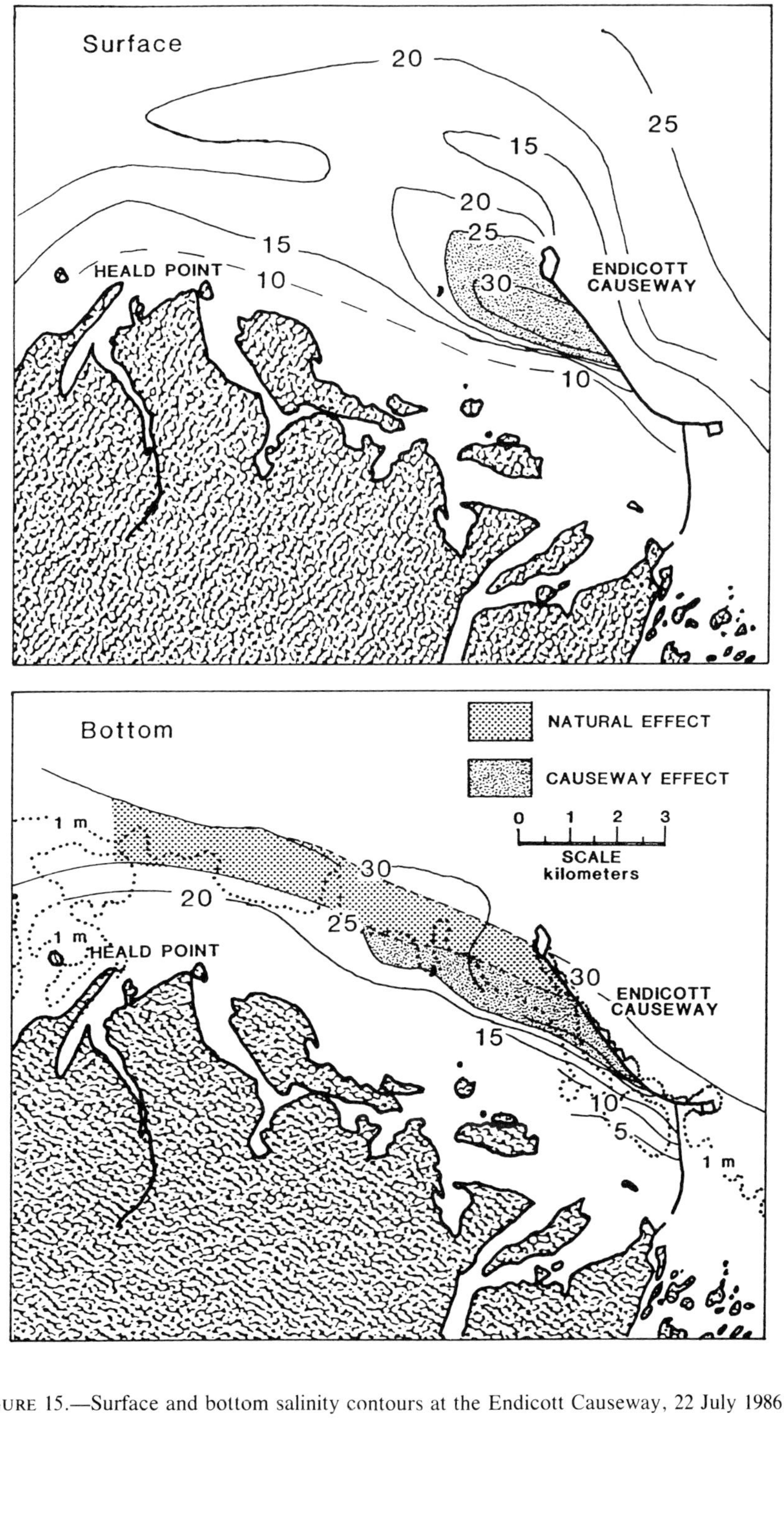

FIGURE 15.—Surface and bottom salinity contours at the Endicott Causeway, 22 July 1986.

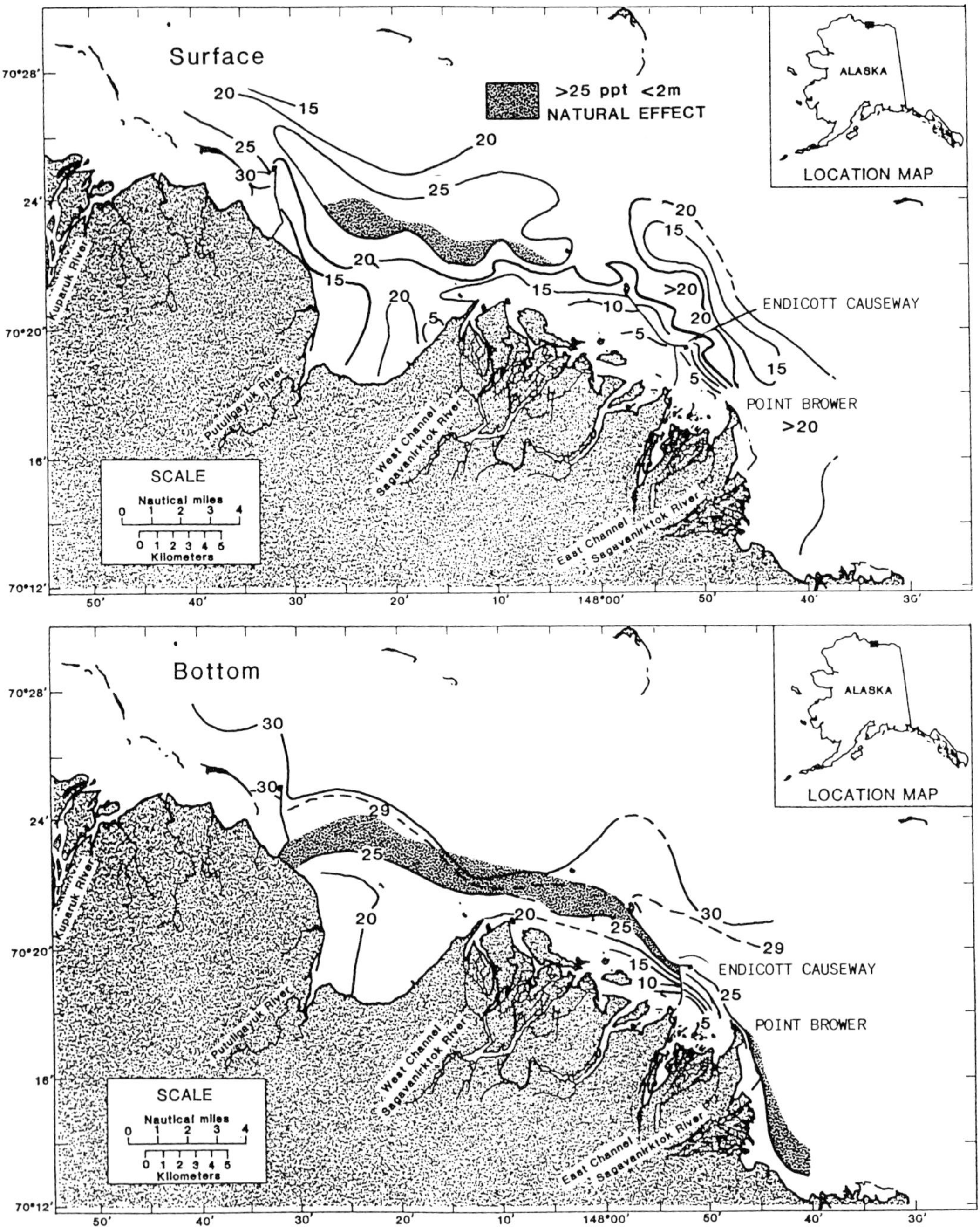

FIGURE 16.—Surface and bottom salinity contours for the Prudhoe Bay region, 15 August 1986.

mixed the waters. About midday of 25 July, winds slackened then turned westerly. A combination of north–south winds brought a mild intrusion into northwest Prudhoe Bay on 26 July, but the effect was restricted spatially and temporally.

15–16 August 1986.—The surface salinity patterns indicated by the CTD sampling on 15 August were complex (Figure 16). A plume of low-salinity water appeared to flow seaward to the northwest away from the study area just east of Point

Brower. West of Point Brower, the 20‰ contour extended across the delta just seaward of the interisland causeway, then across Prudhoe Bay to the end of West Dock. A surface expression of water with salinity greater than 25‰ was apparent across the mouth of Prudhoe Bay, mostly outside the 2-m depth contour (Figure 16).

The 25‰ bottom salinity contour extended along or even inshore of the 1-m depth contour east of the delta in Foggy Island Bay; from there it turned north and westward to the shore at Point Brower, and continued westward from that point to the satellite drilling island (SDI) of the Endicott Causeway (Figure 16). West of the Endicott Causeway, the 25‰ bottom salinity contour wavered along, or just inshore (1 km or less) of the 1-m depth contour to the northwest shore of Prudhoe Bay (Figure 16). The nearshore positioning of this contour east of Point Brower confirms that its nearshore location west of the causeway was not an effect of the causeway. Britch et al. (1983) reported strong thermoclines at water depths of 1–3 m in nearshore waters across the Sagavanirktok River delta before causeway construction. Below the thermocline, salinities were typically at or above 30‰, compared with 14–16‰ above the thermocline (Britch et al. 1983).

West and southwest winds of 16–23 August brought an onshore surge of the high-salinity waters into Prudhoe Bay beginning on 17 August. High-salinity water flowed in from the northwest, and appeared to move counter-clockwise around the bay. Salinities were markedly greater in Prudhoe Bay than on the Sagavanirktok River delta and did not fully dissipate until about 29 August.

Effect on Continuity of the Alongshore Zone

Little or no evidence supports the contention that the effects of the Endicott Causeway on local oceanography have caused high-salinity–low-temperature discontinuities that would block fish movement into or across the low-salinity refugia of the delta. Even under the most severe upwelling–intrusion event of mid-August 1986, a 1.6–3.2-km-wide corridor of low-salinity water extended across the delta. On occasion, the outer breach of the causeway may lie in waters with sufficient salinity to restrict anadromous use. However, under these conditions, which are ephemeral, the nearshore breach always lies in waters of favorable temperature–salinity, which should permit anadromous fish use and passage.

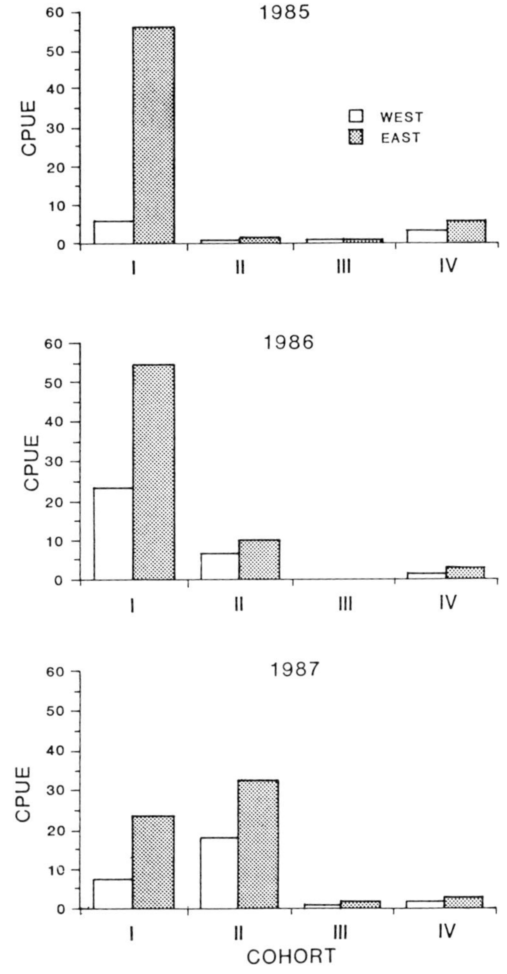

FIGURE 17.—Mean catch per unit effort (CPUE: fish/[cod end · d]) of Arctic ciscoes, size-cohorts I–IV, in the Endicott Project area, 1985–1987. Catch is shown for each side of the net. East-side catches denote westward-moving fish and west-side catches denote eastward movement.

This was the primary reason why the nearshore breach was incorporated in the project design.

During the first 3 years of the monitoring program, meteorological conditions tended to magnify potential effects of the causeway on low-salinity, refuge habitat for anadromous fish, especially 1985 and 1986. Extended periods of light northeast winds each year caused regional-scale upwellings close to the shore. Similar wind patterns had not occurred in the only other year a study of the dimension of the Endicott Monitoring Program had been conducted (1984). In 1984, a mixture of east and west winds produced more

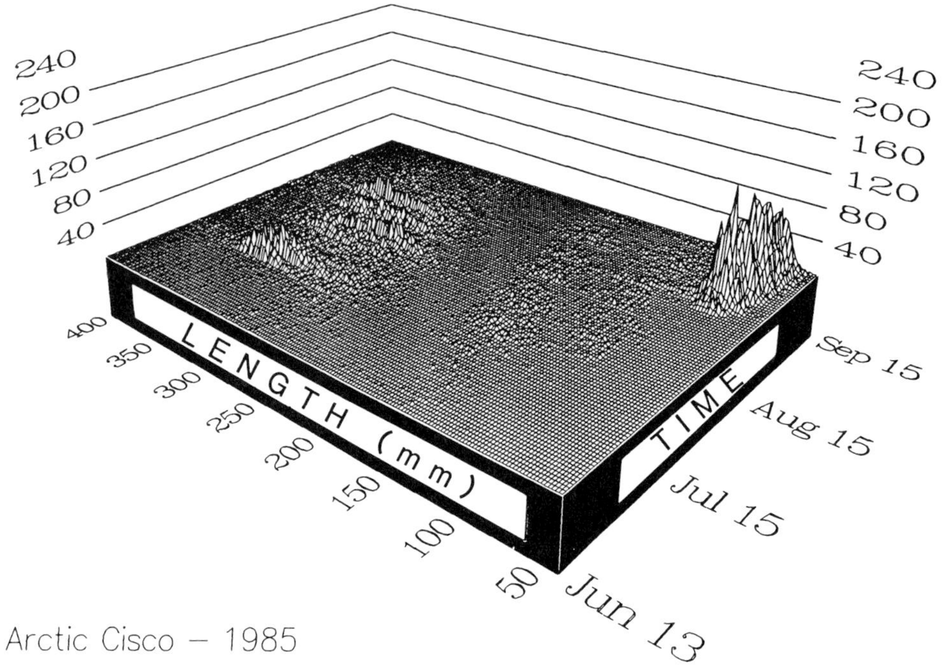

FIGURE 18.—Length frequency of Arctic ciscoes in the Endicott Project area, 1985.

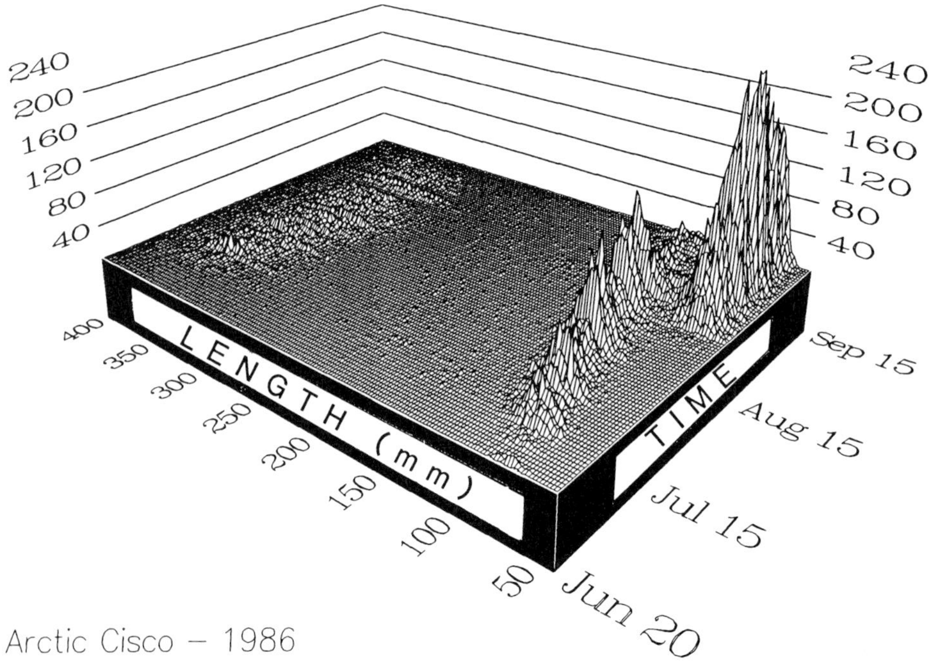

FIGURE 19.—Length frequency of Arctic ciscoes in the Endicott Project area, 1986.

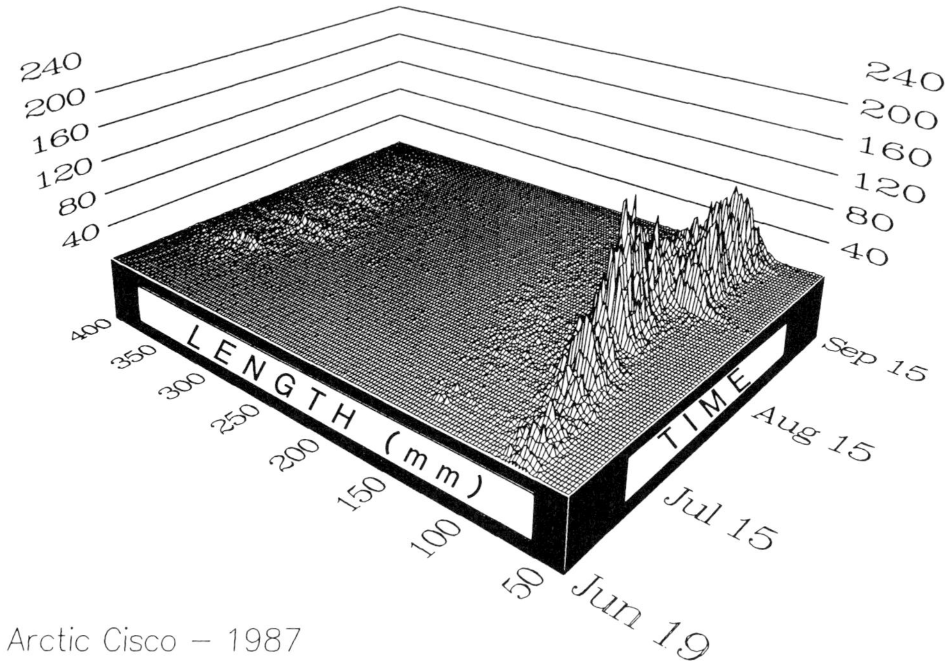

FIGURE 20.—Length frequency of Arctic ciscoes in the Endicott Project area, 1987.

moderate temperature–salinity patterns on a regional basis than those that occurred in 1985–1987. Given that meteorological conditions favorable for the development of habitat discontinuities at the Endicott Causeway occurred during 1985–1987 but no discontinuity was evident, it is unlikely that such events will occur.

Movement Patterns of Young Arctic Ciscoes

Annual abundance and size.—Young-of-the-year Arctic ciscoes (cohort I) were abundant in the project area during each year of 1985–1987 (Figure 17). Each year was characterized by an extended period of east and northeast winds during July–mid-August, as was the case at Barter Island, farther to the east. Young-of-the-year Arctic ciscoes enter the coastal zone at the mouth of the Mackenzie River during early July (Bond 1982). Once in the coastal zone, their movements are thought to be strongly governed by coastal currents, which are driven by winds (Gallaway et al. 1983, 1989). Several studies offer support for this hypothesis: simulation models of fish movement that incorporate temperature preference, salinity levels, and currents (Fechhelm and Gallaway 1983); documentation of actual dispersal patterns of young-of-the-year Arctic ciscoes from the Mackenzie River during 1985–1988 (Moulton 1989; Schmidt et al. 1989); correlation analyses between catch levels in the Colville River fishery and the wind patterns 5 years before (Fechhelm and Fissel 1988); and correlation of year-class strength in Alaska and prevalence of easterly winds in the eastern Alaskan Beaufort Sea during July–15 August of the same year (Fechhelm and Griffiths 1990).

Abundance of young-of-the-year Arctic ciscoes in the project area was higher in 1985 and 1986 than in 1987 (Figure 17), but the size of cohort-I Arctic ciscoes in 1987 was markedly larger than the sizes of cohort-I fish in 1985 and 1986 (Figures 18–20). Length-frequency plots in Figures 18–20 show only data for fish measured, and are not adjusted for the differences in annual effort or extrapolated to the entire collection. They should not be used as an quantitative index to annual abundance. The smallest size-cohort in each graph represents young-of-the-year fish (cohort I), the next smallest size-group represents 1-year-old fish (cohort II) and the third smallest size-group represents 2-year-old fish (cohort III). Cohort IV represents fish 3 years and older.

Young-of-the-year fish were generally present

from mid-August through 15 September, the date when sampling was terminated. Few fish of cohorts II and III were caught and measured in 1985, and the few that were taken were not generally encountered in the project area until about the middle (cohort III) or end of July (cohort II: Figure 18). In 1986, cohort-II fish were present earlier but were relatively rare in samples taken from 20 June through 1 July. Abundance apparently increased after early July, because large sample sizes were measured from July to mid-August. However, few cohort-II fish were measured after young-of-the-year began to dominate the samples (Figure 19).

Two marked differences were apparent in the 1987 length-frequency data (Figure 20), compared with data for 1985 and 1986 (Figures 18, 19). First, although the time of arrival and period of abundance was about the same in 1987 as in 1985 and 1986, the size of the young-of-the-year fish was markedly greater in 1987. Second, cohort-II fish were not rare during the early part of the 1987 season. The first observation suggests that conditions encountered before or during the 1987 migration were more favorable for growth than conditions in 1985 and 1986; and the second suggests that a higher proportion of young-of-the-year fish may have overwintered in the Sagavanirktok River delta in 1986 than in 1985. Schmidt et al. (1989) observed few young-of-the-year Arctic ciscoes overwintering in the Sagavanirktok River delta during 1985, although large specimens were relatively abundant.

The relative abundance of cohort-II fish in 1986 and 1987 versus the abundance and mean CPUE by side of net of cohort-I fish in 1985 and 1986 support the contention that more young-of-the-year Arctic ciscoes overwintered in the Sagavanirktok River delta in the winter of 1986–1987 than in the winter of 1985–1986 (Figure 17). In 1985, mean CPUE by side of net suggested that cohort-I fish moved through the area in a strong east-to-west direction, consistent with the prevalent northeasterly winds (Figures 11, 17). Mean CPUE on the west side of the nets was only about 11% of the mean CPUE on the east side of the nets. West winds occurred on about 20% of the sampling days.

In 1986, mean CPUE of cohort-I fish on the west sides of the nets was higher than in 1985 and was 43% as large as the mean CPUE on the east side of the nets. As will be shown later, most of the west side net catches were made at Heald Point as young-of-the-year fish moved out of Prudhoe Bay into low-salinity waters of the delta in response to marine intrusions on west winds. In 1986, protracted periods of both east and west winds occurred 15 August–15 September; west winds occurred 52% of the time. Mean CPUE of cohort-II fish in the project area was low in 1986 compared with mean CPUE of cohort-II fish in 1987. Although some cohort-II fish may have entered the project area from the Colville River each year, thereby influencing the results, it seems likely that more young-of-the-year Arctic ciscoes overwintered in the Sagavanirktok River delta in the winter of 1986–1987 than in the winter of 1985–1986.

In 1987, the mean CPUE of cohort-I Arctic ciscoes on the west side of nets was 32% as large as the mean east-side CPUE (Figure 17). Similar to 1986, west winds after 15 August were estimated to occur 53% of the time. Overall, the ratio of west-to-east side CPUE (0.11, 0.43, and 0.32; 1985–1987) tended to increase with percent of time the winds in the project area were from the west (20, 52, and 53; 1985–1987).

Temporal and spatial distributions.—In 1985, the mean daily CPUE of cohort-I Arctic ciscoes was low between 15 and 19 August, after which mean daily CPUE increased dramatically, generally above 50 fish/(cod end · d) for the remainder of the season (Figure 21). Marked peaks in the mean daily CPUE levels occurred periodically from 22 to 30 August, after which mean daily CPUE levels were lower and more uniform. During 22–30 August 1985, mean daily CPUE exhibited peaks on 22, 24–25, 28, and 30 August. The peak on 22 August mostly reflected a large catch in Foggy Island Bay about 6 km east of the Sagavanirktok River delta, and the peak on 24 August reflected a high catch at station 7 on the east side of the Endicott Causeway. (See Figure 7 for station sites.) On 25 August, another high CPUE was observed at station 7, but peak CPUE was at station 19 west of the Endicott Causeway across Prudhoe Bay at the east base of the West Dock Causeway. The CPUE peak on 28 August reflected another huge catch in Foggy Island Bay as well as high catches seaward of the Endicott Causeway, at station 19 at the east base of West Dock and at station 25 on the west side of the Kuparuk River delta. On 30 August, the highest catches were made around the Kuparuk River delta west of West Dock and in Prudhoe Bay. Few fish were caught anywhere east of this area around the Sagavanirktok River delta. The data for this period clearly demonstrate movement

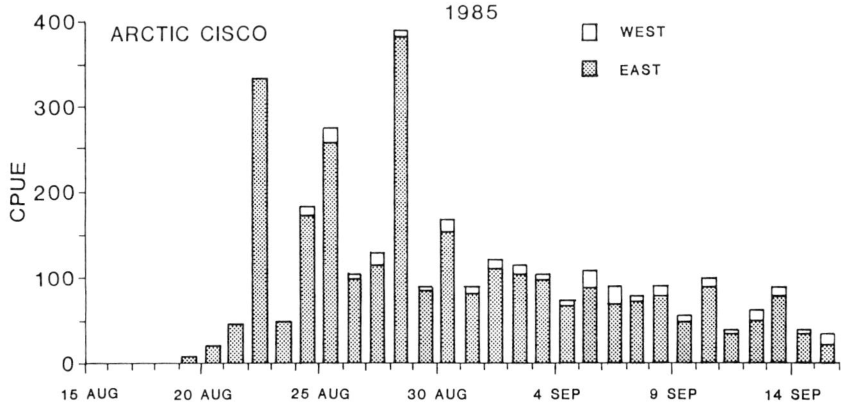

FIGURE 21.—Mean daily catch per unit effort (CPUE: fish/[cod end · d]) for young-of-the-year Arctic ciscoes in the Endicott Project area, 1985. Shaded segment of bars represent catches from the east sides of the nets, and the white segment shows west-side catches.

from east to west across the study area, shown by successive peak east-side-of-net CPUE values from east to west. The same pattern was usually evident for other periods, but was not as easy to discern in the absence of notably large catches.

Summary of the CPUE data on a spatial basis in 1985 did not indicate any thermal trapping effect on the east side of the Endicott Causeway (Figure 22). Such an effect would be indicated by high CPUE values at stations 4, 5, and 6, along with an equitable distribution of catch by side of net resulting from milling behavior of the fish confronted with a barrier. Fish apparently approached the project area along the shoreline from the east, moved around the outer delta seaward of the interisland causeway, and then rejoined the shoreline on the northwestern side of Prudhoe Bay. Relatively high CPUE values at the Kuparuk River delta suggested that the fish then bypassed West Dock and continued westward. This pattern is entirely consistent with the premise of wind-governed dispersal with little impedance from the area causeways.

Mean daily CPUE patterns in 1986 exhibited a pronounced cyclical pattern; peaks occurred during 15–16, 24–26 August, 31 August–2 September, 5–8, 11, and 13–15 September (Figure 23). Even though CPUE values for the east side of the fyke nets were usually higher than those for the west side, marked exceptions to this pattern were observed in early September, especially at some nets.

The CPUE on 15 August reflected high catches made under east winds on the east sides of only

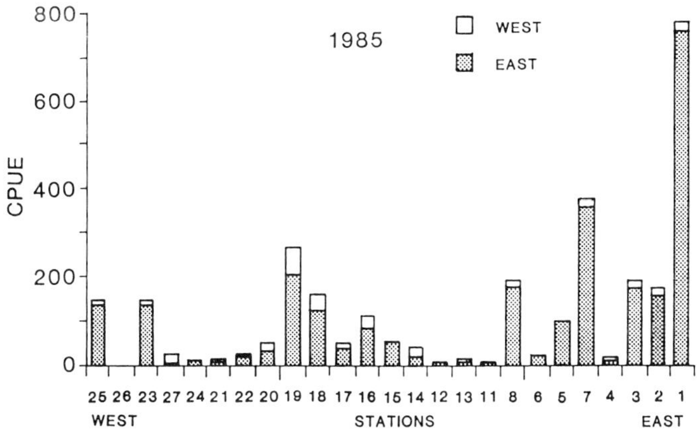

FIGURE 22.—Mean catch per unit effort (CPUE: fish/[cod end · d]) for young-of-the-year Arctic ciscoes at stations sampled in the Endicott Project area, 1985. Shaded and white bars show east- and west-side catches, respectively.

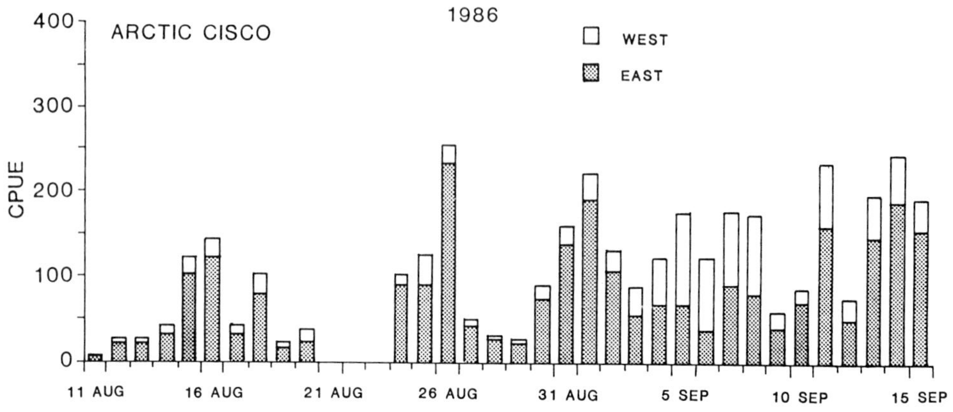

FIGURE 23.—Mean daily catch per unit effort (CPUE: fish/[cod end · d]) for young-of-the-year Arctic ciscoes in the Endicott Project area, 1986. Shaded segment of bars represent catches from the east sides of the nets, and the white segment shows west-side catches.

two nets, one at station 16 on the eastern shore of Prudhoe Bay, and the other at station 23 on the east side of the Kuparuk River delta. Smaller CPUE values were observed at station 20 at the west base of West Dock, mostly from the east side of the net, and at station 25, mostly from the west side of the net. On 16 August, the high mean daily CPUE resulted from a single catch at station 23, on the west side of the Kuparuk River delta. From 16 to 17 August, light westerly winds developed, became increasingly stronger during 18–19 August, and were of storm strength between 19 and 23 August. Although nets could not be tended on 21–23 August, mean daily CPUE had already declined to low levels by 19 and 20 August.

The first mean daily CPUE peak of 1986 at stations 16 and 23 demonstrated that when the fish arrived from Canada they were more offshore than the nets in Foggy Island Bay and at the Endicott Causeway. The implication is that the eastward dispersal was not confined to the shoreline, but rather was associated with a particular water mass. In 1986, as in 1985, the open-water season from 1 July–15 August was characterized by an extended period of northwesterly winds. However, progressive vector plots for this period showed that 1986 was characterized by a stronger eastern (versus northeastern) component than was the case in 1985 (Figure 11). This difference may have affected the location of the water mass in which the fish were moving.

During 23–26 August, the west winds shifted from the east and were relatively light (Figure 11). Mean daily CPUE peaked on the 26th, mainly from a particularly large catch at the east side of station 1, located east of the project area in Foggy Island Bay. However, wind strength greatly increased on the 26th, and remained strong from the east through 29 August. Mean daily CPUE plummeted, and was low through 29 August, possibly because the fish were transported offshore of the nets by Ekman transport or, more likely, because the rough water combined with the low water level caused fish to leave the shoreline (Craig and Haldorson 1981).

The strong east winds decreased on 30 August, then turned southwesterly until 31 August, when moderate east winds prevailed. These moderate east winds ended midday on 2 September, when strong southwest winds began and blew until 6 September (Figure 11). Mean daily CPUE increased from 30 August to 1 September, when a peak was reached, after which a decline to a relative low on 3 September occurred (Figure 23). The highest CPUEs on 30 August were observed at four stations (16, 68, 18, 19) along the southern and western shores of Prudhoe Bay, where the fish were apparently moving eastward. As had occurred during the initial abundance peak, the fish first arrived in Prudhoe Bay from off shore. The increase in mean daily CPUE that occurred on 31 August and 1 September resulted from fish arriving at the shore in areas both east and west of Prudhoe Bay. The subsequent CPUE by side-of-net patterns among stations suggested that most of these fish may have moved westward beyond the area by 3 September (Figure 24).

The sequential directions of the daily wind vectors for 5–12 September were to the northeast (5 September), west (6 September), south (7 Sep-

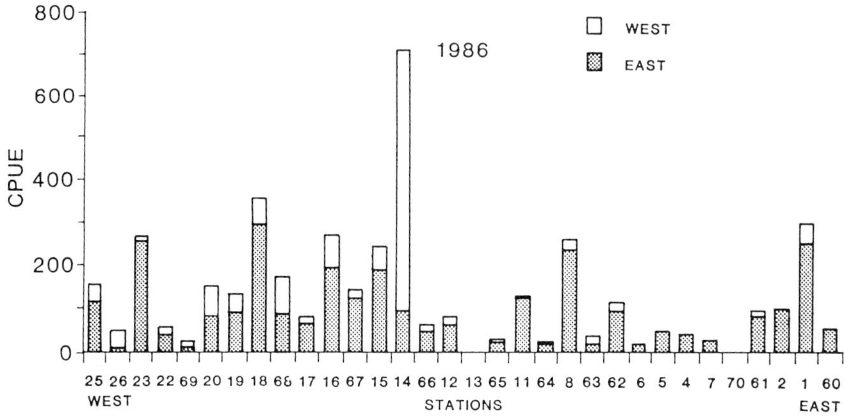

FIGURE 24.—Mean catch per unit effort (CPUE: fish/[cod end · d]) for young-of-the-year Arctic ciscoes at stations sampled in the Endicott Project area, 1986. Shaded and white bars show east- and west-side catches, respectively.

tember), northeast (8–9 September), north (10–11 September), and finally to the east on 12 September (Figure 11). Overall, the general direction of water flow would have been easterly with a shoreward surge, except for the west and north directional components, which would have carried water west and offshore, respectively. Except for 9–10 and 12 September, mean daily CPUE was high during this period, and the average catches on the west side of the nets usually exceeded those on the east side of the nets, suggesting eastward movement of the fish (Figure 23). The most remarkable aspect of the high catches on the west side of nets was that they were made mainly from the Heald Point net (station 14, Figure 24). This suggests that when the fish were near shore on west winds they may have been better able to respond to favorable salinity and temperature conditions than when they were more off shore under east winds.

Moderate east winds resumed on 13 September and continued for the next 2 d. Mean daily CPUE increased from 12 to 13 September and remained high (Figure 23). Most fish were caught on the east side of the nets at all stations where large catches were made, except at Heald Point. At this station (14), large catches on the west side of the net suggested that fish in southern Prudhoe Bay continued to move into the low-salinity waters of the delta.

The CPUE patterns among stations in 1986 (Figure 24) were similar to those observed for 1985 (Figure 22), but with two important differences. The first is the above-mentioned movement of young-of-the-year Arctic ciscoes into the Sagavanirktok River delta from Prudhoe Bay on west winds. The second is the increase in CPUE at stations on the west side of the delta in 1986 compared with 1985. This increase probably resulted from installation of the permanent breaches in the Endicott Causeway in winter 1985–1986. Young-of-the-year Arctic ciscoes apparently moved through the inner breach near station 62 and through the outer delta (stations 11 and 12) as well as around the seaward side of the causeway (stations 7 and 8).

The reduction in sampling effort in 1987 corresponded to a marked decrease in mean daily CPUE. Two of the best stations (7 and 14) in the Prudhoe Bay area in terms of CPUE were not sampled in 1987, nor was station 12. Stations 12 and 14 were both 1982 baseline stations. The loss of stations in Foggy Island Bay and at the Kuparuk River made it difficult to determine the nature of fish movement through the project area. However, late-season sampling near Harrison Bay was conducted and can be used to estimate westward movement.

The mean daily CPUE levels in 1987 seldom exceeded, or even reached, 100 fish/(cod end · d); such events occurred only on 24 August and 7–8 September (Figure 25). Young-of-the-year Arctic ciscoes were first recorded in appreciable numbers on 16 August, a day of north winds, and were taken at only two stations (8 and 70). At station 8, most of the catch was taken on the west side of the net, whereas in the collection east of Point Brower at station 70, nearly all the fish were taken on the east side of the net. Fish were taken at more stations, but in lower numbers, on 17 August, and a very large collection was taken on the east side of the net at station 8 on 18 August.

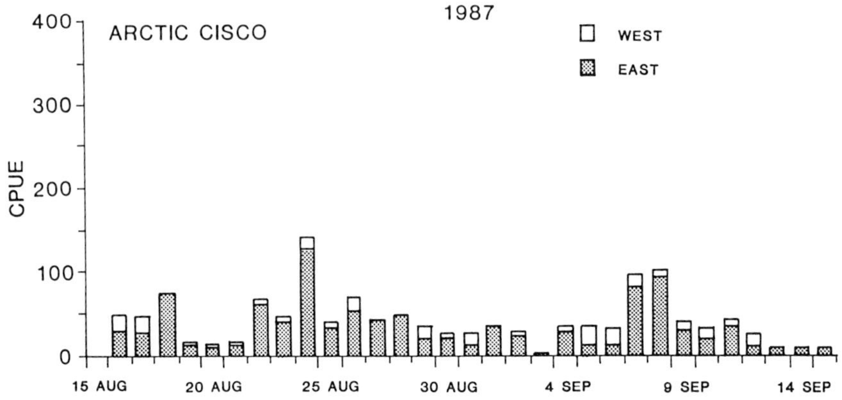

FIGURE 25.—Mean daily catch per unit effort (CPUE: fish/[cod end · d]) for young-of-the-year Arctic ciscoes in the Endicott Project area, 1987. Shaded segment of bars represent catches from the east sides of the nets, and the white segment shows west-side catches.

Winds were from the northeast during 15–18 August. Sampling at Oliktok Point in 1987 did not begin until 19 August and was conducted until 8 September. Few young-of-the-year (size-cohort I) fish were present early; most were caught on 21–22 August. After 22 August, they were seldom encountered until 4 September.

Strong southwest and west winds dominated most of the period from 18 to 31 August. Mean daily CPUE was low throughout this period except for 24 August (Figure 25). Typically, fish were found to be abundant at only a few stations on a given day. On 24 August, high catches were made at three stations, Point Brower (station 70), the nearshore breach (station 62), and seaward of the interisland causeway (station 8). High catches on other days during this period were spotty and restricted to the area between Point Brower and the western side of the Sagavanirktok River delta. Mean daily CPUE on the east side of the nets was generally much larger than on the west side (Figure 25).

Although the side of net catch patterns might be interpreted to suggest that the fish were moving west against the currents during this period, relatively few fish were caught at stations to the west of the Sagavanirktok River delta. For example, on 24 August, the CPUE values for the east side of the nets at Point Brower (station 70), at the inner breach (station 62), and seaward of the interisland causeway (station 8) were 450, 804, and 587 fish/(cod end · d), respectively. West of the delta, the maximum CPUE for the 7-d period 25–31 August was 149 fish/(cod end · d) at station 72 on 26 August. If the catches at station 72 on 26 August reflected the same fish abundant in the delta on 24 August, this would equate to a movement rate to the west of about 10 km/d. At this rate, the fish should have arrived at Oliktok Point in about 5 d, on 31 August. They were not abundant in the Oliktok Point area until 4 September. The clustering of high catches around the Sagavanirktok River delta during the west-wind period 19–31 August thus may have resulted from fish remaining in the area, despite the predominance of catches on the east side of the nets. High CPUE values were observed only at stations in the lee of structures where eddy flow (wake effect) might explain why fish were caught on the east side of the nets even though the fish were not moving beyond the area. Some fish may have also moved into the river during this period.

During 1–4 and 6–8 September, winds were strong and from the northeast; they blew from the north on 4 September and from the southwest on 5 September. Mean daily CPUE at Oliktok Point peaked on 4 September after 4 d of strong northeast winds. In the Sagavanirktok River delta, catches were generally low except for 7–8 September (Figure 25). On 7 September, the high CPUE levels were associated with the east side of the nets at Niakuk Island: 797 fish/(cod end · d) at station 15 and 230 fish/(cod end · d) at station 67. On 8 September, the high CPUE levels were again associated with the east side of only two nets, but they were west of West Dock, at station 22 on Stump Island (1,519 fish/[cod end · d]). This CPUE pattern clearly reflects westward movement. No other high CPUE levels occurred during the rest of the season. Winds during 9–15 Septem-

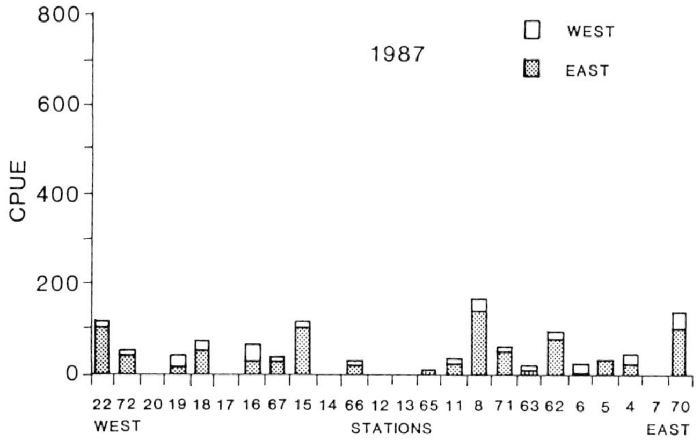

FIGURE 26.—Mean catch per unit effort (CPUE: fish/[cod end · d]) for young-of-the-year Arctic ciscoes at stations sampled in the Endicott Project area, 1987. Shaded and white bars show east- and west-side catches, respectively.

ber were from the southwest. The effects of the few large catches described above on the overall station abundance pattern for 1987 are shown in Figure 26. Highest CPUE was observed at stations 70, 62, 8, 15, and 22.

Assessment.—The 1985–1987 catch patterns of young-of-the-year Arctic ciscoes provide additional support for the hypothesis that the dispersal of young-of-the-year Arctic ciscoes is a current-governed transport phenomenon, modified to some degree by behavioral responses of the fish to environmental gradients. The behavioral responses appeared to be more pronounced under west-wind conditions than under east winds. During west winds, the fish tended to be closer to the shoreline than during east winds, because Ekman transport produces an onshore flow. Currents along the shoreline are more likely to be disrupted than those further off shore because of the many stream mouths, points, spits, and bights characteristic of the shorefront; thus, fish near shore can better respond to more favorable conditions, if present. This is suggested by the data from 1986, when the fish apparently moved into the Sagavanirktok River delta at Heald Point under west winds. Also, a large seine catch of young-of-the-year Arctic ciscoes was made in the Putuligayuk River on 17 August 1986, during a west-wind period when salinities on the west side of Prudhoe Bay were high. Fish moving into such areas may remain there or return to the coastal zone as salinities recede and then be carried to the west when east winds resume.

Such an explanation suggests that the distance fish are transported to the west on east winds might not be markedly reduced on west winds, if the fish can shelter in river mouths. It also suggests that young-of-the-year Arctic ciscoes would likely be found at the mouths of many rivers at the start of winter, and their fate would then depend on the carrying capacity of that habitat.

Two schools of thought have emerged concerning the recruitment of young-of-the-year fish to Alaskan waters, especially the Colville River. One hypothesis is that the fish are genetically predisposed to migrate to the Colville River; that is, a distinct population or stock has evolved that relies upon the Colville River as overwintering habitat and the adjacent coastal area as a summer nursery ground. The opposing hypothesis is that the recruitment is simply a result of wind-governed transport of young-of-the-year fish entrained in coastal currents. The correlation of catch and wind patterns (Fechhelm and Griffiths 1990), as well as the available genetic evidence (Bickham et al. 1989), strongly support the latter premise.

In summary, the Endicott and West Dock area causeways do not appear to affect the westward dispersal of young-of-the-year Arctic ciscoes; this is an expected outcome if dispersal is mainly a transport phenomenon. On west winds, the fish are moved onshore into less marine habitats and apparently move into river mouths if the fish encounter them. The fish may use such areas as refugia and may also overwinter in these habitats.

Condition and Growth Changes

Condition and in-season growth data can provide responsive indices to the health of a population, as well as an integrated measure of the success of fish in acquiring food and converting

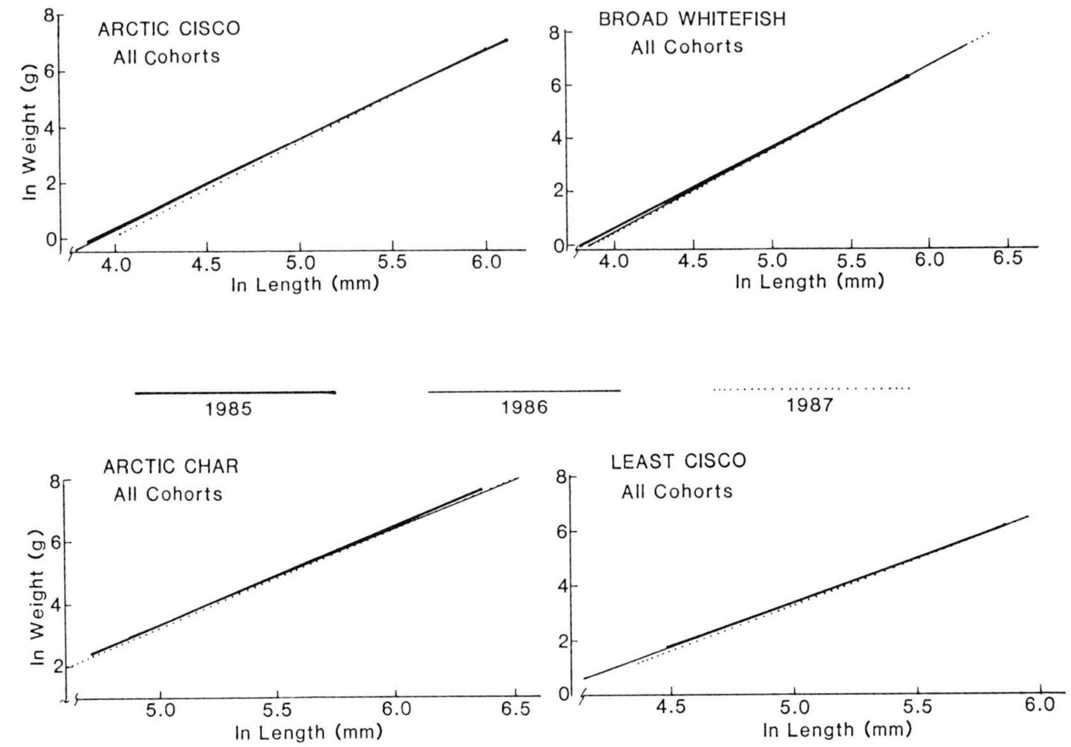

FIGURE 27.—Length–weight regressions for Arctic ciscoes, broad whitefish, Arctic char, and least ciscoes collected in the Endicott Project area, 1985–1987; ln is $\log_e$.

that food to body tissue, if all other factors are equal. Whereas growth analysis has a long history of use, and its utility for detecting sublethal effects should be obvious, length–weight relationships are less frequently used to measure change in population condition. The method is not new, however; Le Cren (1951) demonstrated that comparisons of length–weight regressions by analysis of covariance provide an accurate analysis of condition. Published papers using length–weight relationships and analysis of covariance to compare conditions of fish populations include Le Cren (1951), Held and Peterka (1974), Armstrong et al. (1977), Mohsin and Gallaway (1977), Harrison and Hadley (1979), McAda and Wydoski (1983, 1985), Dolloff (1987), Jessop (1987), Vives (1987), Crossland (1988), Maceina and Murphy (1988), and Osenberg et al. (1988). More recently, Cone (1989) has shown that the calculation of ordinary least-squares regression parameters is the preferred method for evaluation of fish condition.

Results of the length–weight regression analyses that we used to compare condition of study area anadromous fish species among the first three post-construction years are detailed below and in Figure 27. Regressions for the years (1985–1987) joined by the same underline had the same or similar slopes, and no significant differences in weight at a given length were observed in these years.

Arctic cisco	85 86 87
Broad whitefish	85 86 87
Arctic char	85 86 87
Least cisco	85 86 87

Broad whitefish and Arctic char exhibited no significant differences either in the slopes or intercept values of the length–weight regressions over the 3-year period. Arctic and least ciscoes exhibited different length–weight regressions in 1987 compared with 1985 and 1986, but no significant differences in condition were detected between 1985 and 1986. Each of the 1987 regressions for these two species had a significantly steeper slope in 1987 than in 1985 and 1986 (Figure 27).

Results of fitting the length-frequency data for Arctic cisco and broad whitefish to the von Bertalanffy growth curve and testing for differences among the parameters following Galluci and Quinn (1979) showed these growth patterns:

Cohort	Arctic cisco	Broad whitefish
I	85 86 87	85 86 87
II	85 86 87	85 86 87
III	85 86 87	85 86 87

Years (1985–1987) joined by the same underline are years of homogeneous growth. For Arctic cisco cohorts I and II, and all cohorts of broad whitefish, growth either remained the same over 1985–1987 or increased in 1987 (Figures 28, 29). For cohort-III Arctic ciscoes, growth was highest in 1985 and lowest in 1986 (Figure 28).

Results of independent growth and condition analyses for anadromous fish have been reported elsewhere for each year of the Endicott Monitoring Program (e.g., Whitmus et al. 1987). Many comparisons were made among years, regions,

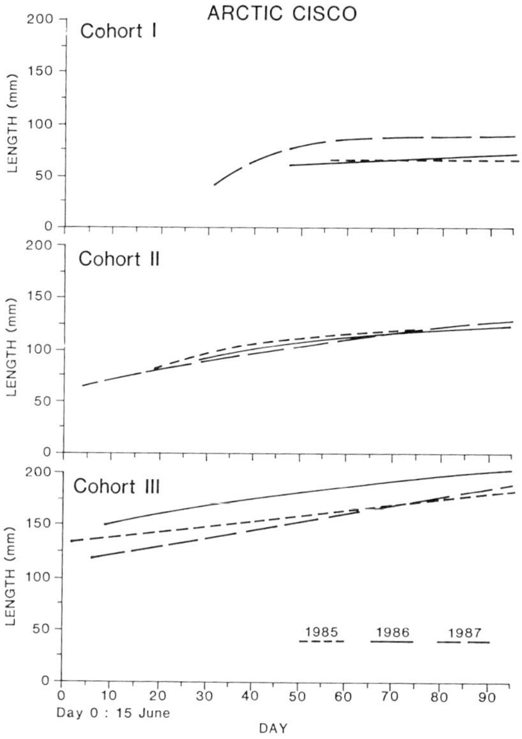

FIGURE 28.—Growth curves for Arctic ciscoes, size-cohorts I–III, collected in the Endicott Project area 1985–1987.

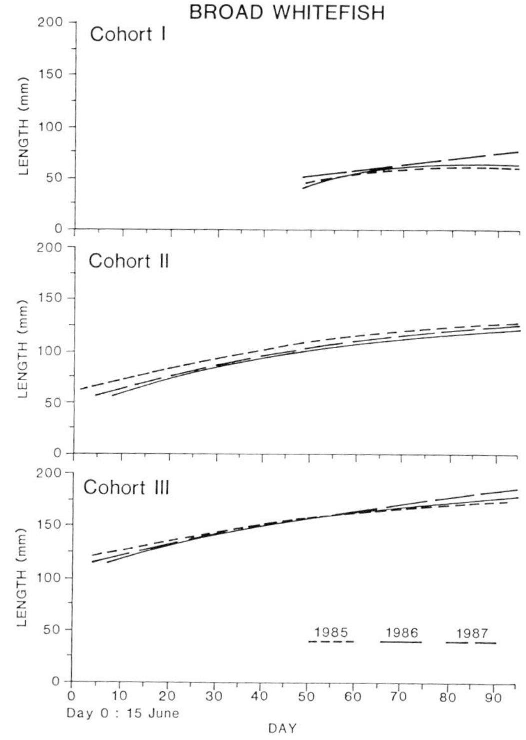

FIGURE 29.—Growth curves for broad whitefish, size-cohorts I–III, in the Endicott Project area, 1985–1987.

and regions by years. The results of these investigations generally concur with the conclusions we have reached. There appear to be few statistically significant differences that would suggest any long- or short-term trend towards reduced growth and condition of anadromous fish using the study area.

If significant changes in environmental quality are occurring and are causing sublethal stress to the fish, reductions in seasonal growth and condition should occur, all other factors being equal. Growth and condition are among the few biological variables that can be measured with accuracy and precision. Detrimental changes (reductions) have yet to be observed, but more detailed analyses need to be performed relating growth to biotic (density of fish and prey) and abiotic (temperature and salinity) factors. Such analyses require data beyond the scope of the 1985–1987 programs, but research to address the relationships of these factors to fish growth was initiated in the 1988 program. Once the relative importance of these factors has been determined and the effects of the causeway on them quantified, a comprehensive assessment can be made.

Population Levels of Small Broad Whitefish and Arctic Cisco

In 1988, mark–recapture data were obtained for small broad whitefish and Arctic ciscoes with the same methods that Griffiths et al. (1983) had used in the area in 1982. In 1988, the estimated population level of small broad whitefish was 272,350 fish (95% confidence interval [CI]: 202,000–378,280 fish). This is an increase over the 1982 population estimate of 179,290 (95% CI: 149,590–216,910). We recalculated the 1982 data from Griffiths et al. (1983) to obtain a figure directly comparable to the 1988 estimate.

The number of small Arctic ciscoes in the study area in 1988 was estimated at 3,162,040 fish (95% CI: 2,828,152–3,547,176), compared with the 1982 estimate of only 1,129,100 (95% CI: 1,040,000–1,228,100) small Arctic ciscoes in the area. In 1988, three year-classes were represented in the population of small Arctic cisco. Only one dominant year-class was represented in the 1982 population (mostly cohort-III fish or the 1979–1980 year-class). These differences in recruitment and population structure explain the increased numbers in 1988.

Summary and Conclusions

During the winter of 1985, the Endicott Causeway was constructed at the mouth of the Sagavanirktok River. The area lies at the eastern end of the Colville–Sagavanirktok Habitat Unit along the central Alaskan Beaufort Sea coast and encompasses an estimated 3,400 km^2. Another causeway, West Dock, is situated nearby. A number of anadromous fish occur in the project area and use different habitats on a seasonal cycle. Broad whitefish use the Sagavanirktok River for spawning and overwintering, and some Arctic ciscoes use the lower reaches of the river channels in the delta for overwintering. Arctic char overwinter and spawn upstream in spring-fed areas of the river. Each of these species uses the coastal waters of the region during summer for feeding, and they are joined by broad whitefish, Arctic char, and least ciscoes from the Colville River that also use the same general coastal habitat. Young-of-the-year Arctic ciscoes from Canada move through the area during the late summer in years when easterly winds are dominant.

Causeways and natural features extending perpendicular to the coast develop an eddy circulation (wake effect) on the lee side of the end of the structure. This circulation causes vertical transport of bottom waters to the surface. On east winds, coastal upwelling occurs due to Ekman transport, which brings cold, marine bottom waters nearshore. If present, these waters can be entrained and advected to the surface by the wake effect of causeways and other structures. Once at the surface, the cold, marine bottom waters continue to mix, and are advected downwind. In some instances these waters can be advected into the nearshore zone, contributing to a less favorable environment for anadromous fish in terms of temperature and salinity. However, this same process may enhance transport of food from the marine to nearshore environment.

The meteorological conditions that were observed during 1985–1987 were favorable for upwelling, thereby increasing the potential for marked effects from the causeway. Marine bottom waters have been advected to the surface by the Endicott Causeway 2–6% of the time that samples have been taken. The higher salinity, colder waters brought to the surface have not been observed to be advected close to shore or to affect much of the nearshore zone (<2 m deep) in the vicinity of the Endicott Causeway.

Recruitment of young-of-the-year Arctic ciscoes from Canada to the Colville River during 1985–1987 has not appeared to have been significantly affected by the Endicott Causeway. The migrations are thought to be driven by winds, and the fish have moved through and around the Endicott and West Dock causeways with little impedance.

None of the anadromous fish populations in the region has exhibited significant reduction in growth or condition. These population characteristics, if anything, have increased over the period 1985–1987. The increases appear to be a response to the progressively longer periods of high temperature that occurred in each successive year. The population estimates in 1988 for small broad whitefish and Arctic ciscoes, which are area residents, are higher than population estimates for 1982.

Collectively, the data indicate that the presence of the Endicott Causeway has not caused significant negative effects on anadromous fish or their habitat. This assessment can also be made for the West Dock Causeway, because both structures occur within the summer range of the fish in question. A possible explanation for the apparent lack of detrimental effects is that the fish are adapted to the environmental variability characteristic of the nearshore Beaufort Sea and are capable of exploiting the most favorable habitats available to them. The range of environmental variability introduced by the causeways does not exceed the natural range, and the area of the affected region has apparently not been extensive enough to change growth and condition of resident or migratory populations. Growth rates and condition levels might have remained high if the populations had been seriously reduced; however, there is no evidence that populations declined.

The above conclusions are mainly preliminary judgments based upon existing data that address a few specific issues. More data are needed to evaluate these and other potential effects, and such data are being collected. Due to the incomplete nature of the available data, a level of uncertainty as to the true effects of the causeways still remains, and other interpretations regarding even the specific issues addressed here are possible. However, if more data are gathered over the next few years, the level of uncertainty will be reduced because fewer explanations that are consistent with the entire body of information will be possible. Even though other explanations are presently possible, there is no evidence that significant adverse effects are occurring; thus, the possibility that fish are suffering severe negative effects from the Endicott Causeway seems low.

Acknowledgments

The overall study from which this work was abstracted was funded by BP Exploration (Alaska), Inc. It was conducted, in part, to fulfill the Endicott Development monitoring requirement pursuant to Title 19 of the North Slope Borough (NSB) Municipal Code. We would like to especially thank members of the Wildlife Management and Planning departments of the NSB for their encouragement and support of the entire 1988–1989 studies.

We also thank the Minerals Management Service and the American Fisheries Society for the opportunity to present results of monitoring studies for large arctic developments to the scientific community in a peer-review forum. The size and importance of the developments in combination with their arctic location requires that decisions be based on the best scientific judgements possible. This opportunity to present findings for peer scrutiny will contribute towards improving the decision process.

References

Aagaard, K., L. Coachman, and E. Cormack. 1981. On the halocline of the Arctic Ocean. Deep-Sea Research 28:529–546.

Armstrong, J. W., C. R. Liston, P. I. Jack, and R. C. Anderson. 1977. Age, growth, maturity, and seasonal food habits of round whitefish, *Prosopium cylindraceum*, in Lake Michigan near Ludington, Michigan. Transactions of the American Fisheries Society 106:151–155.

Barnes, P., E. Reimnitz, D. Drake, and L. Toimil. 1977. Miscellaneous hydrologic and geologic observations on the inner Beaufort Sea shelf, Alaska. U.S. Geological Survey, Open File Report 77-477, Menlo Park, California.

Bernatchez, L., and J. Dodson. 1985. Influence of temperature and current speed on the swimming capacity of lake whitefish (*Coregonus clupeaformis*) and cisco (*C. artedii*). Canadian Journal of Fisheries and Aquatic Sciences 42:1522–1529.

Bickham, J. W., S. M. Carr, B. G. Hanks, D. W. Burton, and B. J. Gallaway. 1989. Genetic analysis of population variation in the Arctic cisco (*Coregonus autumnalis*) using electrophoretic, flow cytometric, and mitochondrial DNA restriction analyses. Biological Papers of the University of Alaska 24:112–122.

Biosonics, Inc. 1984. Prudhoe Bay Waterflood Project fish monitoring program, 1983. Report to U.S. Army Corps of Engineers, Anchorage, Alaska.

Bond, W. 1982. A study of the fishery resources of Tuktoyaktuk Harbor, southern Beaufort Sea coast, with special reference to life histories of anadromous coregonids. Canadian Technical Report of Fisheries and Aquatic Sciences 1119.

Brett, J. R. 1971. Energetic responses of salmon to temperature. A study of some thermal relations in the physiology and freshwater ecology of sockeye salmon (*Oncorhynchus nerka*). American Zoologist 11:99–113.

Britch, R. P., R. C. Miller, J. Downing, T. Petrillo, and M. Veit. 1983. Physical processes. Pages 1–218 *in* B. J. Gallaway and R. D. Britch, editors. Environmental summer studies (1982) for the Endicott Development, volume 2. Report of LGL Ecological Research Associates, Inc., and Northern Technical Services to Sohio Alaska Petroleum Company. Available from BP Exploration (Alaska), Inc., Anchorage, Alaska.

Cannon, T. C., B. A. Adams, D. Glass, and T. Nelson. 1987. Fish distribution and abundance. Pages 1–129 *in* 1985 Final report for the Endicott Environmental Monitoring Program, volume 6. Report of Envirosphere Co. to U.S. Army Corps of Engineers, Anchorage, Alaska.

Colonell, J. M., and B. J. Gallaway. 1990. An assessment of marine environmental impacts of West Dock Causeway. Report of Woodward-Clyde Consultants and LGL Alaska Research Associates, Inc., to Prudhoe Bay Unit Owners. Available from ARCO–Alaska, Inc., Anchorage, Alaska.

Colonell, J. M., and A. W. Niedoroda. 1990. Nearshore oceanographic processes of potential importance to marine and anadromous fishes. Pages 35–42 *in* R. Meyer and T. M. Johnson, editors. Fisheries oceanography—a comprehensive formulation of technical objectives for offshore application in the Arctic: workshop proceedings. Minerals Management Service, Report 88-0042, Anchorage, Alaska.

Colonell, J. M., and T. Weingartner. 1982. Physical oceanography. Pages 2–1 to 2–53 *in* Duck Island development marine environmental studies. Report of Woodward-Clyde Consultants to Exxon USA, Los Angeles.

Cone, R. S. 1989. The need to reconsider use of condition indices in fishery science. Transactions of the American Fisheries Society 118:510–514.

Craig, P. C. 1989. An introduction to anadromous fishes in the Alaskan Arctic. Biological Papers of the University of Alaska 24:27–54.

Craig, P. C., W. Griffiths, L. Haldorson, and H. McElderry. 1985. Distributional patterns of fishes in an Alaskan arctic lagoon. Polar Biology 4:9–18.

Craig, P. C., and L. Haldorson. 1981. Fish. Pages 384–678 *in* Beaufort Sea barrier island–lagoon ecological process studies: final report, Simpson Lagoon. Environmental assessment of the Alaskan continental shelf, final report of the principal investigators, volume 7. Bureau of Land Management and National Oceanic and Atmospheric Administration, Outer Continental Shelf Environmental Assessment Program, Boulder, Colorado.

Crossland, N. O. 1988. A method for evaluating effects of toxic chemicals on the productivity of freshwater ecosystems. Ecotoxicology and Environmental Safety 16:279–292.

Dames and Moore. 1988. Lisburne Offshore Project environmental assessment. Report to ARCO–Alaska, Inc., Exxon USA, and Standard Alaska Production Company. Available from ARCO–Alaska, Inc., Anchorage, Alaska.

Dew, B. 1982. Under-ice survey of overwintering fish in the Sagavanirktok River in the vicinity of the Sag River bridge. Report of Woodward-Clyde Consultants to Exxon USA, Los Angeles.

Dillinger, R. E., Jr. 1989. The taxonomy of the *Coregonus autumnalis* complex in Alaska and analysis of the life histories of coregonine fishes of North America and Europe/Asia. Doctoral dissertation. Memorial University of Newfoundland, St. Johns.

Dolloff, C. A. 1987. Seasonal population characteristics and habitat use by juvenile coho salmon in a small southeast Alaska stream. Transactions of the American Fisheries Society 116:829–838.

Draper, N. R., and H. Smith. 1981. Applied regression analysis, 2nd edition. Wiley, New York.

Dryden, R. L., B. G. Sutherland, and J. N. Stein. 1973. An evaluation of the fish resources of the Mackenzie River valley as related to pipeline development, volume 2. Northern Pipelines Task Force, Environmental–Social Committee, Northern Oil Development Report 73-2, Ottawa.

Envirosphere Company. 1987. 1985 Final report for the Endicott Environmental Monitoring Program, volumes 1–7. Report to U.S. Army Corps of Engineers, Anchorage, Alaska.

Fechhelm, R. G., J. S. Baker, W. B. Griffiths, and D. R. Schmidt. 1989. Localized movement patterns of least cisco (*Coregonus sardinella*) and Arctic cisco (*C. autumnalis*) in the vicinity of a solid-fill causeway. Biological Papers of the University of Alaska 24:75–106.

Fechhelm, R. G., J. D. Fechhelm, C. J. Herlugson, and D. K. Beaubien. 1991. Bioassay tests of acute temperature and salinity shock to Arctic cisco. American Fisheries Society Symposium 11:109–118.

Fechhelm, R. G., and D. Fissel. 1988. Wind-aided recruitment of Canadian Arctic cisco (*Coregonus autumnalis*) into Alaskan waters. Canadian Journal of Fisheries and Aquatic Sciences 45:906–910.

Fechhelm, R. G., and B. J. Gallaway. 1983. Fish modeling. Pages 1–39 *in* B. J. Gallaway and R. P. Britch, editors. Environmental summer studies (1982) for the Endicott Development, volume 6. Report of LGL Alaska Research Associates, Inc., and Northern Technical Services to Sohio Alaska Petroleum Co., Anchorage, Alaska.

Fechhelm, R. G., and W. B. Griffiths. 1990. Effect of wind on the recruitment of Canadian Arctic cisco (*Coregonus autumnalis*) into the central Alaskan Beaufort Sea. Canadian Journal of Fisheries and Aquatic Sciences 47:2164–2171.

Fechhelm, R. G., W. Neill, and B. J. Gallaway. 1983. Temperature preferences of juvenile Arctic cisco from the Alaskan Beaufort Sea. Biological Papers of the University of Alaska 21:24–38.

Gallaway, B. J., and R. P. Britch, editors. 1983. Environmental summer studies (1982) for the Endicott Development. Report of LGL Alaska Research Associates, Inc., and Northern Technical Services to Sohio Alaska Petroleum Co., Anchorage, Alaska.

Gallaway, B. J., W. J. Gazey, and L. L. Moulton. 1989. Population trends for the Arctic cisco (*Coregonus autumnalis*) in the Colville River of Alaska as reflected by the commercial fishery. Biological Papers of the University of Alaska 24:153–165.

Gallaway, B. J., W. Griffiths, P. Craig, W. Gazey, and J. Helmericks. 1983. An assessment of the Colville River delta stock of Arctic cisco (*Coregonus autumnalis*)—migrants from Canada? Biological Papers of the University of Alaska 21:4–23.

Gallucci, V. F., and T. J. Quinn. 1979. Reparameterizing, fitting, and testing a simple growth model. Transactions of the American Fisheries Society 108:14–25.

Gazey, W. J., and M. J. Staley. 1986. Population estimation from mark–recapture experiments using a sequential Bayes algorithm. Ecology 67:941–951.

Griffiths, W. B. 1983. Fish. Pages 176–222 *in* Environmental characterization and biological use of lagoons in the eastern Beaufort Sea. Final report of LGL Ecological Research Associates, Inc., to Minerals Management Service, Bureau of Land Management and National Oceanic and Atmospheric Administration, Outer Continental Shelf Environmental Assessment Program. Available from Minerals Management Service, Anchorage, Alaska.

Griffiths, W. B., P. C. Craig, G. L. Walder, and G. J. Mann. 1975. Fisheries investigations in a coastal region of the Beaufort Sea (Nunaluk Lagoon, Y.T.). Chapter 2 *in* Arctic Gas Biological Report Series 34. Report of Aquatic Environments, Limited, to Canadian Arctic Gas Study, Limited, and Alaska Arctic Gas Study Co. Available from LGL Ecological Research Associates, Inc., Bryan, Texas.

Griffiths, W. B., and R. Dillinger. 1981. Part 5. Invertebrates. Pages 1–198 *in* Beaufort Sea barrier island–lagoon ecological process studies: final report, Simpson Lagoon. Environmental assessment of the Alaskan continental shelf, final report of the principal investigators, volume 8. Bureau of Land Management and National Oceanic and Atmospheric Administration, Outer Continental Shelf Environmental Assessment Program, Boulder, Colorado. Available from Minerals Management Service, Anchorage, Alaska.

Griffiths, W. B., and B. J. Gallaway. 1983. Prudhoe Bay Waterflood Project, fish monitoring program. Report of LGL Alaska Research Associates to Woodward-Clyde Consultants and U.S. Army Corps of Engineers, Anchorage, Alaska.

Griffiths, W. B., and seven coauthors. 1983. Fish ecology. Pages 1–323 *in* B. J. Gallaway and R. P. Britch, editors. Environmental summer studies (1982) for the Endicott Development, volume 3. Report of LGL Alaska Research and Northern Technical Services to Sohio Alaska Petroleum Co. Available from BP Exploration (Alaska), Inc., Anchorage.

Hachmeister, L. E., K. S. Short, K. B. Winnick, G. C. Schrader, and J. W. Johannessen. 1987. Oceanographic monitoring. Pages 1–162 *in* 1985 Final report for the Endicott Environmental Monitoring Program, volume 3. Report of Envirosphere Co. to U.S. Army Corps of Engineers, Anchorage, Alaska.

Harrison, E. J., and W. F. Hadley. 1979. Biology of muskellunge (*Esox masquinongy*) in the upper Niagara River. Transactions of the American Fisheries Society 108:444–451.

Held, J. W., and J. J. Peterka. 1974. Age, growth, and food habits of the fathead minnow, *Pimephales promelas*, in North Dakota saline lakes. Transactions of the American Fisheries Society 103:743–756.

Hummer, P. G. 1987. Meteorology. Pages 1–7 *in* 1985 Final report for the Endicott Environmental Monitoring Program, volume 2. Report of Envirosphere Co. to U.S. Army Corps of Engineers, Anchorage, Alaska.

Jessop, B. M. 1987. Migrating American eels in Nova Scotia. Transactions of the American Fisheries Society 116:161–170.

Kozo, T. L. 1984. Mesoscale wind phenomena along the Alaskan Beaufort Sea coast. Pages 23–45 *in* P. W. Barnes, D. M. Schell, and E. Reiminitz, editors. The Alaskan Beaufort Sea ecosystems and environments. Academic Press, New York.

Le Cren, E. D. 1951. The length–weight relationship and seasonal cycle in gonad weight and condition in the perch (*Perca fluviatilis*). Journal of Animal Ecology 20:201–219.

LGL (LGL Alaska Research Associates, Inc.). 1982. Biological and archaeological investigations in the vicinity of the proposed Duck Island unit pipeline through the Sagavanirktok River delta, Alaska. Report to Exxon USA, Los Angeles.

LGL (LGL Alaska Research Associates, Inc.), Hunter/ESE Environmental Services, Inc., and W. J. Gazey Research. 1990. Analysis of the 1985–1987 data collected by the Endicott Development monitoring program, volumes 1–3. Final report to BP Exploration (Alaska) Inc., Anchorage, and the North Slope Borough, Barrow, Alaska.

MacDonald, P. D. M., and P. E. J. Green. 1985. An interactive program for fitting mixtures of distributions. Ichthus Data Systems, Hamilton, Ontario.

Maceina, M. J., and B. R. Murphy. 1988. Variation in the weight-to-length relationship among Florida and northern largemouth bass and their intraspecific F_1 hybrid. Transactions of the American Fisheries Society 117:232–237.

Mangarella, P. A., J. R. Harper, and T. J. Weingartner. 1982. Physical processes monitoring program from 1981. Pages A1–A247 *in* Prudhoe Bay waterflood project environment monitoring program 1981. Report of Woodward-Clyde Consultants to U.S. Army Corps of Engineers, Anchorage, Alaska.

Marquardt, D. W. 1963. An algorithm for least squares estimation of nonlinear parameters. SIAM (Society for Industrial and Applied Mathematics) Journal 11:431–441.

Matthews, J. B. 1979. Characterization in the nearshore hydrodynamics of an Arctic barrier island system. Pages 57–97 *in* Environmental assessment of the Alaskan continental shelf, annual report, volume 8. National Oceanic and Atmospheric Administration, Outer Continental Shelf Environmental Assessment Program. Available from Minerals Management Service, Anchorage, Alaska.

McAda, C. W., and R. S. Wydoski. 1983. Maturity and fecundity of the bluehead sucker *Catostomus discobolus* Catostomidae, in the upper Colorado River basin, 1975–76. Southwestern Naturalist 28:120–123.

McAda, C. W., and R. S. Wydoski. 1985. Growth and reproduction of the flannelmouth sucker, *Catostomus latipinnis*, in the upper Colorado River basin, 1975–76. Great Basin Naturalist 45:281–286.

Mohsin, A. K. M., and B. J. Gallaway. 1977. Seasonal abundance, distribution, food habitats and condition of the southern brook lamprey, *Ichthyomyzon gagei* Hubbs & Trautman, in an East Texas watershed. Southwestern Naturalist 22:107–114.

Moulton, L. L. 1989. Recruitment of Arctic cisco (*Coregonus autumnalis*) into the Colville Delta, Alaska, in 1985. Biological Papers of the University of Alaska 24:107–111.

Neill, W., R. Fechhelm, B. Gallaway, J. Bryan, and S. Anderson. 1983. Modeling movement and distribution of young Arctic cisco (*Coregonus autumnalis*) relative to temperature salinity regimes of the Beaufort Sea near the waterflood causeway, Prudhoe Bay, Alaska. Biological Papers of the University of Alaska 21:39–61.

Neill, W. H., and B. J. Gallaway. 1989. "Noise" in the distributional responses of fish to environment: an exercise in deterministic modeling motivated by the Beaufort Sea experience. Biological Papers of the University of Alaska 24:123–130.

Nelson, T., T. C. Cannon, W. R. Olmstead, and K. C. Wiley. 1987. Part VI. Surveys of domestic and commercial fisheries in the central and eastern Beaufort Sea. Pages 1–38 *in* 1985 Final report of the Endicott Environmental Monitoring Program, volume 7. Report of Envirosphere Company to U.S. Army Corps of Engineers, Anchorage, Alaska.

Osenberg, C. W., E. E. Werner, G. G. Mittlebach, and D. J. Hall. 1988. Growth patterns in bluegill (*Lepomis macrochirus*) and pumpkinseed (*L. gibbosus*) sunfish: environmental variability and importance of ontogenetic niche shifts. Canadian Journal of Fisheries and Aquatic Sciences 45:17–26.

Savoie, M. A., and D. E. Wilson. 1983. Prudhoe Bay Waterflood Project physical processes monitoring program for 1982. Pages 3-1 to 3-122 *in* Prudhoe Bay Waterflood Project Environmental Monitoring Program, 1982, volume 2. Report of Kinnetic Laboratories, Inc., to U.S. Army Corps of Engineers, Anchorage, Alaska.

Savoie, M. A., and D. E. Wilson. 1984. Prudhoe Bay Waterflood Project physical oceanographic processes monitoring program for 1983. Pages 2-1 to 2-124 *in* Prudhoe Bay Waterflood Project Environmental Monitoring Program, 1983, volume 2. Report of Kinnetic Laboratories, Inc., to U.S. Army Corps of Engineers, Anchorage, Alaska.

Savoie, M. A., and D. E. Wilson. 1986. Prudhoe Bay Waterflood Project physical oceanographic processes monitoring program for 1984. Pages 2-1 to 2-155 *in* Prudhoe Bay Waterflood Project Environmental Monitoring Program, 1984, volume 1. Report of Kinnetic Laboratories, Inc., to U.S. Army Corps of Engineers, Anchorage, Alaska.

Schmidt, D. R., W. B. Griffiths, and L. R. Martin. 1989. Overwintering biology of anadromous fish in the Sagavanirktok River delta, Alaska. Biological Papers of the University of Alaska 24:55–74.

Slaybaugh, D. K., B. J. Gallaway, and J. S. Baker. 1989. The databank for Arctic anadromous fish: description and overview. Biological Papers of the University of Alaska 24:4–26.

Thomson, D., D. Fissel, J. Marko, R. Davis, and G. Borstad. 1986. Distribution of bowhead whales in relation to hydro-meteorological events in the Beaufort Sea. Canada Department of Indian and Northern Affairs, Environmental Studies Revolving Funds Report 028, Ottawa.

Truett, J. C. 1981. Synthesis, impact analysis and a monitoring strategy. Pages 259–359 *in* Beaufort Sea barrier island–lagoon ecological process studies: final report, Simpson Lagoon. Environmental assessment of the Alaskan continental shelf, final report biological studies, volume 8. Bureau of Land Management and National Oceanic and Atmospheric Administration, Outer Continental Shelf Environmental Assessment Program, Boulder, Colorado.

USACE (U.S. Army Corps of Engineers). 1984. Endicott Development Project. Final environmental impact statement. USACE, Anchorage, Alaska.

Vives, S. P. 1987. Aspects of the life history of the slender madtom *Noturus exilis* in northeastern Oklahoma (Pisces: Ictaluridae). American Midland Naturalist 117:167–176.

Walker, D. A., K. R. Everett, P. J. Webber, and J. Brown. 1980. Geobotanical atlas of the Prudhoe Bay region, Alaska. U.S. Army Corps of Engineers, Cold Regions Research and Engineering Laboratory, Hanover, New Hampshire.

Whitmus, C. J., T. C. Cannon, and S. S. Parker. 1987. Age, growth, and condition of anadromous fish. Pages 1–34 *in* 1985 Final report for the Endicott Environmental Monitoring Program. Report of Envirosphere Company to U.S. Army Corps of Engineers, Anchorage, Alaska.

Woodward-Clyde Consultants. 1982. Duck Island development: marine environmental studies. Report to Exxon USA, Los Angeles.

Woodward-Clyde Consultants. 1983. Lisburne development area: 1983 environmental studies—final report. Report to ARCO-Alaska, Inc., Anchorage, Alaska.

American Fisheries Society Symposium 11:81–96, 1991

Effects of Solid-Fill Gravel Causeways on the Coastal Central Beaufort Sea Environment

L. E. HACHMEISTER, D. R. GLASS, AND T. C. CANNON

Ebasco Environmental
10900 NE 8th Street, Bellevue, Washington 98004, USA

Abstract.—The West Dock and Endicott causeways extend several kilometers from the coast into the Beaufort Sea near Prudhoe Bay, Alaska. The effects of these two solid-fill gravel causeways on the physical and biological environment of the coastal central Beaufort Sea of Alaska were assessed as part of a 3-year postconstruction monitoring program. The conclusions of this program are summarized in this paper. Physical oceanography and fisheries monitoring data were used to assess the effects of the causeways on summer migrating and feeding habitat of anadromous fishes. The causeways were found to alter sea ice formation and breakup, currents, coastal mixing, and water temperature and salinity distribution patterns. The causeways locally enhanced upwelling of cold, marine waters and deflected river plumes offshore, thereby reducing the nearshore extent of warm, low-salinity fish habitat. Analyses of trends in 3 years of precauseway and 3 years of postcauseway data indicate that growth of juvenile Arctic ciscoes *Coregonus autumnalis* and broad whitefish *C. nasus* in the study area during the short summer growing season is positively related to the amount of warm, low-salinity habitat available in time and space to the fish.

Oil and gas development on the North Slope of Alaska at Prudhoe Bay on the Beaufort Sea coast has included the construction of solid-fill gravel causeways for docks, waterflood intakes, and access to drilling–production islands. Two major causeways have been constructed in the Prudhoe Bay area of the central Beaufort Sea coast, the West Dock (1974–1981) and Endicott (1985–1986) causeways. The Endicott Causeway is 6 km long and extends several kilometers offshore from the Sagavanirktok River delta into the Beaufort Sea to two drilling–production islands (Figure 1). Two breaches, one spanning 61 m and the other 152 m, were included in the causeway design to allow fish passage and cross-delta water movement. The West Dock Causeway is a combination dock and waterflood intake facility. It is 4 km long and also extends several kilometers offshore (Figure 1). West Dock has one 15.2-m breach between the second and outer legs for fish passage.

The causeways were permitted and constructed in the Prudhoe Bay area following an extensive evaluation of their potential effects that included preparation of environmental impact statements (EISs) by the U.S. Army Corps of Engineers (USACE). In each case, effects of the causeways on the coastal physical and biological marine environment were predicted to be localized and minor (USACE 1980, 1984). The principal issue addressed in the EISs was the effect of the causeways on the summer feeding and migratory habitat of anadromous fishes, which support sport, commercial, and subsistence fisheries.

Permits from the USACE for the causeways stipulated environmental monitoring programs to evaluate the predictions in the EISs. Since 1981, intensive study of the physical and biological marine environment of coastal waters in the Prudhoe Bay area has provided information necessary to evaluate the EIS predictions. Emphasis has been on coordinated oceanography and fisheries studies. West Dock studies were conducted from 1981 through 1984 (Woodward-Clyde Consultants 1982; Envirosphere 1983, 1984, 1985). Endicott studies were conducted by the authors from 1985 through 1987 (Envirosphere 1987, 1990, and in press). This paper summarizes the conclusions of the authors as presented in the Endicott Monitoring Program final assessment report (Cannon et al. 1990).

The Endicott Monitoring Program

The primary objectives of the Endicott Monitoring Program were (1) to assess the effects of the Endicott Causeway and cumulative effects of the Endicott and West Dock causeways on the marine environment, (2) to determine the effectiveness of approximately 200 m of causeway breaching in ameliorating the effects of the Endicott Causeway on fish movement and habitat, and (3) to validate predictions of environmental effects in the USACE EIS.

Physical oceanographic and fisheries studies

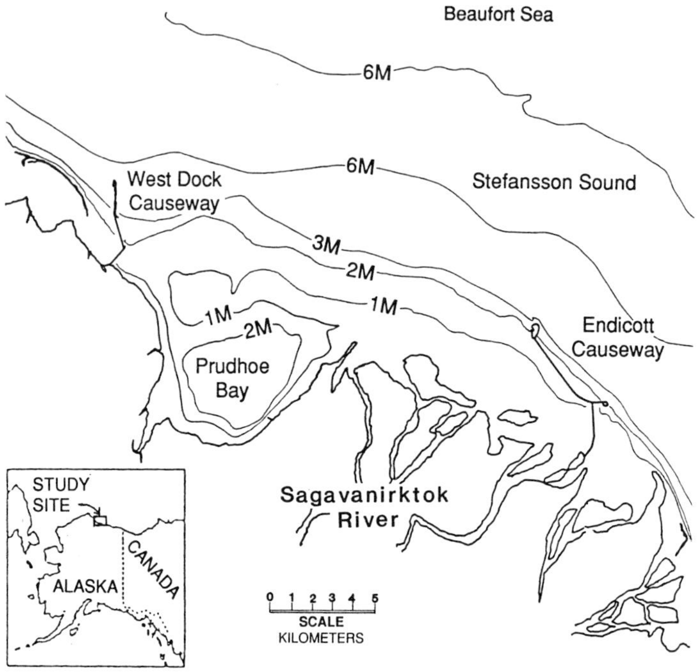

FIGURE 1.—Central Beaufort Sea study area and location of the West Dock and Endicott causeways.

were implemented to address these objectives. The oceanography program included surveys of local and regional patterns of weather, river discharge, sedimentation and erosion, ice processes, currents, plume tracking, water temperature, salinity, and turbidity. These surveys provided data necessary to characterize physical habitat conditions of the coastal marine environment and to assess the effects of the causeways on those habitat conditions.

Fisheries studies were undertaken to evaluate the effects of the causeways on the distribution, movement, and the quantity and quality of habitat that supports the regional populations of fish important to commercial and subsistence fisheries. Field data were also collected on fish feeding habits, fish prey abundance and distribution, and fish population characteristics (e.g., age structure, growth, and condition).

A secondary goal of the monitoring program was to assess the potential changes in the anadromous fish populations resulting from causeway effects on fish habitat and migration. Little baseline information was available on population dynamics of the regional fish populations. In recognition of this limitation, it was understood that this program goal could effectively detect only large-scale catastrophic effects over the short term, and that many years, perhaps as many as 20, would be necessary to detect significant long-term effects of low-level impacts to the populations.

The Coastal Central Beaufort Sea Environment

The study area is on the coast of Alaska's Beaufort Sea adjacent to Prudhoe Bay (Figure 1). This area of the Beaufort Sea coast has numerous shallow bays and river deltas that have estuarine characteristics due to the influx of fresh water from ice melt and river discharge.

Physical Oceanography and Ice Processes

Ice dominates coastal shelf waters of the Beaufort Sea for most of the year, except for several months of open water during the summer, when solar radiation, warm (10–15°C) river discharge, and winds keep the nearshore waters relatively ice free. The Sagavanirktok, Kuparuk, and Putuligayuk rivers flow into the study area. Their average discharge rates in summer are 245, 40, and 1.2 m^3/s, respectively (Johannessen 1987).

Relatively warm, brackish waters dominate the

coastal central Beaufort Sea during the summer (Hufford and Bowman 1974; Barnes et al. 1977; Aagaard 1978; Matthews 1978; Chin 1979; Mangarella et al. 1982; Britch et al. 1983; Savoie and Wilson 1984; Berry and Colonell 1985; Hachmeister et al. 1987). The fresh water from ice melt and river discharge dominates coastal surface waters (2–4 m) in early summer (July). Water temperatures are warm (6–12°C) and salinities low (1–15‰) in alongshore and coastal surface waters. By midsummer (August), river discharge declines, and warm, low-salinity waters are confined to protected bays, lagoons, and river plumes. By late summer (September), water temperatures fall below 5°C, river discharge falls as river headwaters freeze, and salinities reach moderate-to-high levels (16–32‰) along the coast (Short et al. 1990).

Wind-induced circulation controls coastal oceanographic conditions during the summer open-water season (Hufford and Bowman 1974; Barnes et al. 1977; Aagaard 1978; Matthews 1978; Hachmeister et al. 1987). Prevailing easterly winds set up westerly flowing currents. Warm, low-salinity surface waters along the coast are pushed toward the west and off shore. The offshore component is due to Ekman transport associated with the Coriolis effect. The offshore deflection of warm, coastal water causes an onshore movement of cold, bottom marine water, commonly referred to as upwelling. West winds have the opposite effect, because surface waters are driven to the east and onshore, and bottom marine waters move offshore (downwelling). Summers dominated by east winds are typically characterized by upwelling and cooler, higher-salinity waters along the coast (Hachmeister et al. 1987). Summers in which west winds are frequent are generally characterized by warmer, lower-salinity coastal waters. West winds are also usually associated with low-pressure systems and precipitation (Hummer 1990). Midsummer and late-summer precipitation is infrequent, but can have a marked effect on river discharge, which in turn may measurably affect the amount of warm, low-salinity water along the coast (Hachmeister et al. 1987; Cannon et al., in press).

Fish Ecology

Many marine, estuarine, anadromous, and freshwater fish use the coastal waters of the Beaufort Sea during the summer. Four species of anadromous fish, Arctic cisco *Coregonus autumnalis*, least cisco *C. sardinella*, broad whitefish *C. nasus*, and Arctic char *Salvelinus alpinus*,[1] are among the most abundant fish in the warm brackish coastal waters during the summer (Furniss 1975; Bendock 1979; Craig and Haldorson 1981; Griffiths et al. 1983; Craig 1984; Moulton et al. 1986; Cannon et al. 1987; Glass et al. 1990). All four species support subsistence fisheries at villages along the North Slope.

Arctic ciscoes spawn in Canada in the Mackenzie River (Gallaway et al. 1983) in late summer or early fall and emerge as fry in late winter and spring. Fry move downriver and disperse into coastal waters. An unknown portion of the young disperse westward into the Alaskan Beaufort Sea during their first summer and eventually overwinter in larger rivers such as the Colville and Sagavanirktok rivers. At maturity (ages 7–9), they return to the Mackenzie River to spawn. Each summer they migrate into coastal waters to feed (Craig and Haldorson 1981; Griffiths et al. 1983).

Least ciscoes spawn and overwinter in the Colville River and disperse into coastal waters during the summer to feed (Craig and Haldorson 1981). Juvenile fish remain near the Colville delta, while adults disperse widely along the coast. Large numbers of adults disperse eastward into the Prudhoe Bay area after ice breakup.

Many stocks of Arctic char use the coastal waters of the central Beaufort Sea during the summer to feed (Craig 1984). Char spawn and overwinter in spring areas of the upper rivers, usually in the foothills of the Brooks Range (Bendock 1979). Char spend their first two summers in the rivers and migrate to coastal waters to feed beginning in their third summer. Juvenile and adult char migrate extensively along the coast during the summer; spawners usually return to the rivers in midsummer.

Broad whitefish have populations in many of the coastal rivers (Craig 1984). They too disperse into coastal waters during the summer to feed (Craig and Haldorson 1981). Adults disperse widely during early summer, while juvenile fish remain near the river deltas.

Two marine species common to nearshore wa-

[1]The taxonomy of the Beaufort Sea char is in question. Historically, fish of the genus *Salvelinus* in the Beaufort Sea have been identified as Arctic char (*S. alpinus*); however, Morrow (1980) and Behnke (1980, 1984) believe this char to be the northern form of Dolly Varden char (*S. malma*). Arctic char continues to be used in most Beaufort Sea studies to maintain consistency with previous work.

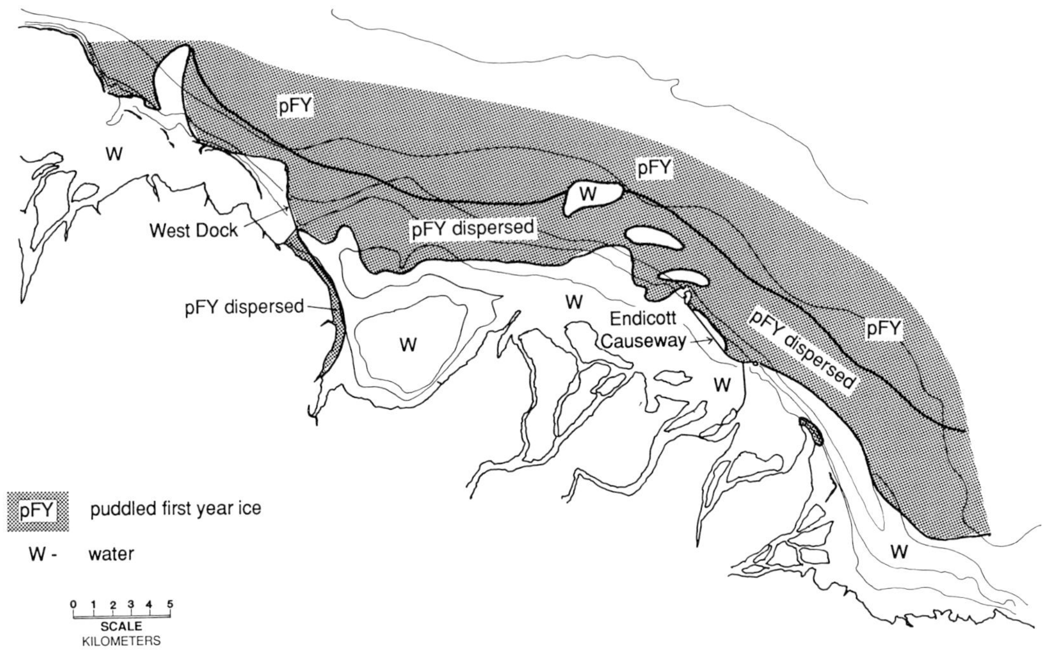

FIGURE 2.—Breakup of ice near the West Dock and Endicott causeways, July 8, 1987.

ters during the summer are the Arctic cod *Boreogadus saida* and fourhorn sculpin *Myoxocephalus quadricornis* (Craig and Haldorson 1981). Arctic cod young are very abundant in coastal waters during the summer. Fourhorn sculpin, a glacial marine relict to many fresh and marine waters of northern North America, is also very abundant in coastal waters during the summer.

Following breakup, anadromous fishes spread along the coast from overwintering rivers to feed (Craig 1984). Most remain near shore, but Arctic char and adult ciscoes are also found in offshore waters (Craig and Haldorson 1981). They feed on abundant crustaceans, particularly mysids and amphipods, which concentrate in intermediate temperature and salinity waters where the warm river waters mix with cold marine waters (Cannon et al. 1991b, this volume).

By midsummer, anadromous fish concentrate in warm low-to-moderate salinity habitats (Craig and Haldorson 1981). During this period, the distribution of fish varies by species and age-group within species because of varied habitat preferences (Glass and Cannon 1990). The fish return to the deltas of rivers in which they overwinter in late summer as water temperatures decline below 5°C along the coast (Cannon et al. 1987). Spawning adults migrate into the rivers as early as late July (Craig and Haldorson 1981; Griffiths et al. 1983; Cannon et al. 1987) after spending only a month or less feeding in coastal waters. Juvenile and nonspawning adults often remain in the river plumes at the mouths of the rivers through late summer to feed (Cannon et al. 1987). They too move into the rivers as river discharges decline, usually by late September or early October (Cannon et al. 1987; Glass et al. 1990).

Effect of the Causeways on the Physical Environment

Sea Ice

Causeways influence sea ice in the study area by affecting the freeze-up and breakup processes (Cannon et al. 1990; Schrader and Hachmeister 1990). Because the causeways provide protection from wind and waves, freeze-up begins earlier near the causeways than in open waters. The causeways also capture slush ice and fresh water flowing from the rivers, which accelerate freeze-up in waters around the causeways. Accumulations of slush and fresh water on the windward sides of the causeways result in an early freeze-up and locally increased ice depths. This buildup of ice around the causeways hinders the breakup and dispersal of ice the following summer.

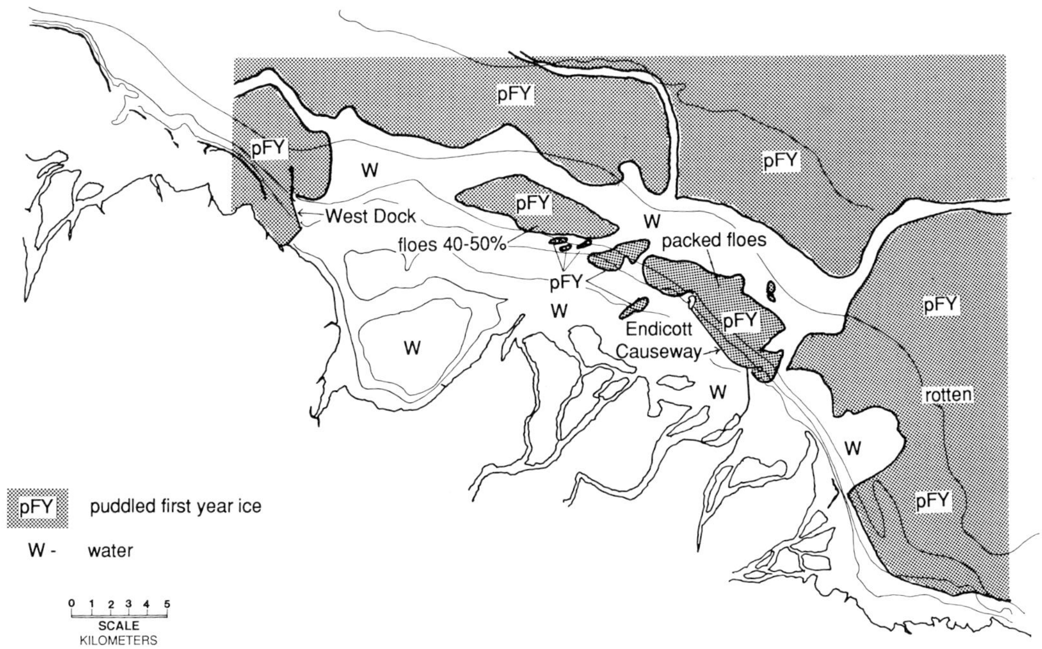

FIGURE 3.—Breakup of floating, fast ice near the West Dock and Endicott causeways, July 9, 1986.

The causeways also hinder the summer breakup of ice by (1) changing the dispersal of flood waters over the ice, (2) providing an anchor for the ice, (3) protecting the ice from wind-induced waves and currents, and (4) allowing broken ice to collect on their windward sides (Cannon et al. 1990). Generally, the causeways protect the ice from breaking up (Figures 2, 3). The causeways and adjacent accumulations of ice form obstacles to the alongshore flow of early-summer, warm, low-salinity waters. Wind-induced waves and currents are hindered by the ice, and coastal waters within the ice floes, and downwind, are colder because of the ice (Cannon et al. 1990, in press).

Physical Oceanographic Processes

After the ice recedes from inshore waters in early summer, the Endicott and West Dock causeways directly affect coastal water circulation by deflecting alongshore currents (Figures 4, 5). These effects on circulation change the mixing and water distribution processes and ultimately local and regional water temperature and salinity regimes (Hachmeister et al. 1987). Warm, low-salinity, coastal waters and river plumes are deflected offshore around the causeways, which results in the loss of plume waters from the nearshore waters and the accelerated degradation of the plume by mixing with offshore marine waters (Hachmeister et al. 1987; Cannon et al. 1990). By interrupting the alongshore flow of water and deflecting river plumes offshore, the causeways allow colder, higher-salinity waters to move onshore and replace warmer, low-salinity plume waters in the lee of the causeways. In extreme cases, cold, high-salinity water can upwell into shallow waters in the lee of the causeway (Figure 6).

Under east winds, the two causeways enhance the natural upwelling of bottom waters that occurs along the coast (Matthews 1978; Mangarella et al. 1982; Hachmeister et al. 1987; Cannon et al. 1990). East winds drive surface waters offshore (to the right of the wind) due to Ekman transport from the Coriolis force. Colder, higher-salinity bottom waters replace the river plumes displaced offshore. The influence of the Sagavanirktok River plume on the west Sagavanirktok River delta and Prudhoe Bay is reduced by the offshore displacement of a portion of the plume by the Endicott Causeway (Hachmeister et al. 1987; Cannon et al. 1990). The Endicott and West Dock causeways both contribute to a reduction in the influence of the Sagavanirktok River plume on the coastal waters west of the West Dock Causeway (Figure 7).

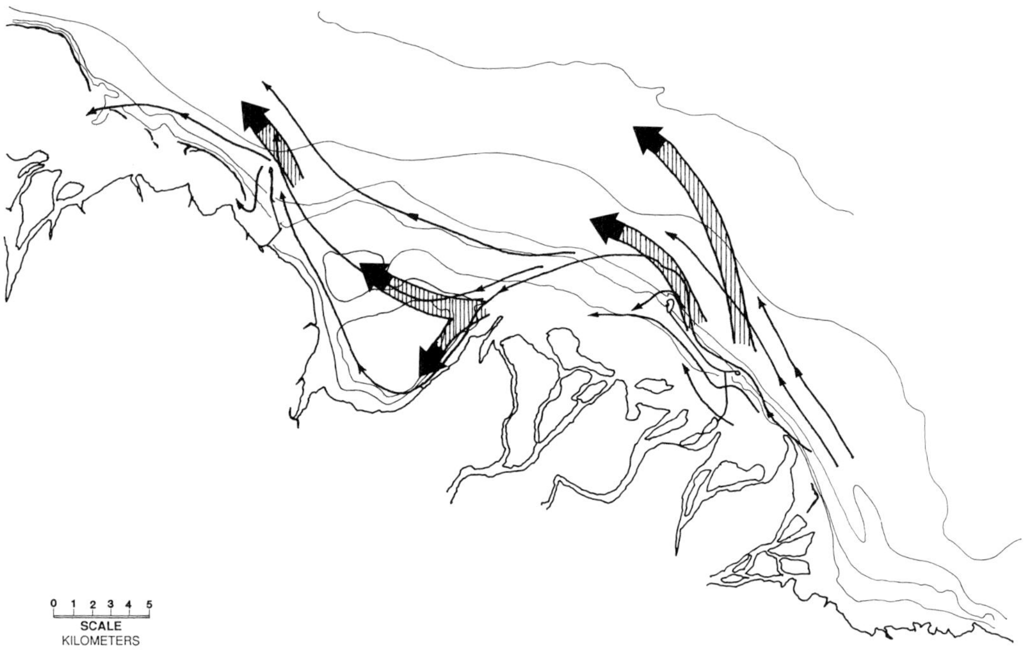

FIGURE 4.—Circulation in the area of the West Dock and Endicott causeways under east winds. Solid lines denote streamlines determined from current meter data. Crosshatched arrows are flow patterns inferred from hydrographic and satellite data.

Under west winds, warm, low-salinity, coastal waters are also affected by the causeways. West winds force surface waters slightly onshore and bottom waters offshore (Matthews 1978; Mangarella et al. 1982; Hachmeister et al. 1987; Short et al. 1990). The effect of periods of west winds on habitat conditions varies and depends on the preceding wind patterns (Hachmeister et al. 1987). The most extreme change in habitat conditions under west winds occurs following a period of east winds and upwelling of cold, high-salinity water west of the causeway (Cannon et al. 1990). Cold, upwelled water from the west side of West Dock is blown east around West Dock into Prudhoe Bay. Figure 8 depicts a cross-section of Prudhoe Bay before and after a shift to westerly wind. After several days of west winds, the cold, marine water may reach east Prudhoe Bay and the west end of the Sagavanirktok River delta. Upon filling Prudhoe Bay, the dense, high-salinity water is difficult to flush from the bay, and thus often remains for most of the summer.

The extent of the effects of the causeways on environmental conditions varies within and among years, depending on the meteorologic and oceanographic conditions that precede any specific event or period (Hachmeister et al. 1987; Cannon et al. 1990). The frequency and duration of west winds are the primary factors that regulate the extent of causeway effects. The greater the frequency of west winds, the more warmer, lower-salinity water remains in inshore waters, the greater the homogeneity of the water masses, and the lesser the effects of the causeways. The amount of river discharge is another important factor that influences interannual variability. Greater discharge means a greater volume of warm, brackish water overlying the marine water along the coast. Midsummer and late-summer rainstorms can markedly alter the size of the river plumes along the coast (Cannon et al., in press).

Effects of the Endicott Causeway on Anadromous Fishes

Changes in sea ice and oceanographic conditions affect migrating and feeding habitat of anadromous fishes during the summer open-water season. Such effects may in turn lead to changes in the dynamics of the fish populations.

Movement and Migration

The causeways affect the movement and migration of anadromous fish to and from the rivers in

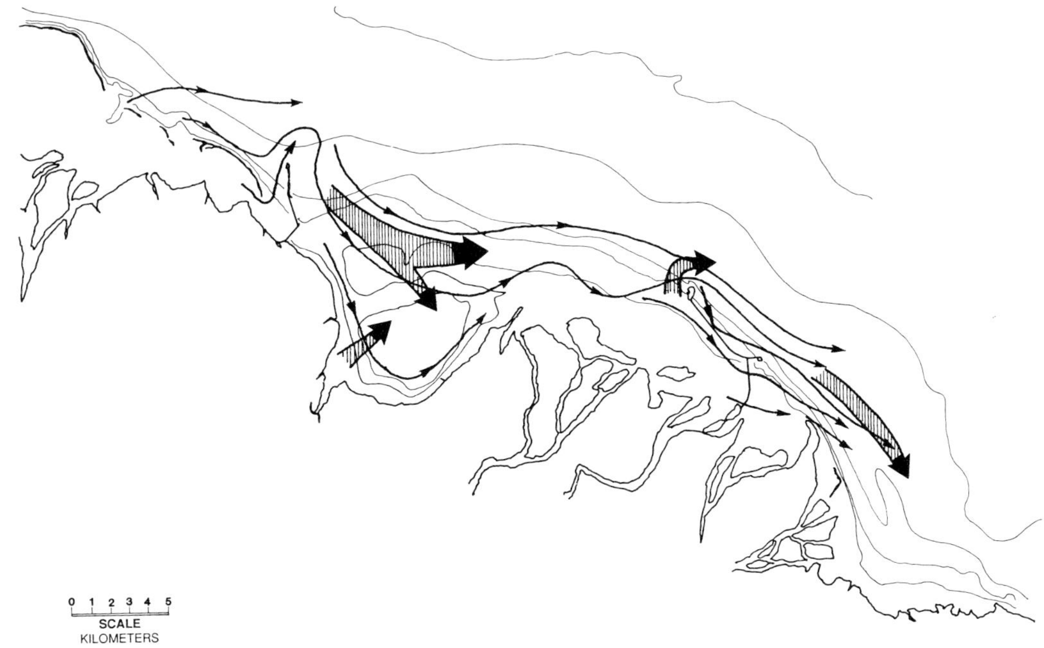

FIGURE 5.—Circulation in the area of the West Dock and Endicott causeways under west winds. Solid lines denote streamlines determined from current meter data. Crosshatched arrows are flow patterns inferred from hydrographic and satellite data.

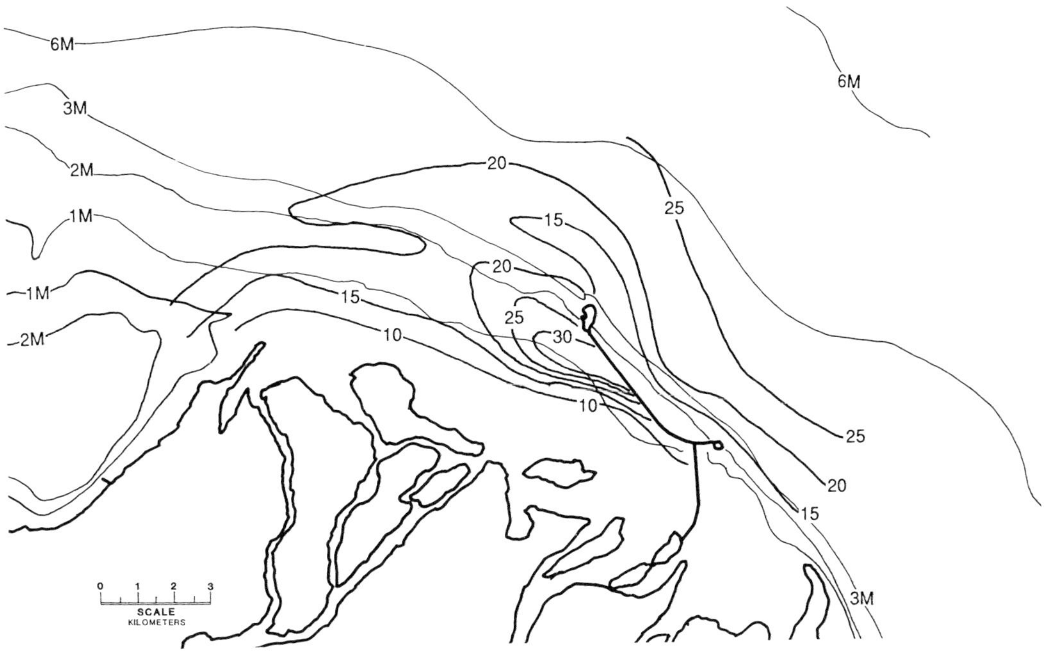

FIGURE 6.—Surface salinity (‰) distribution near the Endicott Causeway, July 22, 1986.

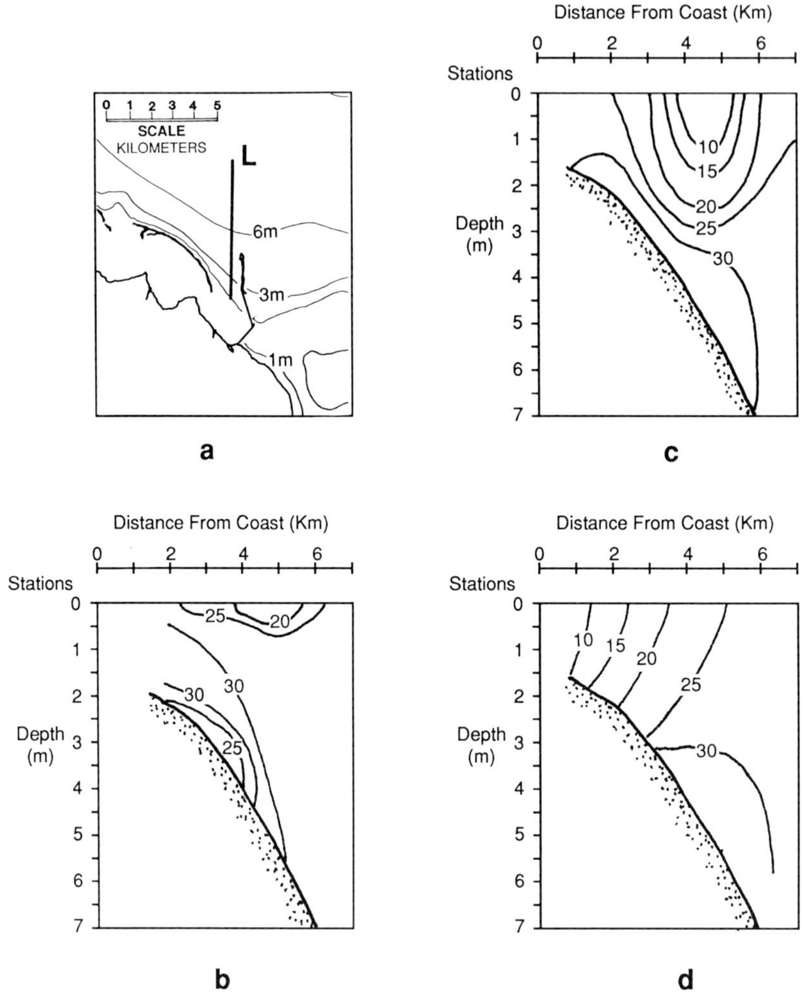

FIGURE 7.—Vertical salinity (‰) distribution along a north–south cross section (L) near West Dock, September 1, 1986: **(a)** location of section, **(b)** measured (present) conditions with West Dock and Endicott causeways, **(c)** hypothesized pre-Endicott Causeway conditions, and **(d)** hypothesized pre-West Dock conditions.

which they overwinter. In early summer, retention of ice along the east sides of the causeways under prevailing easterly winds and the resulting cold-water barriers around and downwind of the causeways hinder and delay dispersal of Colville River Arctic and least ciscoes eastward across the study area (Cannon et al. 1990, in press). Large concentrations of Colville ciscoes in Gwydyr Bay west of West Dock typically do not extend eastward past West Dock until much of the ice disperses from around West Dock. Similarly, juvenile Arctic char that overwinter in the Sagavanirktok River concentrate in Prudhoe Bay during breakup and do not move west past West Dock until the ice disperses. The effect of the Endicott Causeway on movement in early summer appears less than that of West Dock because the discontinuity in habitat conditions is not as great at the Endicott Causeway. Portions of the Sagavanirktok River plume that enter the delta on both sides of the Endicott causeway reduce its effect on habitat. Ciscoes do build up temporarily along the west side of the Endicott Causeway due to ice retention on the east side of the Endicott Causeway; however, once the ice has dispersed, fish appear to move readily around the causeway.

In late July after ice has gone, the intrusion of upwelled marine water along the west side of West Dock further hinders dispersal of fish from the Colville River into waters east of Gwydyr Bay

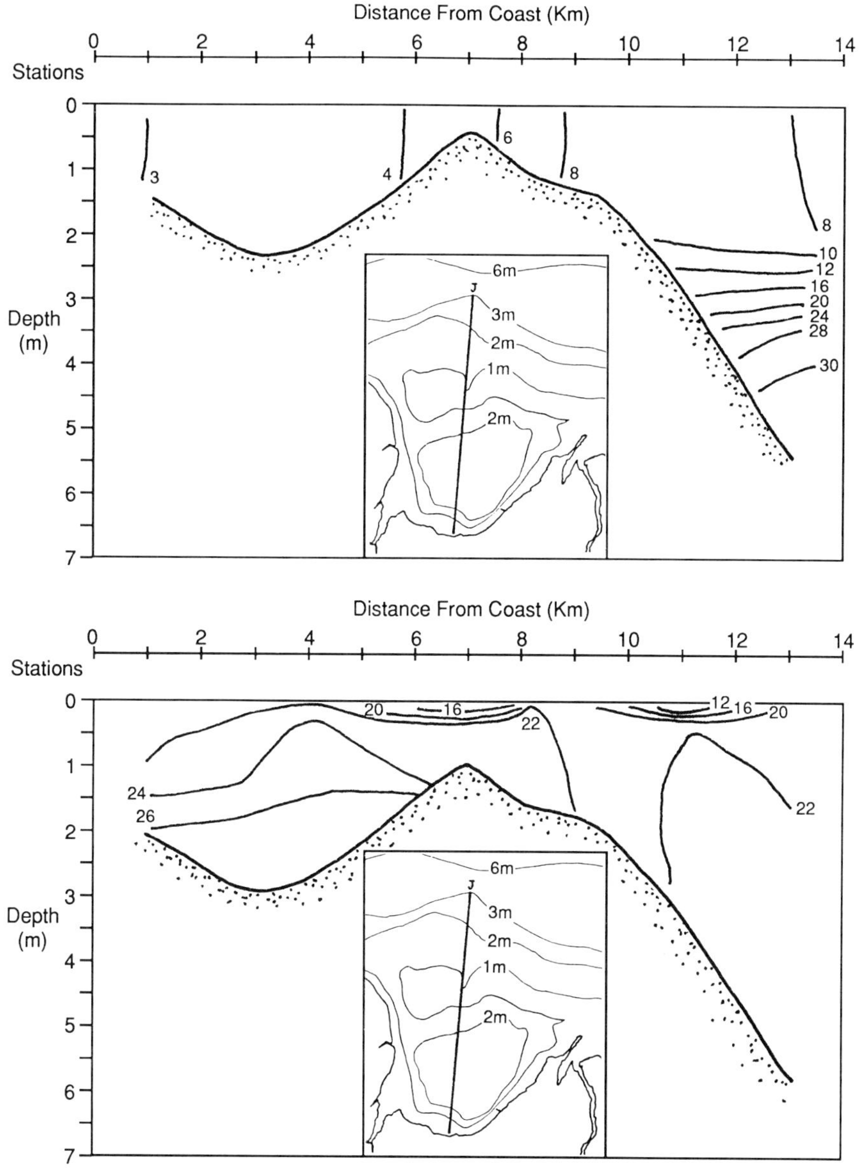

FIGURE 8.—Vertical salinity (‰) distribution along a north–south cross section (J) of Prudhoe Bay under east winds, July 18, 1987 (top panel) and after a shift to west winds, July 22, 1987 (bottom panel).

(Cannon et al. 1990, in press). In east wind upwelling years, high concentrations of Colville ciscoes and, to a lesser extent, broad whitefish are found in the Kuparuk River plume in western Gwydyr Bay. The cold, high-salinity water that upwells behind West Dock and intrudes into the east end of Gwydyr Bay forms a barrier to the fish. When the barrier of upwelled marine water is broken down by a shift to west winds or very strong east winds, movements of ciscoes and broad whitefish into eastern Gwydyr Bay and eastward around West Dock into Prudhoe Bay and the Sagavanirktok River delta usually occur. Subsequent shifts to east winds reestablish the marine water barrier west of West Dock, forcing ciscoes and broad whitefish from the east end of Gwydyr Bay lagoon.

The majority of the juvenile Arctic ciscoes that overwinter in the Sagavanirktok River are concentrated between the two causeways during mid-

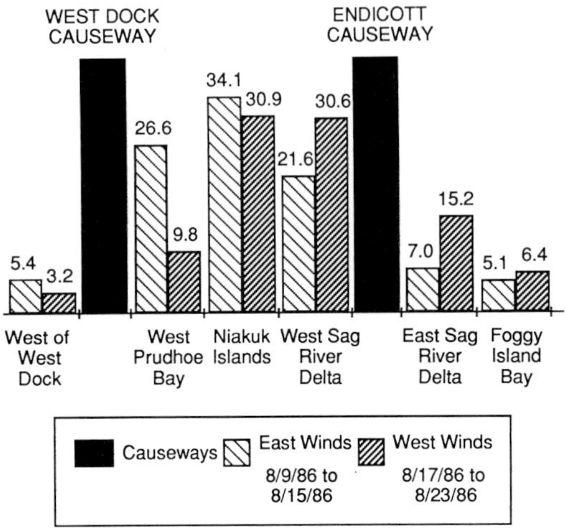

FIGURE 9.—Distribution of juvenile (ages 1 and 2) Arctic ciscoes under east and west winds, August 9–23, 1986. Numerical values represent the mean daily catch in each region under east or west winds, based on daily fyke net catches.

summer, as they were in 1986 (Figure 9). Their distribution is driven by habitat conditions (Cannon et al. 1990). Under east winds, these fish extend westward to West Dock with the warm, low-salinity plume of the Sagavanirktok River. Under west winds, these fish move eastward with the retracting river plume and the intrusion of cold, high-salinity water eastward around West Dock and into Prudhoe Bay.

Most of the anadromous fish that feed during the summer in the Prudhoe Bay region begin their migrations toward overwintering rivers when regional water temperatures fall to around 5°C (Cannon et al. 1987; Glass et al. 1990). During westward return migrations to the Colville River, fish (primarily large Arctic and least ciscoes) often accumulate along the east side of West Dock (Figure 10).

The physical presence of West Dock, the upwelled water in its lee, and the reduced volume of the Sagavanirktok River plume combine to delay migrations (Cannon et al. 1990). Continued movement past West Dock most frequently occurs when west winds carry Colville River water towards West Dock. In the absence of west winds, many fish may move westward following the remnants of the Sagavanirktok River plume as it passes over the West Dock Causeway and along the seaward side of the barrier islands. Catch patterns and tag returns indicate that most of these fish eventually return from the Prudhoe Bay region to the Colville River (Moulton et al. 1986; Cannon et al. 1987; Glass et al. 1990). However, during their migration these fish are exposed to colder, more marine waters than they would otherwise encounter in the absence of the causeways.

Juvenile Arctic char returning in late summer to the Sagavanirktok River from summer feeding habitat west of West Dock are also hindered by West Dock (Cannon et al. 1990). Figure 11 shows the catch distribution of Arctic char in late summer 1986. The higher catch on the west side of the causeway is a consistent late-summer pattern from year to year.

Age-0 Arctic ciscoes moving westward from their natal Mackenzie River arrive in the Prudhoe Bay region anywhere from early August to mid-September, moving within a cooler, moderate-salinity coastal plume (Cannon et al. 1987; Glass et al. 1990). Their migrations through the area usually continue until freeze-up in late September or October, and may continue under the ice following freeze-up. These fish appear to have a greater tolerance of marine conditions than their older counterparts, and indeed, most of the anadromous fish in the area (Cannon et al. 1987). Although large numbers may build up temporarily on the east sides of the causeways, they also pass around the causeways in large numbers (Cannon et al. 1987; Glass et al. 1990).

The extent of the causeway effects on the movements and migrations of anadromous fish in the region is highly variable between and within years (Cannon et al. 1990, in press). As discussed earlier, the frequency of westerlies and the volume of water discharged from the rivers have a large influence on the extent of causeway effects on the aquatic environment relative to conditions believed to have occurred prior to causeway construction. The effects of the causeways are greatest in years of infrequent west winds. In years when west winds are frequent, the effects of causeways on fish movements and migrations are greatly reduced.

Summer Feeding Habitat

The effect of the causeways on the warm, brackish-water feeding habitat of anadromous fish varies with season and year. In years when east winds predominate, the causeways play a greater role, due to enhanced upwelling and a greater offshore deflection of the river plumes (Cannon et al. 1990, in press). In early summer, the causeways have little effect on habitat, due to the large

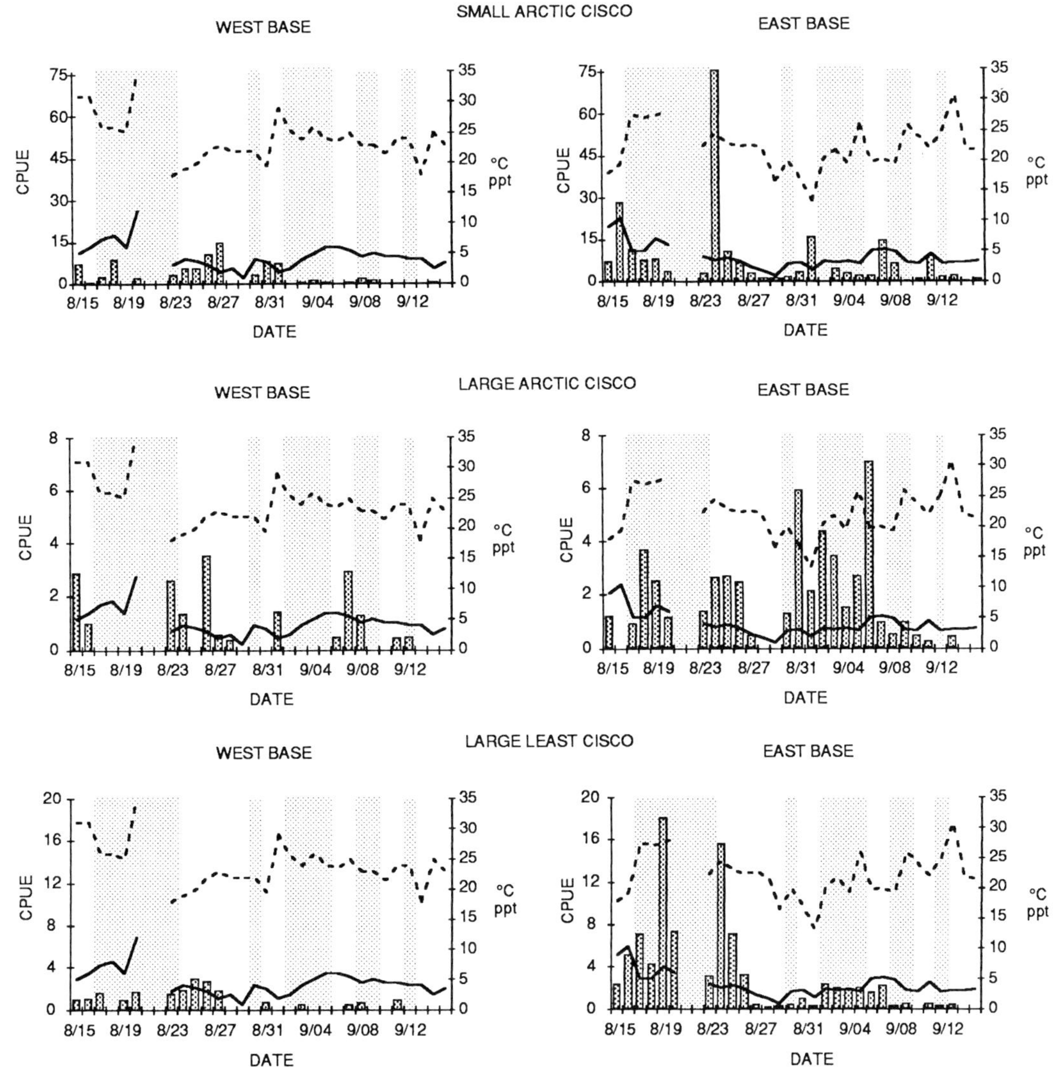

FIGURE 10.—Catch per unit effort (CPUE: fish/[fyke net · d]) of small (age 1) Arctic ciscoes, large (age 3 and older) Arctic ciscoes, and large (>250 mm) least ciscoes, adjacent to the east and west sides of the West Dock Causeway, August 15–September 15, 1986. Surface water temperature (°C: solid line) and surface salinity in per mille (ppt: dashed line) are also shown. Shaded areas indicate periods of west winds.

amount of warm, fresh water along the coast, with one exception: ice accumulated along the causeways reduces water temperatures near the causeways and downwind. In middle and late summer, enhanced upwelling and the offshore deflection of river plumes can greatly reduce the amount of warm, low-salinity, anadromous fish habitat in the study area.

Juvenile broad whitefish and Arctic cisco summer habitat in and adjacent to the Sagavanirktok River delta is affected to the greatest extent (Cannon et al. 1990). These fish concentrate in the warm, low-salinity waters of the river plume (Glass and Cannon 1990). The Endicott Causeway degrades this habitat by deflecting a portion of the plume offshore; the plume is then replaced by colder, higher salinity offshore waters, as described previously.

Larger ciscoes and Arctic char use a wider range of habitat conditions (Glass and Cannon 1990); thus, changes in salinity and temperature in the region have a lesser effect on their summer

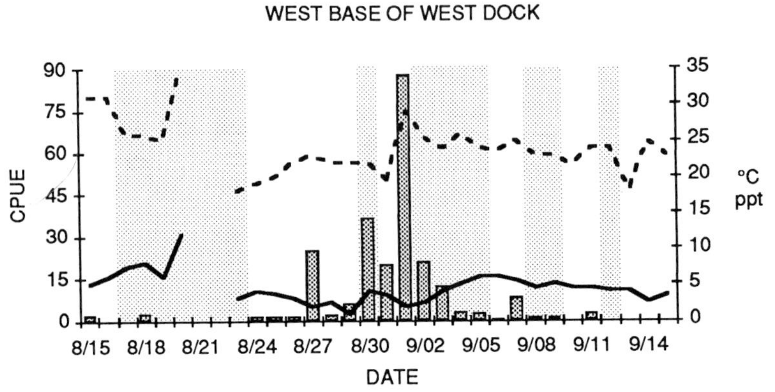

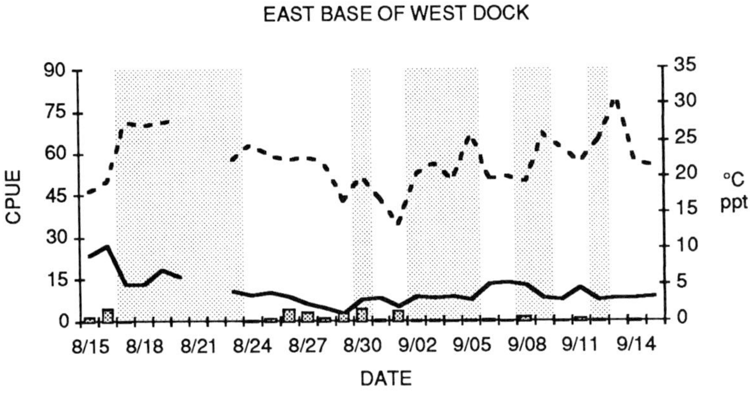

FIGURE 11.—Catch per unit effort (CPUE: fish/[fyke net · d]) of small (<250 mm) Arctic char adjacent to the east and west sides of the West Dock causeway, August 15–September 15, 1986. Surface water temperature (°C: solid line) and surface salinity in per mille (ppt: dashed line) are also shown. Shaded areas indicate periods of west winds.

habitat. During east winds, however, the coastal area between West Dock and western Gwydyr Bay supports few ciscoes, broad whitefish, or Arctic char, due to the presence of upwelled marine water in the lee of West Dock (Cannon et al. 1990).

Enhanced upwelling of cold marine water into coastal waters by the causeways in middle and late summer also affects the summer distribution of juvenile Arctic cod, the most abundant fish in coastal Beaufort Sea waters. Juvenile Arctic cod are most abundant in waters of intermediate temperature and salinity along the coast and avoid the warmer waters of the river plumes, bays, and lagoons, as well as cold, marine water (See Cannon et al. 1991a, this volume). Causeway enhancement of the onshore movement of marine waters brings large numbers of Arctic cod into the shallow coastal waters, thus increasing competition among the fishes for food and space in the remaining intermediate temperature and salinity habitat of the study area (Cannon et al. 1990).

Population-Level Effects

Because modifications in the quality of fish habitats are known to affect the productivity of the species using those habitats (e.g., Waters 1977), the monitoring program was designed to emphasize how causeways affect fish migration and feeding habitat. Program planners concentrated on habitat and prescribed only the simplest monitoring of growth rates, fish condition, and population age structure, in the belief that such monitoring would detect any catastrophic effects of the causeways (Cannon et al. 1990). The sparse data collected on population parameters during the monitoring program are poorly suited to address questions of population effects. The paucity of baseline data further limits the potential for determining long-term trends or short-term changes that may have occurred due to the causeways.

Despite these deficiencies, questions regarding population effects have gained in importance in recent years as effects to fish habitat have been documented. We have used the data available to reveal clues to the potential population effects of causeway-induced changes to habitat (Cannon et al. 1990). Analyses of growth rates of Arctic ciscoes and broad whitefish based on length-frequency data have indicated a relationship between the frequency of west winds and river discharge, and growth of these fish (Figure 12). Although only 6 years of data are available, results suggest that variations in environmental conditions within the range observed in the region can affect growth of juvenile Arctic ciscoes and broad whitefish. Because the difference in environmental conditions found in "east wind" and "west wind" years is amplified by the presence of the causeways, it is likely that some of the differences in observed growth rates can be attributed to causeway effects on habitat.

Analyses of age structure of the broad whitefish population of the Sagavanirktok River (Cannon et al. 1990) suggest poor survival of juveniles in recent years. By 1987, four year-classes, each well represented at age 0, were virtually absent from the population (Figure 13). Although the cause of this trend has not been identified, the trend is suggestive of an unstable population that could be vulnerable to habitat changes.

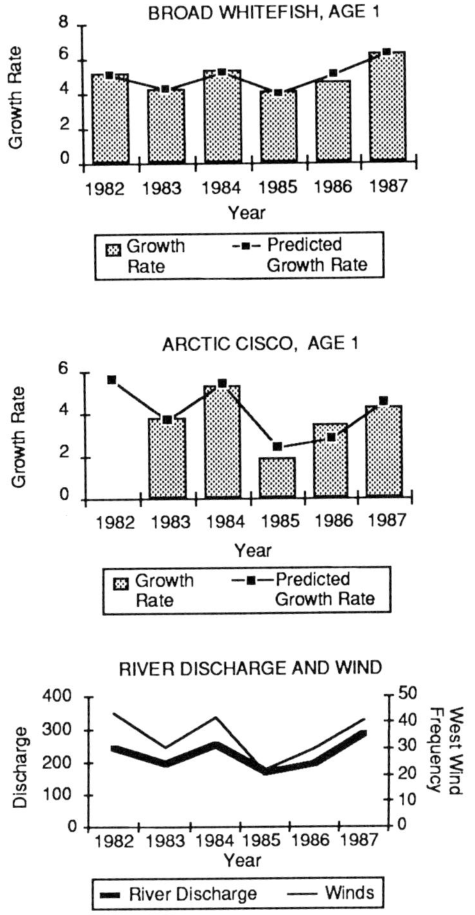

FIGURE 12.—Mean summer growth rate (mm/week) of age-1 broad whitefish and age-1 Arctic ciscoes calculated from length-frequency data (bars) and predicted from a regression equation with river discharge and frequency of west winds (solid lines) for summers 1982–1987. River discharge (m^3/s) and frequency (%) of west winds are also shown.

Limitations and Difficulties Encountered

The greatest difficulty associated with assessing the effects of the causeways on fish and fish habitat has been the inadequacy of the baseline data. Prior to the construction of West Dock, the data base consisted of sampling conducted in a region considerably to the west of the causeway area and numerous small studies that were not adequate to characterize precauseway conditions. Assessment of the effects of the West Dock Causeway has therefore required extrapolation of preexisting conditions. Resource agency planners decided that numerical models would not be used to estimate those conditions; thus, extrapolations were based upon detailed conceptual models developed over the course of the monitoring programs.

The baseline data for the assessment of the Endicott Causeway were somewhat better. It included 4 years of extensive fish studies in the region, although the Sagavanirktok River area was extensively sampled in only one of those years. Unfortunately, the Endicott baseline data collection did not include extensive oceanographic or fish population studies. In addition, the 4 years of data (1981–1984) collected prior to the construction of the Endicott causeway included 3 years in which west winds were frequent and 1 year in which west winds were frequent for half of the season. No data were collected in the precauseway years under conditions characteristic of low-frequency west winds, which are historically the norm and which occurred in the postcauseway monitoring years (1985–1987). The high-frequency west wind years have been determined to be years

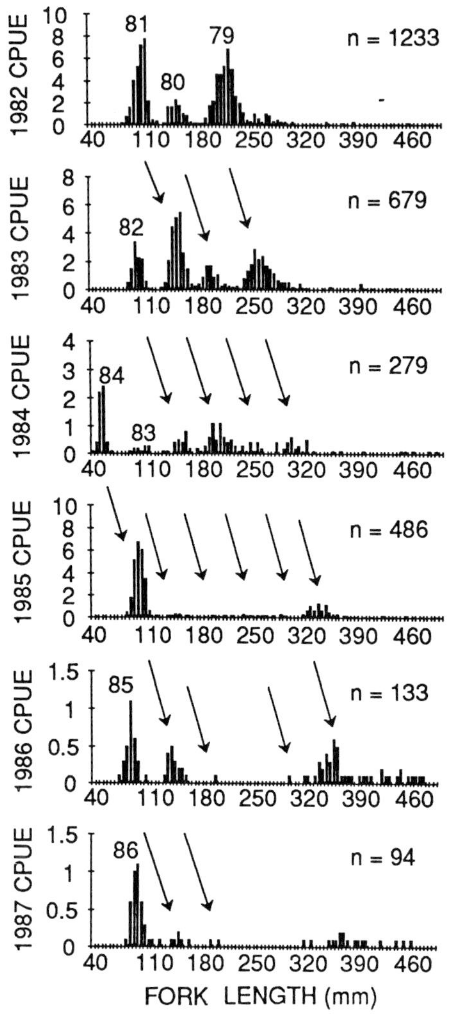

FIGURE 13.—Length-frequency distribution of broad whitefish captured from the Sagavanirktok River delta, summers 1982–1987. CPUE is average catch per 24-hour fyke-net sample. Year-classes are noted above peaks in distribution. Arrows track year-classes to subsequent years; n is sample size.

of low causeway effects, and low-frequency west wind years have been determined to be years of higher causeway effects. The absence of pre-causeway, low-frequency west wind fish and oceanographic data has necessitated the extrapolation of these values from conceptual models.

The second greatest difficulty in assessing causeway effects was the dynamic character of the arctic system, including not only large inter-annual differences in environmental conditions, but also large daily and seasonal changes in conditions. In addition to environmental variability and the paucity of data, large variations in the year-classes of various species obscure the identification of population-level effects.

Conclusions

After 3 years of study, it is apparent that the Endicott Causeway has reduced anadromous fish habitat in the study area, and that this effect compounds the already existing effects of the West Dock Causeway. In addition, the causeways' effects on habitat coincide with variations in growth of juvenile Arctic ciscoes and broad whitefish. The long-term consequences of causeway effects to these fish populations are unknown. At this time, no long-term population studies have commenced or are planned.

The USACE and state of Alaska permits for the causeways specify that mitigation measures, specifically larger breaches in the causeways, must be implemented if effects on the fish populations and their habitat are found to be unacceptable by the District Engineer or the State. In early 1991, the decision was made to enlarge breaches in both the Endicott and West Dock causeways by 230 m.

References

Aagaard, K. 1978. Physical oceanography and meteorology. *In* Interim synthesis: Beaufort/Chukchi, chapter 2. National Oceanic and Atmospheric Administration, Outer Continental Shelf Environmental Assessment Program, Boulder, Colorado.

Barnes, P., E. Reimnitz, and G. Smith. 1977. Some coastal oceanographic observations—Beaufort Sea, Alaska. *In* Miscellaneous hydrologic and geologic observations on the inner Beaufort Sea shelf, Alaska, part A. U.S. Geological Survey Open File Report 77-477, Anchorage, Alaska.

Behnke, R. J. 1980. A systematic review of the genus *Salvelinus*. Pages 441–480 *in* E. K. Balon, editor. Charrs: salmonid fishes of the genus *Salvelinus*. Dr. W. Junk, The Hague, Netherlands.

Behnke, R. J. 1984. Organizing the diversity of the Arctic charr complex. Pages 3–21 *in* L. Johnson and B. L. Burns, editors. Biology of the Arctic charr. Proceedings of the international symposium on Arctic charr. University of Manitoba Press, Winnepeg.

Bendock, T. 1979. Beaufort Sea estuarine fish study. Pages 670–729 *in* Environmental assessment of the Alaska Outer Continental Shelf. Annual report of the principal investigators. Bureau of Land Management and National Oceanic and Atmospheric Administration, Outer Continental Shelf Environmental Assessment Program, Boulder, Colorado.

Berry, A. D., and J. M. Colonell. 1985. Oceanography. *In* Lisburne Development environmental studies: 1984, volume 2. Report of Woodward-Clyde Consultants to ARCO–Alaska, Inc., Anchorage, Alaska.

Britch, R. P., R. C. Miller, J. Dowing, T. Petrillo, and M. Veit. 1983. Physical processes. Pages 1–218 *in* B. J. Gallaway and R. D. Britch, editors. Environmental summer studies (1982) for the Endicott Development, volume 2. Report of LGL Ecological Research Associates, Inc., and Northern Technical Services, Inc., to Sohio Alaska Petroleum Co., Anchorage, Alaska.

Cannon, T. C., B. Adams, D. Glass, and T. Nelson. 1987. Fish distribution and abundance. Pages 1–129 *in* 1985 Final report of the Endicott Environmental Monitoring Program, volume 6. Report of Envirosphere Company to U.S. Army Corps of Engineers, Anchorage, Alaska.

Cannon, T. C., D. R. Glass, and C. M. Prewitt. 1991a. Habitat use patterns of Arctic cod in the coastal Beaufort Sea near Prudhoe Bay, Alaska. American Fisheries Society Symposium 11:157–162.

Cannon, T., L. Hachmeister, and D. Glass. 1990. Integration and assessment. *In* 1986 Final report of the Endicott Environmental Monitoring Program, volume 1. Report of Envirosphere Company to U.S. Army Corps of Engineers, Anchorage, Alaska.

Cannon, T. C., L. H. Hachmeister, and D. R. Glass. In press. Integration and assessment. *In* 1987 Final report of the Endicott Environmental Monitoring Program, volume 2. Report of Envirosphere Company to U.S. Army Corps of Engineers, Anchorage, Alaska.

Cannon, T. C., J. A. Knutzen, and D. R. Glass. 1991b. Habitat use by epibenthic and planktonic invertebrates along the coast of the central Beaufort Sea, Alaska. American Fisheries Society Symposium 11:163–172.

Chin, H. 1979. Physical oceanography. *In* Environmental studies associated with the Prudhoe Bay Dock, final report, chapter 2. Report of Woodward-Clyde Consultants to ARCO–Alaska, Inc., Anchorage, Alaska.

Craig, P. C. 1984. Fish use of coastal waters of the Alaskan Beaufort Sea: a review. Transactions of the American Fisheries Society 113:265–282.

Craig, P. C., and L. Haldorson. 1981. Fish. Pages 384–678 *in* Beaufort Sea barrier island–lagoon ecological process studies: final report, Simpson Lagoon. Environmental assessment of the Alaskan outer continental shelf program, final report of the principal investigators, volume 7. Bureau of Land Management and National Oceanic and Atmospheric Administration, Outer Continental Shelf Environmental Assessment Program, Boulder, Colorado.

Envirosphere Company. 1983. Final report 1982 Prudhoe Bay Waterflood Environmental Monitoring Program. Report to U.S. Army Corps of Engineers, Anchorage, Alaska.

Envirosphere Company. 1984. Final report 1983 Prudhoe Bay Waterflood Environmental Monitoring Program. Report to U.S. Army Corps of Engineers, Anchorage, Alaska.

Envirosphere Company. 1985. Final report 1984 Prudhoe Bay Waterflood Environmental Monitoring Program. Report to U.S. Army Corps of Engineers, Anchorage, Alaska.

Envirosphere Company. 1987. Final report 1985 Endicott Environmental Monitoring Program. Report to U.S. Army Corps of Engineers, Anchorage, Alaska.

Envirosphere Company. 1990. Final report 1986 Endicott Environmental Monitoring Program. U.S. Army Corps of Engineers, Anchorage, Alaska.

Envirosphere Company. In press. Final report 1987 Endicott Environmental Monitoring Program. U.S. Army Corps of Engineers, Anchorage, Alaska.

Furniss, R. A. 1975. Inventory and cataloging of Arctic area waters. Alaska Department of Fish and Game, Annual Report 16:1–47, Juneau.

Gallaway, B. J., W. B. Griffiths, P. C. Craig, W. J. Gazey, and J. W. Helmericks. 1983. An assessment of the Colville River delta stock of Arctic cisco (*Coregonus autumnalis*)—migrants from Canada? Biological Papers of the University of Alaska 21:4–23.

Glass, D. R., and T. C. Cannon. 1990. Fish habitat suitability estimate. *In* 1986 Final report for the Endicott Environmental Monitoring Program, volume 5. Report of Envirosphere Company to U.S. Army Corps of Engineers, Anchorage, Alaska.

Glass, D. R., C. Whitmus, and C. M. Prewitt. 1990. Fish distribution and abundance. *In* 1986 Final report of the Endicott Environmental Monitoring Program, volume 6. Report of Envirosphere Company to U.S. Army Corps of Engineers, Anchorage, Alaska.

Griffiths, W. B., and seven coauthors. 1983. Fish ecology. Pages 1–323 *in* B. J. Gallaway and R. P. Britch, editors. Environmental summer studies (1982) for the Endicott Development, volume 3. Report of LGL Alaska Research Associates and Northern Technical Services to Sohio Alaska Petroleum Co., Anchorage, Alaska.

Hachmeister, L. E., K. S. Short, G. C. Schrader, K. T. Winnick, and J. W. Johannessen. 1987. Oceanographic monitoring. *In* 1985 Final report for the Endicott Environmental Monitoring Program, volume 3. Report of Envirosphere Company to U.S. Army Corps of Engineers, Anchorage, Alaska.

Hufford, G. L., and R. D. Bowman. 1974. Airborne temperature survey of Harrison Bay. Arctic 27:69–70.

Hummer, P. G. 1990. Meteorology. *In* 1986 Final report for the Endicott Environmental Monitoring Program, volume 2. Report of Envirosphere Company to U.S. Army Corps of Engineers, Anchorage, Alaska.

Johannessen, J. W. 1987. River discharge. *In* 1985 Final report for the Endicott Environmental Monitoring Program, volume 2. Report of Envirosphere Company to U.S. Army Corps of Engineers, Anchorage, Alaska.

Mangarella, P. A., J. R. Harper, and T. J. Weingartner. 1982. Physical processes monitoring program. Final report of 1981 Prudhoe Bay Waterflood Environmental Monitoring Program, volume 2. Report of Woodward-Clyde Consultants to U.S. Army Corps of Engineers, Anchorage, Alaska.

Matthews, J. B. 1978. Characterization of the nearshore hydrodynamics of an Arctic barrier island–lagoon system. Pages 607–619 *in* Environmental assessment of the Alaskan continental shelf, annual report, volume 8. National Oceanic and Atmospheric Administration, Outer Continental Shelf Environmental Assessment Program, Boulder, Colorado.

Morrow, J. E. 1980. Analysis of the Dolly Varden char, *Salvelinus malma*, of northwestern North America and southeastern Siberia. Pages 323–338 *in* E. K. Balon, editor. Charrs: salmonid fishes of the genus *Salvelinus*. Dr. W. Junk, The Hague, Netherlands.

Moulton, L. L., and seven coauthors. 1986. 1984 Central Beaufort Sea fish study. Pages 1–300 *in* Prudhoe Bay Waterflood Environmental Monitoring Program 1984, volume 2. Final report of Woodward-Clyde Consultants, Entrix Inc., and LGL Alaska Research Associates to U.S. Army Corps of Engineers, Anchorage, Alaska.

Savoie, M. A., and D. E. Wilson. 1984. Physical processes. *In* Prudhoe Bay Waterflood Environmental Monitoring Program 1983, volume 1. Final report of Kinnetic Laboratories, Inc., to U.S. Army Corps of Engineers, Anchorage, Alaska.

Schrader, G. C., and L. E. Hachmeister. 1990. Ice breakup/freezeup monitoring. *In* 1986 Final report for the Endicott Environmental Monitoring Program, volume 2. Report of Envirosphere Company to U.S. Army Corps of Engineers, Anchorage, Alaska.

Short, K. S., G. C. Schrader, L. E. Hachmeister, and C. J. VanZee. 1990. Oceanography. *In* 1986 Final report of the Endicott Environmental Monitoring Program, volume 2. Report of Envirosphere Company to U.S. Army Corps of Engineers, Anchorage, Alaska.

USACE (U.S. Army Corps of Engineers). 1980. Prudhoe Bay Waterflood Development Project final environmental impact statement. USACE, Anchorage, Alaska.

USACE (U.S. Army Corps of Engineers). 1984. Endicott Development Project final environmental impact statement. USACE, Anchorage, Alaska.

Woodward-Clyde Consultants. 1982. Final report 1981 Prudhoe Bay Waterflood Environmental Monitoring Program. Report to U.S. Army Corps of Engineers, Anchorage, Alaska.

Waters, T. F. 1977. Secondary production in inland waters. Advances in Ecological Research 10:91–164.

American Fisheries Society Symposium 11:97–108, 1991

Habitats versus Populations: Approaches for Assessing Impacts on Fish

SCOTT B. ROBERTSON
ARCO Alaska, Inc.
Post Office Box 100360, Anchorage, Alaska 99510, USA

Abstract.—Predictions of the effects of industrial activities on fish populations are often based on anticipated alteration or degradation of their habitat. Loss of critical habitat (habitat necessary for at least part of a fish's life history) can reduce the health and survival of a population. In the absence of data on various attributes of the population, assessment of possible population changes can be estimated through speculation (typically necessary in an environmental impact statement) or measurement (possible after the project is in place) of changes in habitat attributes. The alternative is to monitor the population attributes themselves. When these data are available it should no longer be necessary to "assess" effects on populations by extrapolating from measured changes in habitat. These alternative methods of evaluating impacts to fish populations are considered in the context of the environmental effects of solid-fill causeways that support the petroleum industry on Alaska's Beaufort Sea coast.

The importance of habitat to every species is an unquestioned tenet of environmental science. Indeed, the entire field of ecology examines the relationships of organisms with habitat in its broadest sense. Habitat can be described by its biotic components, such as prey and competition, and by its physical components, such as turbidity, temperature, and salinity. From the many habitat characteristics available for study, the interests and resources of investigators dictate the completeness of the list of habitat characteristics selected. Habitats for freshwater fish can be described by a few or many specific parameters such as turbidity, temperature, current velocity, and substratum (Pajak and Neves 1987); depth, current velocity, substratum, cross-stream and vertical position, and turbulence (Gorman 1988); temperature, cover, and current velocity (Taylor 1988); structure (Johnson et al. 1988; Moore and Gregory 1988); food and cover (Mesick 1988); or by more general descriptors such as forest, meadow, or slough tributary (Dolloff 1987). Of course, these kinds of habitat descriptors can be used for other biota as well; for example, phytoplankton habitat has been described by water residence time, depth, clarity, drainage area, and flow (Søballe and Kimmel 1987). A wide variety of inventory and description techniques for aquatic habitats were described at a 1981 symposium (Armantrout 1982). In generalizations of important habitat characteristics for freshwater and marine fish communities, Pitcher and Hart (1982) used river form and flow, current types, vegetation, temperature, oxygen, depth, substratum, and water color.

The U.S. federal government, through enactment of a variety of laws, has recognized that habitat protection is one clear method of reducing the negative effects of human activity on the environment. The National Environmental Policy Act of 1969 (NEPA: Public Law [PL] 91-190), requires a detailed statement on environmental effects of "major Federal actions significantly affecting the quality of the human environment." These proposed actions may be actual federal projects or the granting of a permit allowing other parties to proceed with an activity. A statement, called an environmental impact statement (EIS), must evaluate the expected consequences of both the proposed action and a set of practicable alternative actions (e.g., a smaller-scale project or no project at all). Environmental consequences of major projects and their alternatives are generally habitat related. Habitat protection is one of the goals of the Coastal Zone Management Act of 1972 (PL 92-583). Preservation of "critical habitat" is one of the cornerstones of the Endangered Species Act of 1973 (PL 93-205). The Clean Water Act of 1977 (PL 95-217) addressed water quality aspects of habitat. The Marine Protection, Research, and Sanctuaries Act of 1972 (PL 92-532) was established to control ocean dumping and protect special areas.

Predictions of the effects of development projects are frequently based on anticipated alteration or degradation of habitat. A common feature

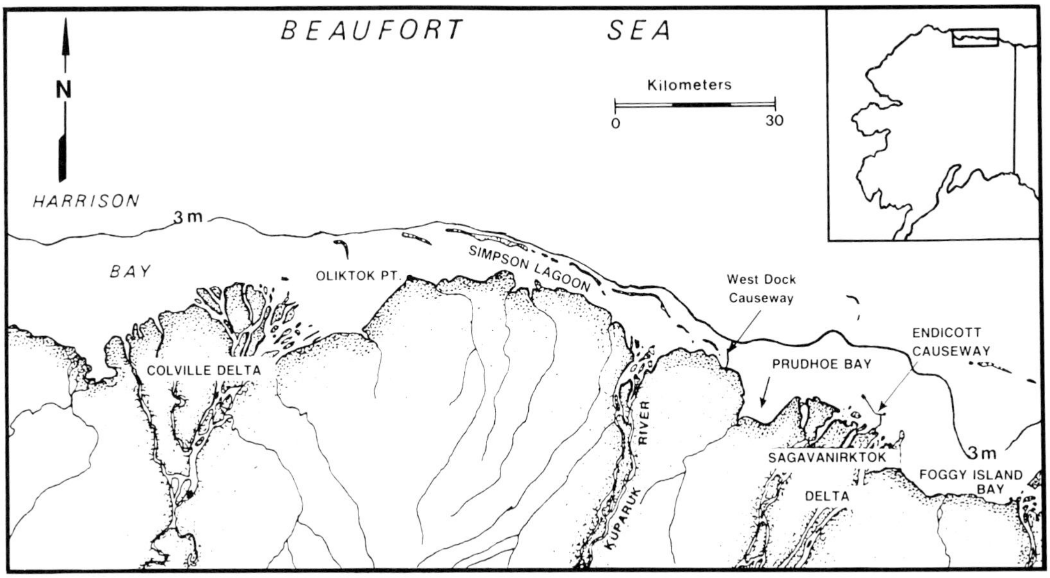

FIGURE 1.—The central Beaufort Sea coast and West Dock and Endicott causeways, Alaska. (From Fechhelm et al. 1989.)

of EISs is the description of possible changes to the affected physical environment and the expected consequences to the human and biological communities that use that environment. Habitat evaluation procedures have been developed by the U.S. Fish and Wildlife Service as a planning tool for assessing potential effects of projects on selected fish and wildlife species (Division of Ecological Services 1980a, 1980b). Additionally, other methods, including energy flow and population estimation, are available for making predictive impact assessments (Division of Ecological Services 1980a). These and other methods are also available to determine effects on populations *after* development projects are in place. However, assessments of effects on populations inferred from observed habitat changes may be less meaningful than assessments based on direct measurements of various population parameters. The following example, of the effects of petroleum industry causeways on arctic fishes, is provided to illustrate this point.

Case Study: Causeways in the Beaufort Sea

Physical and Biological Features

The central Beaufort Sea coast is bounded to the south by a large plain, the Arctic Coastal Plain of Alaska. The area has been described previously (Craig 1984, 1989) and is summarized briefly here. Several small tundra streams feed into the coastal area, but summer nearshore oceanography is more strongly influenced by the Colville and Sagavanirktok rivers, west and east of Prudhoe Bay (Figure 1). Between the Colville River and Prudhoe Bay, a string of barrier islands separates the shallow sea from Simpson Lagoon.

The open-water period along the coast is usually from mid-July through the latter part of September. Due to the shallowness of the nearshore seas, current patterns are primarily a function of local winds, which are bimodal and blow predominantly from the east-northeast and west. During winter, ice freezes to a depth of 2 m and ocean currents are low.

The greatest discharge of rivers along the Arctic Coastal Plain occurs early in the summer. The early open-water period (breakup) is strongly influenced by this high volume of meltwater. The nearshore surface water is quite fresh and mixes little with the deeper marine layer. Lower discharge rates and increased mixing by storms cause the nearshore waters to become more marine as summer progresses (Colonell and Niedoroda 1990).

Nearshore upwelling is induced by easterly winds. The winds move surface waters offshore, which results in the onshore and upward movement of deeper marine water, occasionally to less than 2 m from the surface. Along shore, brackish water (10–15‰) is maintained at the surface by continued input from rivers (Savoie and Wilson 1986; Colonell and Niedoroda 1990). Although tidal range is low in this region (<0.2 m), wind-driven positive and negative surges can change water levels more than 1 m (Savoie and Wilson 1986). A nearshore transition to a more marine condition occurs in midsummer upon one of the switches from east to west winds that occur throughout the summer. West winds can move the upwelled marine water further onshore by creating a positive surge. Late-summer nearshore surface salinities range from 15 to 30‰ (Savoie and Wilson 1986).

The distribution and characteristics of fishes in the coastal Beaufort Sea have been described by Craig (1984, 1989). The most abundant nearshore fishes are two marine species (Arctic cod *Boreogadus saida* and fourhorn sculpin *Myoxocephalus quadricornis*) and four anadromous salmonid species (Dolly Varden *Salvelinus malma*, broad whitefish *Coregonus nasus*, Arctic cisco *C. autumnalis*, and least cisco *C. sardinella*). The Dolly Varden in this area has been historically misidentified as Arctic char (*S. alpinus*), but analysis of the meristics favors the *S. malma* designation (Morrow 1980; Behnke 1984). Recent reports have begun to recognize this (Moulton and Field 1988; Griffiths et al. 1990). To minimize confusion, in this paper the fish will simply be called char. The four anadromous fishes overwinter in the local rivers and move into the nearshore brackish–marine environment to feed during the summer.

Fyke net catch data from a 1985 monitoring study illustrate the typical relative abundances of these fishes (Figure 2). The two marine species were numerically dominant. Least cisco and char were each approximately one-fifth as abundant as the marine species; Arctic cisco (not including young-of-the-year) and broad whitefish were somewhat lower. The remaining 2% of the summer's catch consisted of 16 other marine, anadromous, and freshwater species (Cannon et al. 1987). Arctic cisco young-of-the-year represent a special situation due to the variability of their wind-aided recruitment to the area from the Mackenzie River (Gallaway et al. 1983; Fechhelm and Fissel 1988; Schmidt et al. 1991, this volume). In some years, such as 1985, they may be very abundant late in the summer; in other years, such as 1984 (Moulton et al. 1986), they may be almost absent.

Habitat description for these nearshore fishes

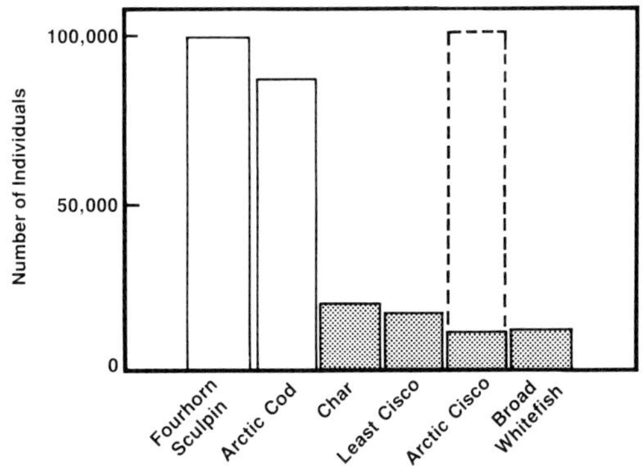

FIGURE 2.—Relative abundances of nearshore central Beaufort Sea fishes as reflected in the total catch of a summer 1985 study from Gwydyr Bay to Foggy Island Bay east of Sagavanirktok River (27 stations, 53 km of coastline). Marine fishes are represented by open columns and anadromous fish by grey columns. Arctic cisco young-of-the-year migrants from Mackenzie River, Canada, are shown by the dashed column. (Data from Cannon et al. 1987.)

has been based primarily on salinity and temperature. As previously mentioned, salinity can vary widely as the summer progresses, from quite fresh at breakup to almost marine at freeze-up. There is some bias in the habitat use data for these six fishes because of the limited distribution of the sampling stations. The stations, ranging from east of the Sagavanirktok River delta west to Gwydyr Bay in Simpson Lagoon, were located along mainland and island shorelines, except for a few offshore in shallow water. Fishes in deeper water (and probably more marine conditions) thus avoided sampling and were not included in the habitat use analyses. The fish catch data came from daily (approximately 24 h) fyke net samples, while the temperature and salinity data were generally instantaneous measurements taken at the end of each 24-h catch period (Moulton et al. 1986). To lessen the possibility that these physical data were not representative of the conditions under which the fish were captured, only data within specified ranges of change were considered (salinity change < 4‰; temperature change < 2°C: Whitmus et al. 1987; Dames and Moore 1988; Houghton et al. 1990).

The habitat use descriptions that follow represent the range wherein most (but not all) of the individuals of a given species were captured. Minor differences in habitat use by different cohorts of fish are not reviewed here. Arctic cod in the study area were found in water of moderate

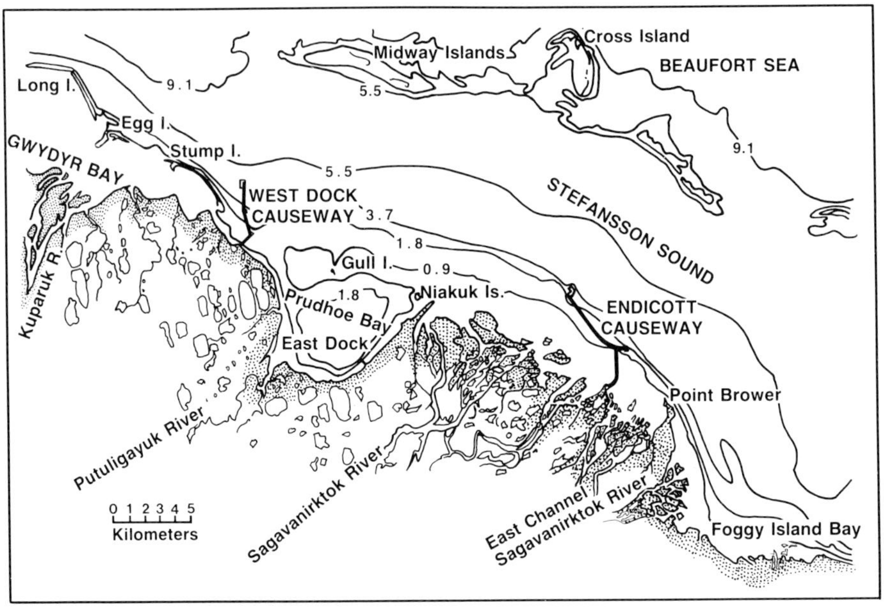

FIGURE 3.—Causeways and bathymetry (m) in the Prudhoe Bay region, Alaska.

salinity (15–25‰) and cool temperatures (2–6°C). Fourhorn sculpin were found in shallow waters of low-to-moderate salinities and intermediate temperatures. Of the four anadromous fishes, broad whitefish were the least tolerant of marine conditions. They generally used the warm (>12°C), low-salinity (<8‰) water in the vicinity of river deltas. Young char, in their first two summers away from fresh water (ages 3 and 4: Craig 1989) used the low-salinity, cool-to-moderate temperature (4–10°C) water typical of the early summer. Of these anadromous species, char age 5 and older were the most tolerant of marine conditions. Arctic ciscoes were found in moderate temperatures (4–10°C) and low-to-intermediate salinities (<16‰). Curiously, under the ice during the fall, they were closely associated with higher salinities (24–26‰: Moulton and Field 1988). Least ciscoes were captured throughout a wide salinity range (0–24‰). Fish up to 3 years of age used water of intermediate temperatures (8–10°C), while older least cisco were found in a broader, cooler temperature range (4–10°C).

Prey distribution and abundance is another aspect of habitat that has been examined in the coastal Beaufort Sea. Mysids and amphipods, which dominate the fishes' food, originate in offshore marine waters and are transported by currents to the nearshore habitat. Not unexpectedly, the distribution of prey was patchy, but abundance was quite high and was not considered to be a limiting factor to the fishes (Craig and Haldorson 1981; Craig et al. 1984; Moulton et al. 1986).

There are two solid-fill gravel causeways in the Prudhoe Bay area that have the potential to influence the populations of nearshore fishes in the region (Figure 3). The 4.3-km West Dock Causeway is west of Prudhoe Bay and east of Stump Island at the end of Simpson Lagoon. A seawater treatment plant, which prepares water for waterflooding the oil reservoir, is located on its terminus at the 4-m contour. Two-thirds of the way along its length is a dock for off-loading the large modules that are used to construct the oilfield facilities. A 16-m breach, intended to facilitate fish passage, is north of the dock. The Endicott Causeway serves two artificial, gravel, drilling and production islands for the Endicott oil reservoir. The interisland portion of the causeway parallels the 2-m depth contour and extends out to 4.5 km from shore; a mainland portion attaches the causeway to the middle of the Sagavanirktok River delta. Total causeway length is about 9 km. Two breaches, 152 and 61 m in width, are on the mainland portion of the causeway.

Habitat Alteration

These causeways affect the local habitat of fishes by altering the distribution of the water

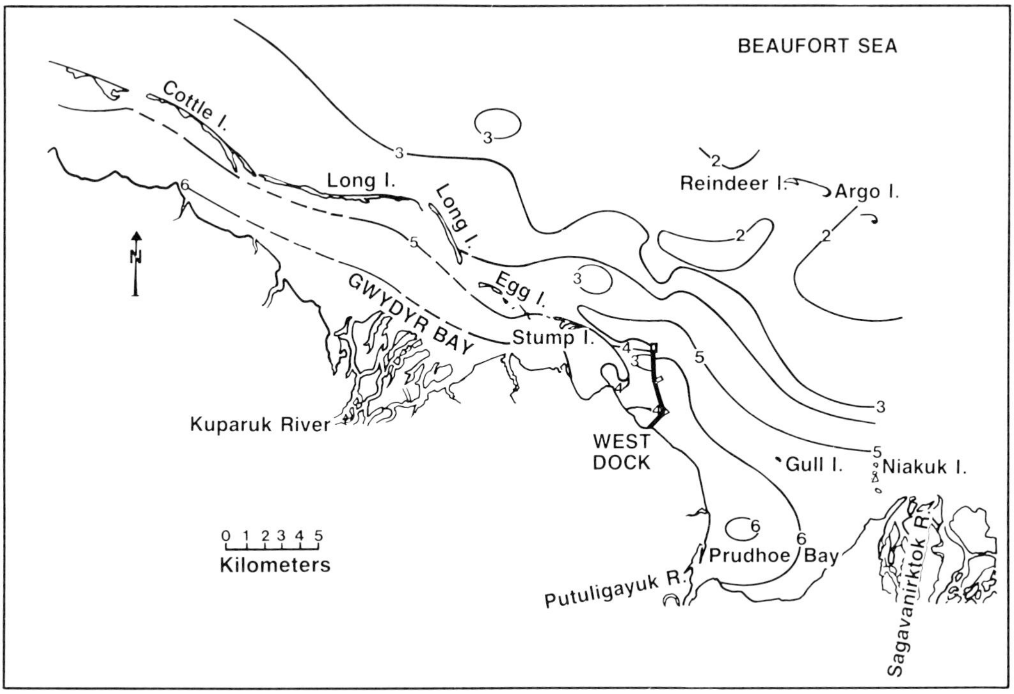

FIGURE 4.—Sea-surface temperature (°C) in the central Beaufort Sea during east winds, July 29, 1983 (Savoie and Wilson 1984).

masses in the area. Several factors cause the effects of the Endicott Causeway to be less than those of West Dock: the Endicott Causeway is in shallower water, is not located at the end of a barrier island lagoon system, and has freshwater input on both sides of it (the east and west channels of the Sagavanirktok River). For this reason, West Dock will be the focus of the examples below. It must be realized, of course, that any changes noted in the fish populations would most likely be the result of the cumulative effects of both causeways. Additional, more detailed information on causeway effects can be found in Gallaway et al. (1991, this volume).

West Dock influences the nearshore distribution of water in two ways: it disrupts the alongshore flow of brackish water, and, in the lee of the tip of the causeway, surface divergence can cause bottom water to be brought to the surface. These phenomena can be manifested under east or west winds by a plume of relatively warmer water downwind from the tip of the causeway and relatively colder water in the lee of the entire causeway (Figures 4, 5). During midsummer the water column is stratified and the bottom water is colder and more saline than the water it replaces. This, then, is the period of the greatest extent of the causeway's influence: up to 10–13 km downwind of the structure in extreme cases. Temperatures may occasionally fall to 5°C below ambient levels and salinity may reach 15‰ above ambient levels in the immediate vicinity of the causeway as a result of the upward mixing of more marine water. The intensity and extent of these effects are functions of wind speed and duration as well as of the magnitude of the stratification. As the wind switches direction between east and west, the effects of the previous event may be completely flushed away, again depending on the duration of the wind. During east winds, natural upwelling through the barrier island inlets provides an additional source of marine water to the lagoon (Savoie and Wilson 1986). This source of natural variability and discontinuity of the coastal band of brackish water also existed prior to construction of the last leg of West Dock (Matthews 1979; Mungall et al. 1979) and before any causeways were in place (Barnes et al. 1977). These natural and causeway-induced hydrological effects significantly alter nearshore waters only in midsummer, because in early summer the nearshore is dominated by fresh water, and in late

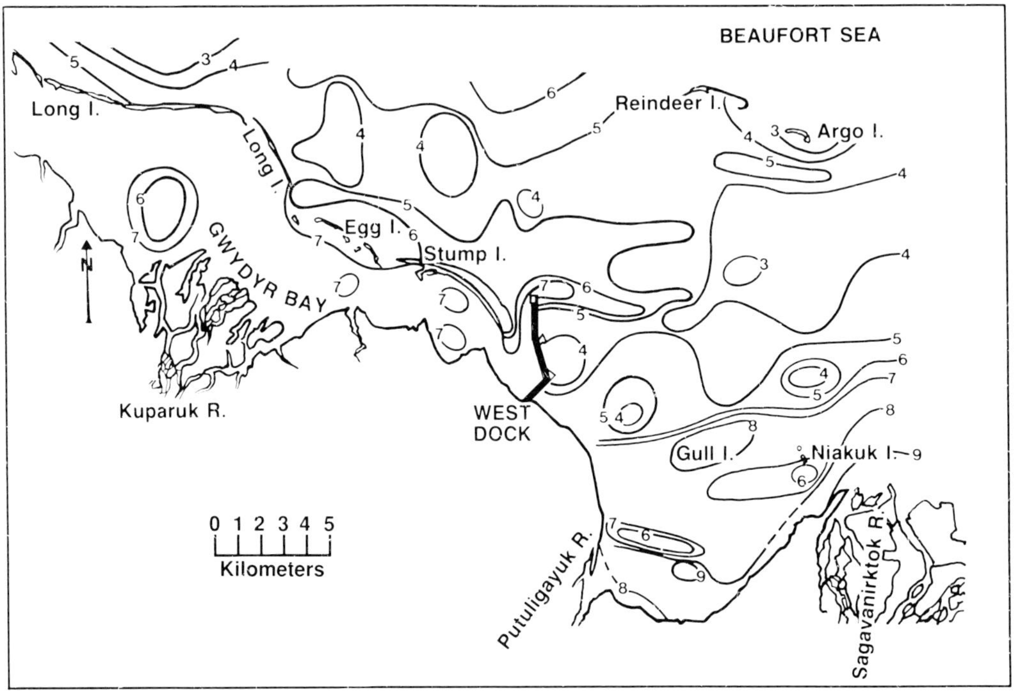

FIGURE 5.—Sea-surface temperatures (°C) in the central Beaufort Sea during west winds, August 8, 1983 (Savoie and Wilson 1984.)

summer conditions are naturally more uniformly marine.

The barrier islands west of the causeway can restrict mixing with offshore water and thus maintain the temporal and areal extent of an effect; that is, the characteristics of the water in the lagoon can exist for a longer period and over a larger area than would be possible in a more open area. If a strong east wind persists for 3–4 d, "marine" water may flow inshore of Stump Island ("marine" here indicates brackish water of higher salinity than would otherwise be found at the surface; the water rarely reaches truly marine salinities, i.e., >33‰). Upwelled marine water also flows through channels between the barrier islands (Savoie and Wilson 1986).

Warmer, lower salinity water flows back to the eastern lagoon during a midsummer west wind (the freshwater input is from the Kuparuk and Colville rivers) and is deflected offshore when it reaches the causeway (Figure 5). As water level rises during the switch from a negative to positive surge caused by the change from east to west winds, upwelled "marine" water can enter Prudhoe Bay by flowing along the eastern side of the causeway and by flowing over the sill of Prudhoe Bay (Figure 3; Savoie and Wilson 1986). The latter event is a natural one that occurs as part of a region-wide seasonal transition from a stratified condition with low-to-intermediate surface salinities to a less stratified condition with more marine surface water.

Anadromous fish distributions can be altered by the causeway in two ways. The causeway could physically block fish if they were unwilling to swim around it. In addition, causeway-induced inputs of "marine" water into the nearshore habitat could periodically create areas of high salinities and low temperatures beyond the bounds of normal fish use (Fechhelm et al. 1989). Tagging studies demonstrated that large fish (>250 mm) did not have difficulty swimming around the causeway (Critchlow 1983; Moulton et al. 1986). These studies also showed that small fish could swim around the causeway, but perhaps with lower frequencies, depending on the species. Small char did not appear to be blocked by the causeway. Small broad whitefish were rarely in the vicinity of West Dock because the low-salinity portion of the Sagavanirktok River plume seldom reached that far west. While small Arctic and least ciscoes could swim around the causeway, these species were more apt to be occasionally inhibited by the character of the water near the causeway than the other two species. Young-of-the-year Arctic ciscoes from the Mackenzie River, transported to the area by east winds, were able to travel past both causeways and reach the Colville River, where there is sufficient overwintering habitat to sustain them (Moulton 1989; Schmidt et al. 1991). In summary, for Arctic and least ciscoes the causeway can cause periodic decreases in available habitat, either through the introduction of upwelled water into the area or by such water masses temporarily delaying or blocking movement to other areas.

Prey, the other important feature of habitat, is probably enhanced by the additional input of this new source of upwelled water. The mysids and amphipods that the nearshore fishes eat are transported from offshore. Their abundance in Simpson Lagoon is about 50 times more than that required by the local consumers, both the fishes and oldsquaw ducks *Clangula hyemalis* (which are greater consumers than the fish, Craig et al. 1984). The continued replenishment of prey depends, in part, on exchange with offshore or adjacent coastal waters. Areas with limited exchange would be poorer feeding areas (Craig et al. 1984). Enhancement of prey abundance in the vicinity of the causeway has been consistently documented (Envirosphere Company 1983; Moulton et al. 1986).

Because the importance of habitat is clear, and because occasionally some of the habitat in the vicinity of Prudhoe Bay is degraded or access to it is impaired as a result of the causeway, it is possible to speculate that the fish populations could suffer negative effects. Fish unable to swim around the causeway might become overcrowded, which could overtax the food supply and reduce feeding efficiency. Fish might be delayed or prevented from reaching preferred overwintering habitat and forced to use a less desirable river. Fish forced to reside in water that is colder or more salty than usual might suffer metabolic consequences that could reduce food assimilation efficiency or extract higher energy costs. This would result in a lower condition that could then reduce reproductive potential or winter survival.

The extent to which these consequences would occur is largely a function of the extent to which the summer habitat of these fishes is limiting; and, if it is limiting, what portion of their habitat is altered. This latter question is a function of fish distribution, which depends on both the age and

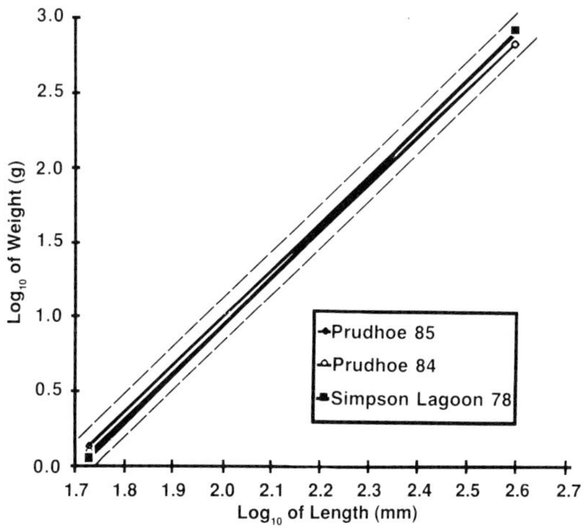

FIGURE 6.—Condition of Arctic ciscoes (length versus weight) from 1978, 1984, and 1985 in central Beaufort Sea region (Whitmus et al. 1987). Dashed lines represent the 95% confidence interval for predictions based on the regression of the 1985 data from Prudhoe Bay (Griffiths et al. 1990).

species of fish. Craig (1989) made a strong case for overwintering habitat as the limiting factor for these arctic anadromous fishes.

Population Parameters

An alternative way to assess the effect of a project is to examine the populations of organisms potentially affected by it, in this case the fishes. The summer health or condition of the fishes will influence their growth rate and potentially their population size. Of course, the size of the population can also be influenced by other factors such as winter mortality, which may depend on a variety of factors including summer health. Researchers have examined condition, growth rate, and population size to determine if reductions in these fish population attributes have occurred.

Fish condition is described by the length–weight relationships of the fish. Comparisons of the condition of Arctic ciscoes in different years (Figure 6) and in different areas for the same year (Figure 7) show no differences (Whitmus et al. 1987). The data from the other years and areas fall within the 95% confidence interval for predictions based on the 1985 Prudhoe Bay regression (Griffiths et al. 1990). Similar comparisons for least cisco data reveal similar findings. Least ciscoes tagged on the east and west sides of West Dock during the summer showed no difference in condition when recovered in the Colville River in the

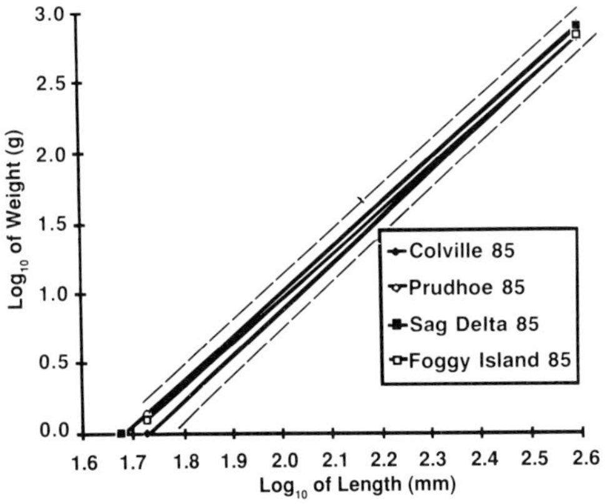

FIGURE 7.—Condition of Arctic ciscoes (length versus weight) from different areas during the same year (1985) in central Beaufort Sea region (Whitmus et al. 1987). Dashed lines represent the 95% confidence interval for predictions based on the regression of the 1985 data from Prudhoe Bay (Griffiths et al. 1990).

fall (Fechhelm et al. 1989). Data from subsequent work have continued to show no reduction in fish condition in the vicinity of the causeways (Glass et al. 1988; Griffiths et al. 1990).

Researchers have evaluated growth rates of the local fishes by comparing fish lengths at different ages. The age–length relationships from several studies of Arctic cisco before and after causeway construction show no noticeable change in fish condition after construction (Figure 8; Whitmus et al. 1987). The wide range of variability in these data is probably due to a variety of factors that, statistically, are forms of sampling error. Apparent age–length relationships of a population can be influenced by the aging method, period of summer, and relative representativeness of the sample subpopulation (a function of sample size and location). Differences in techniques and sample locations make direct comparisons between data sets difficult, but recent work has continued to indicate that the length-at-age distribution falls within the range of historical variability (Glass et al. 1988).

Mark-and-recapture techniques have provided population estimates of varying quality for these fishes. Unfortunately, for some species necessary assumptions were violated and estimates for some years were hindered by small sample sizes. Least cisco estimates were based on captures in the Colville River commercial gill-net fishery (Moulton and Field 1988); and, coming from an apparently closed (though anadromous) system, are probably fairly reasonable: 300,000–500,000 catchable fish in most years. These estimates began in 1976, which coincided with the construction of the second leg of the West Dock Causeway but does not provide a strong precauseway baseline. Because the primary overwintering habitat for Arctic and least ciscoes is the Colville River, a surrogate

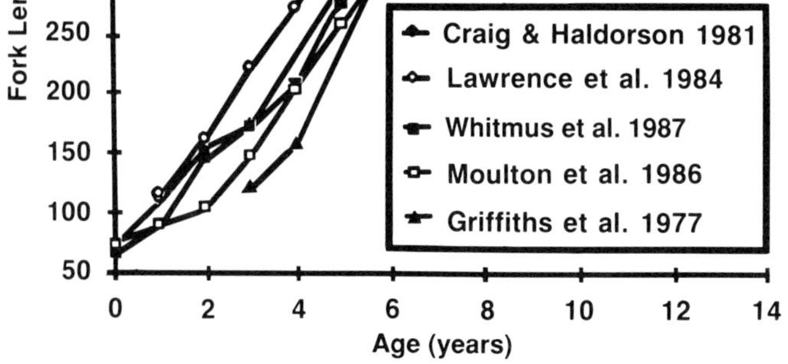

FIGURE 8.—Growth curves of Arctic ciscoes from various studies and locations along central Beaufort Sea coast. (From Whitmus et al. 1987.)

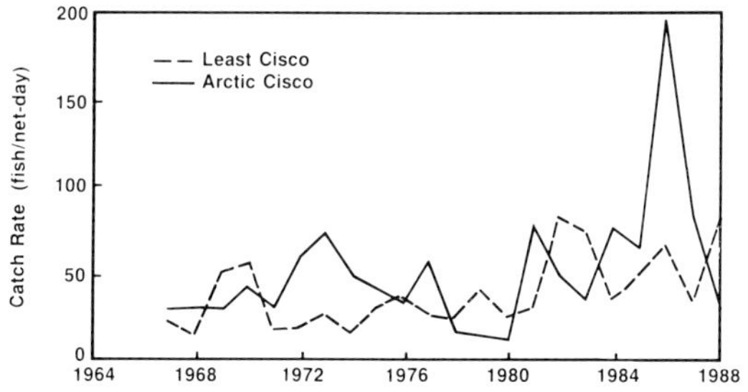

FIGURE 9.—Catch per unit effort (fish/net·d) of least and Arctic ciscoes during the fall, under-ice, commercial fishery on the Colville River, Alaska, 1967–1988 (Moulton and Field 1988; Moulton et al. 1990).

for relative change in population size exists in the 23 years of catch statistics from the commercial fishery on the Colville River delta (Figure 9). During October and November, commercial fishermen and subsistence fishermen from Nuiqsut fish with gill nets under the ice. Over the past 10 years, the combined annual catch has averaged approximately 48,000 Arctic ciscoes and 36,000 least ciscoes (Moulton and Field 1988).

In one respect the Arctic cisco data are difficult to use as an indicator of the impacts of the causeway on population size because their abundance is a function of the wind-aided recruitment from the Mackenzie River. On the other hand, based on the knowledge of previous years' wind conditions and success of recruitments to the area, predictions can be made regarding the size of subsequent catches. The good 1986 and 1987 catches of the very abundant 1979–1980 year-classes indicated that the causeways were not harming the local population. The fact that catch statistics closely match predicted trends (Gallaway et al. 1989; Moulton et al. 1991, this volume) is another indicator that the causeways were not detrimental to these populations.

Least ciscoes both spawn and overwinter in the Colville River; therefore, evaluation of the catch data of this species is more straightforward. Despite the variability in catch data over the 23 years of record, there is a significant positive slope to the regression of catch rate through time (Moulton and Field 1988); thus, the presence of the causeways has obviously not reduced the ability of this population to grow to catchable size.

Conclusion

The West Dock Causeway was built out to the dockhead in 1976; construction of the final leg, out to the seawater treatment plant, occurred in 1981. The Endicott Causeway was built in 1985. These structures can cause local, transient changes to the habitat of the local fishes; these changes vary with wind speed, direction, and duration. Several population-level effects can be hypothesized if these habitat changes are critical to the fishes. If these causeways are affecting the local fish populations, one would suspect that sufficient time has passed to begin to see effects beyond local, ephemeral, distribution changes. Poor body condition, as a result of poor feeding or higher-than-usual metabolic costs, is one of the possible consequences to the fish. There is no direct evidence, as measured by condition factor, that there has been any change in the health of these fishes. Growth rate may be affected by poor health itself, or, in order to maintain a certain condition in poor habitat, a reduction in growth could occur without an observed effect on condition. The data do not indicate that causeway-induced changes in growth have occurred. Another indirect measure of health is the ability to survive the winter. Mortality could be due to poor, site-specific, winter conditions or to general inability to cope with winter stresses because of poor body condition at the end of summer. While mortality has been observed due to oxygen depletion from overcrowding in an ice-closed overwintering site (Schmidt et al. 1989), commercial fishing success indicates that there has not been a general increase in overwintering mortality since construction of the causeways.

In the case of causeways and the nearshore fishes of the central Beaufort Sea, hypothesized population effects based on habitat changes have not been found by actual measurement. Given that the causeway-induced habitat changes are similar to observed, natural variations in these fishes' environment, and that these fishes have evolved in a rigorous, highly variable environment (both among and within years), it should perhaps be expected that the fishes would not be affected by this new source of variability. Gorman (1988) has noted, "A high degree of habitat specialization may not be an appropriate ecological trait for species persistence in fluctuating stream environments with changing community composition." It would seem that the same is true for arctic coastal environments.

In conclusion, while impact *predictions* may be appropriately based on expected changes to habitat, impact *assessments* are best made by study of the populations involved.

References

Armantrout, N. B., editor. 1982. Acquisition and utilization of aquatic habitat inventory information. American Fisheries Society, Western Division, Bethesda, Maryland.

Barnes, P., E. Reimnitz, and G. Smith. 1977. Some coastal oceanographic observations—Beaufort Sea, Alaska. Part A. Pages A1–A19 *in* Miscellaneous hydrologic and geologic observations on the inner Beaufort Sea shelf, Alaska. U.S. Geological Survey Open-File Report 77-477, Menlo Park, California.

Behnke, R. J. 1984. Organizing the diversity of the Arctic charr complex. Pages 3–21 *in* L. Johnson and B. L. Burns, editors. Biology of the Arctic charr: proceedings of the international symposium on Arctic charr. University of Manitoba Press, Winnipeg.

Cannon, T. C., B. Adams, D. Glass, and T. Nelson. 1987. Fish distribution and abundance. Pages 1–129 *in* 1985 Final report for the Endicott Environmental Monitoring Program, volume 6. Report of Envirosphere Company to U.S. Army Corps of Engineers, Alaska District, Anchorage.

Colonell, J. M., and A. W. Niedoroda. 1990. Nearshore oceanographic processes of potential importance to marine and anadromous fishes. Pages 35–42 *in* R. M. Meyer and T. M. Johnson, editors. Fisheries oceanography—a comprehensive formulation of technical objectives for offshore application in the Arctic. U.S. Minerals Management Service, OCS Study MMS 88-0042, Anchorage, Alaska.

Craig, P. C. 1984. Fish use of coastal waters of the Alaska Beaufort Sea: a review. Transactions of the American Fisheries Society 113:265–282.

Craig, P. C. 1989. An introduction to anadromous fishes in the Alaskan Arctic. Biological Papers of the University of Alaska 24:27–54.

Craig, P. C., W. B. Griffiths, S. R. Johnson, and D. M. Schell. 1984. Trophic dynamics in an Arctic lagoon. Pages 347–380 *in* P. Barnes, D. Schell, and E. Reimnitz, editors. The Alaskan Beaufort Sea: ecosystems and environments. Academic Press, New York.

Craig, P. C., and L. Haldorson. 1981. Fish. Pages 384–678 *in* Beaufort Sea barrier island–lagoon ecological process studies: final report, Simpson Lagoon. Environmental assessment of the Alaskan continental shelf, final report of the principal investigators, volume 7. National Oceanic and Atmospheric Administration and Bureau of Land Management, Outer Continental Shelf Environmental Assessment Program, Boulder, Colorado.

Critchlow, K. R. 1983. Prudhoe Bay Waterflood Project fish monitoring program, 1982. Pages 5-1 to 5-219 *in* Prudhoe Bay Waterflood Environmental Monitoring Program, 1982, volume 3. Report of Woodward-Clyde Consultants to U.S. Army Corps of Engineers, Alaska District, Anchorage.

Dames and Moore. 1988. Anadromous fish habitat usage. Appendix B *in* Lisburne Offshore Project environmental assessment. Report to ARCO Alaska, Inc., Exxon USA, and Standard Alaska Production Company. Available from ARCO Alaska, Anchorage.

Division of Ecological Services. 1980a. Habitat as a basis for environment assessment. U.S. Fish and Wildlife Service, Ecological Services Manual, part 101, Washington, D.C.

Division of Ecological Services. 1980b. Habitat evaluation procedures (HEP). U.S. Fish and Wildlife Service, Ecological Services Manual, part 102, Washington, D.C.

Dolloff, C. A. 1987. Seasonal population characteristics and habitat use by juvenile coho salmon in a small southeast Alaska stream. Transactions of the American Fisheries Society 116:829–838.

Envirosphere Company. 1983. Marine invertebrates. Pages 4-1 to 4-101 *in* Prudhoe Bay Waterflood Environmental Monitoring Program, 1982, volume 2. Report to U.S. Army Corps of Engineers, Alaska District, Anchorage.

Fechhelm, R. G., J. S. Baker, W. B. Griffiths, and D. R. Schmidt. 1989. Localized movement patterns of least cisco (*Coregonus sardinella*) and Arctic cisco (*C. autumnalis*) in the vicinity of a solid-fill causeway. Biological Papers of the University of Alaska 24:75–106.

Fechhelm, R. G., and D. B. Fissel. 1988. Wind-aided recruitment of Canadian Arctic cisco (*Coregonus autumnalis*) into Alaskan waters. Canadian Journal of Fisheries and Aquatic Sciences 45:906–910.

Gallaway, B. J., W. J. Gazey, J. M. Colonell, A. W. Niedoroda, and C. J. Herlugson. 1990. The Endicott Development Project—preliminary assessment of impacts from the first major offshore oil development in the Alaskan arctic. American Fisheries Society Symposium 11:42–80.

Gallaway, B. J., W. J. Gazey, and L. L. Moulton. 1989. Population trends for the Arctic cisco (*Coregonus autumnalis*) in the Colville River of Alaska as reflected by the commercial fishery. Biological Papers of the University of Alaska 24:153–165.

Gallaway, B., W. Griffiths, P. Craig, W. Gazey, and J. Helmericks. 1983. An assessment of the Colville River delta stock of arctic cisco (*Coregonus autumnalis*)—migrants from Canada? Biological Papers of the University of Alaska 21:4–23.

Glass, D. R., L. Russell, and S. S. Parker. 1988. Age, growth, and condition of anadromous fish. Draft report for the 1987 Endicott monitoring program. Report of Envirosphere Company to U.S. Army Corps of Engineers, Alaska District, Anchorage.

Gorman, O. T. 1988. The dynamics of habitat use in a guild of Ozark minnows. Ecological Monographs 58:1–18.

Griffiths, W. B., J. Den Beste, and P. C. Craig. 1977. Fisheries investigations in a coastal region of the Beaufort Sea (Kaktovik Lagoon, Barter Island, Alaska). Pages 1–190 *in* P. McCart, editor. Arctic Gas Biological Report Series, volume 40, chapter 2.

Report of Aquatic Environments Limited to Canadian Arctic Gas Pipeline, Limited.

Griffiths, W. B., B. J. Gallaway, R. J. Fechhelm, D. R. Schmidt, and W. J. Gazey. 1990. Species accounts describing anadromous fish use of the area, 1985–1987. Pages 117–192 *in* Analysis of 1985–1987 data collected by the Endicott Development Monitoring Program, volume 2: results and preliminary impact assessment. Report of LGL Alaska Research Associates, Inc., Hunter/ESE Environmental Services, Inc., and W. J. Gazey Research to BP Exploration (Alaska) Inc., Anchorage, and the North Slope Borough, Barrow, Alaska. Available from BP Exploration (Alaska) Inc., Anchorage.

Houghton, J. P., C. J. Whitmus, and A. W. Maki. 1990. Habitat relationships of Beaufort Sea anadromous fish: integration of oceanographic information and fish-catch data. Pages 73–88 *in* R. M. Meyer and T. M. Johnson, editors. Fisheries oceanography—a comprehensive formulation of technical objectives for offshore application in the Arctic. U.S. Minerals Management Service, OCS Study MMS88-0042, Anchorage, Alaska.

Johnson, D. L., R. A. Beaumier, and W. E. Lynch, Jr. 1988. Selection of habitat structure interstice size by bluegills and largemouth bass in ponds. Transactions of the American Fisheries Society 117:171–179.

Lawrence, M. J., G. Lacho, and S. Davies. 1984. A survey of the coastal fishes of the southeastern Beaufort Sea. Canadian Technical Report of Fisheries and Aquatic Sciences 1220.

Matthews, J. B. 1979. Characterization of the nearshore hydrodynamics of an Arctic barrier island–lagoon system. Pages 57–97 *in* Environmental assessment of the Alaskan continental shelf, annual report of the principal investigators, volume 8. Bureau of Land Management and National Oceanic and Atmospheric Administration, Outer Continental Shelf Environmental Assessment Program, Boulder, Colorado.

Mesick, C. F. 1988. Effects of food and cover on numbers of Apache and brown trout establishing residency in artificial stream channels. Transactions of the American Fisheries Society 117:421–431.

Moore, K. M. S., and S. V. Gregory. 1988. Response of young-of-the-year cutthroat trout to manipulation of habitat structure in a small stream. Transactions of the American Fisheries Society 117:162–170.

Morrow, J. E. 1980. Analysis of the Dolly Varden char, *Salvelinus malma*, of northwestern North America and northeastern Siberia. Pages 323–338 *in* E. K. Balon, editor. Charrs: salmonid fishes of the genus *Salvelinus*. Dr. W. Junk, The Hague, Netherlands.

Moulton, L. L. 1989. Recruitment of Arctic cisco into the Colville Delta in 1985. Biological Papers of the University of Alaska 24:107–111.

Moulton, L. L., and L. J. Field. 1988. Assessment of the Colville River fall fishery, 1985–1987. Final report of Environmental Sciences and Engineering, Inc., to ARCO Alaska, Inc., North Slope Borough, and the City of Nuiqsut. Available from ARCO Alaska, Anchorage.

Moulton, L. L., L. J. Field, and R. Kovalsky. 1990. The 1988 fall gill net fishery for ciscos in the Colville River, Alaska. Pages 1–64 *in* The 1988 Endicott Development fish monitoring program, volume 4. Report of LGL Alaska Research Associates, Inc., to BP Exploration (Alaska) Inc., Anchorage, and the North Slope Borough, Barrow, Alaska. Available from BP Exploration (Alaska) Inc., Anchorage.

Moulton, L. L., L. J. Field, and R. Kovalsky. 1991. Predictability in the catch of Arctic cisco in the Colville River, Alaska. American Fisheries Society Symposium 11:145–156.

Moulton, L. L., and seven coauthors. 1986. 1984 central Beaufort Sea fish study. Pages 1-1 to 6-8 *in* Prudhoe Bay Waterflood Project Environmental Monitoring Program, 1984, volume 2. Final report of Entrix, Inc., LGL Ecological Research Associates, and Woodward-Clyde Consultants to U.S. Army Corps of Engineers, Alaska District, Anchorage.

Mungall, J. C. H., R. E. Whitaker, and S. D. Pace. 1979. Oceanographic processes in a Beaufort Sea barrier island–lagoon system: numerical modeling and current measurements. Pages 182–287 *in* Environmental assessment of the Alaskan continental shelf, annual reports of principal investigators, volume 8. Bureau of Land Management and National Oceanic and Atmospheric Administration, Outer Continental Shelf Environmental Assessment Program, Boulder, Colorado.

Pajak, P., and R. J. Neves. 1987. Habitat suitability and fish production: a model evaluation for rock bass in two Virginia streams. Transactions of the American Fisheries Society 116:839–850.

Pitcher, T. J., and P. J. B. Hart. 1982. Fisheries ecology. AVI, Westport, Connecticut.

Savoie, M. A., and D. E. Wilson. 1984. 1983 physical processes monitoring program. Pages 2-1 to 2-124 *in* Prudhoe Bay Waterflood Environmental Monitoring Program, 1983, volume 2. Report of Kinnetic Laboratories, Inc., to U.S. Army Corps of Engineers, Alaska District, Anchorage.

Savoie, M. A., and D. E. Wilson. 1986. Physical marine processes monitoring program 1984, final report. Pages 2-1 to 2-195 *in* Prudhoe Bay Waterflood Environmental Monitoring Program, 1984, volume 1. Report of Kinnetic Laboratories, Inc., to U.S. Army Corps of Engineers, Alaska District, Anchorage.

Schmidt, D. R., W. B. Griffiths, D. K. Beaubien, and C. J. Herlugson. 1991. Movement of young-of-the-year Arctic ciscoes across the Beaufort Sea coast, 1985–1988. American Fisheries Society Symposium 11:132–144.

Schmidt, D. R., W. B. Griffiths, and L. R. Martin. 1989.

Overwintering biology of anadromous fish in the Sagavanirktok River delta, Alaska. Biological Papers of the University of Alaska 24:55–74.

Søballe, D. M., and B. L. Kimmel. 1987. A large-scale comparison of factors influencing phytoplankton abundance in rivers, lakes, and impoundments. Ecology 68:1943–1954.

Taylor, E. B. 1988. Water temperature and velocity as determinants of microhabitats of juvenile chinook and coho salmon in a laboratory stream channel. Transactions of the American Fisheries Society 117:22–28.

Whitmus, C. J., T. C. Cannon, and S. S. Parker. 1987. Age, growth, and condition of anadromous fish. Pages 1–34 *in* 1985 Final report for the Endicott Environmental Monitoring Program, volume 7. Report of Envirosphere Company to U.S. Army Corps of Engineers, Alaska District, Anchorage.

American Fisheries Society Symposium 11:109–118, 1991

Bioassay Tests of Acute Temperature and Salinity Shock to Arctic Cisco

ROBERT G. FECHHELM
LGL Ecological Research Associates, Inc.
1410 Cavitt Street, Bryan, Texas 77801, USA

JANICE D. FECHHELM
Department of Wildlife and Fisheries Sciences
Texas A&M University, College Station, Texas 77803, USA

CHRISTOPHER J. HERLUGSON AND DEBRA K. BEAUBIEN
BP Exploration (Alaska), Inc.
900 East Benson Boulevard, Anchorage, Alaska 99519, USA

Abstract.—Bioassay laboratory studies were performed on 2- to 4-year-old Arctic ciscoes *Coregonus autumnalis* to determine short-term (96 h) stress and mortality levels associated with sharp declines in temperature and increases in salinity. There was no mortality when fish were subjected to a maximum thermal shock of 12 to 3°C and simultaneous salinity shocks of 0 to 22‰ and 0 to 24‰. Eight additional experiments involving salinities of 25‰ or less elicited no significant mortality or overt signs of stress, regardless of the associated temperature shock. Conversely, mortalities of 5–80% were observed in all seven experiments involving test salinities of 30–31‰, two of which involved no temperature shock at all. Background mortality in groups slowly acclimated to 32‰ salinity was higher than that for fish acclimated to lower salinities. In 22 tests, the only case of mortality in a control group occurred at a salinity of 32‰. Results at test salinities of 26–28‰ varied depending on the magnitude of the salinity shock and of the associated temperature shock. Results are discussed in terms of environmental conditions typically encountered in the Beaufort Sea during summer.

The advent of petroleum exploration and production in the Prudhoe Bay region of Alaska has brought with it extensive environmental monitoring programs intended to assess the oil industry's impact on local fish populations. A primary concern has centered on the construction of solid-fill causeways along the Beaufort Sea coast and their effect on the summer feeding habits of local anadromous fish species such as the Arctic cisco *Coregonus autumnalis*, least cisco *C. sardinella*, and broad whitefish *C. nasus* (Figure 1). These fishes, which overwinter beneath the ice in rivers and river deltas for nearly 9 months of the year, move into coastal waters and feed on marine invertebrates during the open-water period from mid-June through September (Moulton et al. 1986). It is during this summer feeding period that fish achieve most of their yearly growth and accumulate fat reserves necessary to survive the arctic winter. Because these species generally avoid marine (cold, saline) conditions, their summer feeding excursion is limited to nearshore areas where waters remain warmer and less saline (Moulton et al. 1986; Cannon et al. 1987; Glass et al. 1988). Causeways can, under certain environmental conditions, alter coastal circulation patterns such that offshore marine water intrudes into the nearshore zone. These marine intrusions may, at times, block alongshore movement of fish such that they are excluded from segments of their coastal feeding habitat (Griffiths and Galla-way 1982; Critchlow 1983; Griffiths et al. 1983; Biosonics, Inc. 1984; Moulton et al. 1986; Cannon et al. 1987; Glass et al. 1988; Fechhelm et al. 1989).

If the alteration of coastal hydrographic features does adversely affect anadromous fishes, the effects could be manifested in several ways. Current studies are attempting to determine if there have been detectable changes in the condition or population sizes of key species. More elemental is the question of acute effects upon fish—can causeways modify temperature and salinity conditions to levels that cause fish kills? To address this question, temperature and salinity bioassay studies were conducted on 2- to 4-year-old Arctic ciscoes during the summer of 1988. The Arctic cisco was chosen as a test subject because it is one of the most numerically abundant anadromous species in the Alaskan–Canadian Beaufort Sea and is an important component of commercial and native subsistence fisheries.

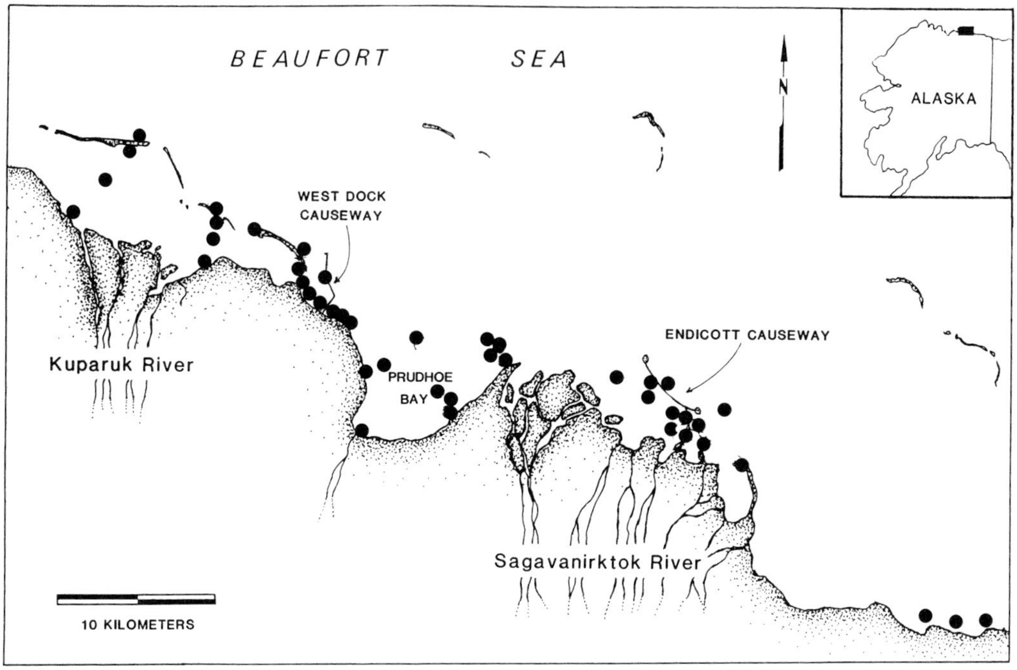

FIGURE 1.—Causeways in the Beaufort Sea, Alaska. Black dots indicate fyke nets sampled at least once during the 1981–1988 monitoring studies.

Methods

Arctic ciscoes ranging in length from 120 to 170 mm fork length (2- to 4-year-olds) were collected in early summer from fyke nets in the Prudhoe Bay area and transported to laboratory facilities on the main production island of the Endicott Causeway (Figure 1). Fish were placed in outdoor, aerated, 341-L (90 gallon) holding tanks under temperature and salinity conditions similar to those recorded in the field at the time of capture. Fish were kept in these tanks for a minimum of 3 d to allow handling mortality to run its course. Water temperature was allowed to fluctuate naturally during this initial holding period and generally remained within a range of 3–10°C.

After the initial holding period, fish were transferred to indoor tanks and acclimated at one of three temperature regimes (12, 7, and 3°C) and one of three salinities (0, 15, and 32‰). Approximately 50 fish were placed in each acclimation holding tank at a temperature and salinity consistent with outdoor holding conditions. Temperature was then changed at a rate of 2°C/d and salinity at a rate of 5‰/d until the desired acclimation conditions were attained. Although temperatures in individual acclimation and test tanks varied up to 2°C from these values (Table 1), the three temperature regimes are henceforth referred to as 12, 7, and 3°C for ease of discussion.

Acclimation tanks were inert, 341-L (90 gallon) commercial Tetra totes® fitted with undergravel filtration systems. Water temperature was controlled by a mechanism that pumped cooled ethylene glycol (−10–0°C) through each tank via a closed system of inert Tygon® plastic tubing and by 75-watt commercial aquarium heaters. This cooling and heating system permitted temperatures to be maintained within approximately 1°C of the designated temperature. All temperatures were monitored with in-glass mercury thermometers calibrated against an ERTCO standard thermometer (FC 63903).

Dechlorinated tap water (treated with a commercial, aquarium water treatment—Novaqua®) was used exclusively in all acclimation and test tanks. Salinity (±1‰) was adjusted with Instant Oceans® commercial aquarium sea salts. Salinities were monitored with Goldberg T/C refractometers (full-scale salinity index from 0 to 160‰) calibrated a minimum of every 3 d against distilled water and 35‰ standard sea water. Refractome-

TABLE 1.—Acclimation and experimental temperature and salinity conditions for the 22 bioassay tests on Arctic ciscoes. Values are means (weighted means for test and control), standard deviations (in parentheses), and ranges.

	Acclimation		Test		Control	
Bioassay	Temperature, °C	Salinity, ‰	Temperature, °C	Salinity, ‰	Temperature, °C	Salinity, ‰
1	12.80 (0.47)	0	3.29 (0.51)	31.47 (0.74)	12.00 (0.08)	0
	11.5–13.2		2.5–4.0	31–34	11.8–12.1	
2	12.23 (0.54)	32.37 (1.01)	3.15 (0.19)	31.16 (0.40)	11.68 (0.27)	30.88 (0.29)
	11.3–13.1	30–34	3.0–3.5	31–33	11.3–12.5	30–32
3	6.99 (0.59)	15.17 (0.62)	3.22 (0.19)	15.04 (0.24)	6.77 (0.31)	14.14 (0.33)
	6.0–7.9	14–16	3.0–3.5	14–16	6.0–7.2	14–15
4	11.98 (0.51)	0	6.73 (0.49)	0	12.29 (0.26)	0
	11.0–13.0		3.0–7.2		12.0–12.8	
5	12.11 (0.55)	14.56 (0.81)	11.84 (0.38)	31.04 (0.02)	12.06 (0.27)	14.48 (0.05)
	11.4–13.0	12–15	11.3–12.8	31–32	11.2–12.4	14–15
6	6.99 (0.53)	15.35 (0.49)	3.83 (0.34)	31.00 (0.00)	7.00 (0.23)	15.00 (0.00)
	6.0–8.0	15–16	3.0–4.2	31–31	6.2–7.4	15–15
7	12.07 (0.67)	0.61 (0.98)	3.80 (0.32)	1.00 (0.00)	11.82 (0.36)	0
	10.9–13.7	0–3	3.1–4.3	1–1	11.0–12.5	
8	12.17 (0.58)	15.06 (0.24)	3.70 (0.28)	15.00 (0.00)	11.90 (0.30)	14.00 (0.00)
	11.2–13.1	15–16	3.0–4.1	15–15	11.2–12.4	14–14
9	12.26 (0.56)	15.35 (0.67)	11.89 (0.48)	25.00 (0.00)	11.85 (0.04)	15.00 (0.00)
	11.2–13.0	14–16	11.1–12.8	25–25	11.0–12.8	15–15
10	11.94 (0.41)	0.75 (1.07)	12.08 (0.11)	24.86 (0.31)	11.39 (0.77)	0
	11.2–12.7	0–3	11.8–12.4	24–25	9.6–12.3	
11	12.29 (0.48)	14.90 (0.64)	6.95 (0.23)	15.00 (0.00)	12.02 (0.12)	15.00 (0.00)
	11.4–13.0	13–16	6.2–7.2	15–15	11.7–12.2	15–15
12	6.99 (0.59)	15.15 (0.59)	2.94 (0.55)	25.00 (0.00)	7.07 (0.08)	15.00 (0.00)
	6.0–7.9	14–16	1.1–3.5	25–25	6.8–7.2	15–15
13	11.80 (0.89)	0	3.28 (0.34)	26.00 (0.00)	11.46 (0.39)	0
	10.3–13.7		3.0–4.3	26–26	10.2–12.2	
14	12.19 (0.56)	0.29 (0.61)	3.05 (0.63)	28.00 (0.00)	11.14 (0.57)	0
	11.3–13.2	0–2	2.0–4.1	28–28	10.2–12.2	
15	7.17 (0.69)	0.50 (0.80)	7.35 (0.51)	28.00 (0.00)	6.96 (0.58)	0
	6.5–8.8	0–2	6.3–8.0	28–28	6.0–7.9	
16	11.79 (0.45)	0.64 (0.74)	2.62 (0.46)	24.00 (0.00)	11.64 (0.39)	0
	11.0–12.8	0–2	2.0–3.7	24–24	11.0–12.3	
17	7.07 (0.48)	15.00 (0.00)	2.39 (0.50)	28.00 (0.00)	6.78 (0.43)	15.00 (0.00)
	6.3–8.2	15–15	2.0–4.5	28–28	6.0–7.6	15–15
18	7.78 (0.62)	15.00 (0.00)	2.37 (0.35)	30.00 (0.00)	7.45 (0.38)	15.00 (0.00)
	7.0–9.0	15–15	1.9–3.2	30–30	6.8–7.9	15–15
19	11.68 (0.36)	0.57 (0.76)	2.01 (0.17)	22.00 (0.00)	11.79 (0.08)	0
	10.8–12.1	0–2	1.6–2.7	22–22	11.6–11.9	
20	11.95 (0.80)	0	1.78 (0.60)	30.00 (0.00)	11.34 (0.29)	0
	10.3–13.7		1.0–2.7	30–30	11.0–12.0	
21	7.26 (0.48)	0.67 (0.90)	7.04 (0.32)	26.00 (0.00)	7.15 (0.49)	0
	6.8–8.5	0–2	6.3–7.4	26–26	6.2–7.9	
22	7.14 (0.56)	0.30 (0.66)	6.88 (0.39)	30.00 (0.00)	6.54 (0.08)	0
	6.5–8.8	0–2	6.1–7.6	30–30	6.4–6.9	

ters were rinsed with distilled water before every measurement. All laboratory tanks were aerated. Tanks were also fitted with translucent polyethylene covers to shield fish from visual disturbances while still permitting light penetration. Fish were maintained under constant light conditions throughout the experimentation period. During the acclimation period, fish were fed to satiation once daily with commercial freeze-dried white shrimp.

Although acclimation tanks were equipped with undergravel filtration systems, the low temperatures and large numbers of fish resulted in steady increases in water nitrate levels (measured with a commercial HACH Water Quality Test Kit). Accordingly, about 90% of the water was changed in every acclimation holding tank every 3–5 d. Test tanks were completely emptied and rinsed after every experiment.

Fish were maintained at their acclimation conditions for a minimum of 14 d prior to testing. At the start of an experiment, 20 individuals (test group) were transferred from their acclimation tank to a test tank. An additional 20 fish (control group) were removed from the acclimation tank and then redeposited. In this way, both the test

TABLE 2.—Mortality of Arctic ciscoes after 96-h temperature–salinity bioassays. N = number of fish.

Bioassay conditions				Bioassay results			
Acclimation		Experimental		Control fish		Test fish	
Temperature, °C	Salinity, ‰	Temperature, °C	Salinity, ‰	N	Mortality, %	N	Mortality, %
12	0	7	0	20	0	20	0
		3	0	20	0	20	0
		3	22	23	0	20	0
		3	24	20	0	20	0
		3	26	19	0	20	15
		3	28	21	0	20	15
		3	30	19	0[a]	20	60[a]
		3	31	20	0	20	80
12	0	12	25	22	0	20	5
7	15	3	15	20	0	20	0
		3	25	20	0	20	0
		3	28	20	0	20	0
		3	30	20	0	20	5
		3	31	20	0	20	25
7	0	7	26	20	0	20	0
		7	28	20	0	20	5
		7	30	20	0[b]	20	20[b]
12	15	3	15	20	0	23	4
		7	15	21	0	20	0
		12	25	20	0	20	0
		12	31	20	0	20	15
12	32	3	31	20	10	20	5

[a]Experiment terminated after 75 h.
[b]Experiment terminated after 72 h.

and control groups were subjected to the same degree of handling stress. The acclimation tank then served as the control tank. All acclimation tanks were subjected to a final water change of about 90% within 3 h of the start of a test. Neither the control nor test groups were fed once the experiment began.

Test and control tanks were monitored for 96 h. Temperature and salinity levels were recorded, as was the time to mortality for any fish. Death was defined as the point at which a fish did not respond to gentle prodding (Peltier and Weber 1984). All test and control fish were weighed and measured at the completion of the experiment. Dissolved oxygen levels in the acclimation, test, and control tanks were monitored with a Yellow Springs Instrument Model 51B Oxygen Meter.

In total, 22 bioassays were completed (Table 1). Because of the 4-d duration of each experiment, test fish were checked with decreasing regularity as the bioassay progressed, and the frequency of observation generally conforming to a geometric progression (Peltier and Weber 1984). Accordingly, the temperature and salinity levels for tests and controls in Table 1 were calculated using a weighted mean standardized to 1-h intervals (Ostle and Mensing 1975).

Results

Under the three temperature regimes (3, 7, and 12°C), the series of tests that subjected fish to the greatest instantaneous temperature shock involved acclimation to 12°C/0‰ and testing at 3°C and salinities of 0, 22, 24, 26, 28, 30, and 31‰ (Table 2). There were no mortalities in the 0, 22, and 24‰ tests; 15% mortality at salinities of 26 and 28‰; 60% mortality at 30‰; and 80% mortality at 31‰. There were no control mortalities in any of these extreme temperature shock experiments. Of the 10 bioassays at salinities less than 26‰ (0, 0, 15, 15, 15, 22, 24, 25, 25, and 25‰), there were only two test mortalities: one fish died in a 12°C/0‰ to 12°C/25‰ test, the other in a 12°C/15‰ to 3°C/15‰ test. Both of these cases equate to a test mortality of 5% or less and therefore fall below the U.S. Environmental Protection Agency (EPA) criteria of significance for toxicity (Peltier and Weber 1984).

Conversely, tests indicated that salinities of 30‰ and higher are very stressful to Arctic ciscoes. Test mortalities ranged from 15 to 80% in five of seven experiments involving salinities of 30–31‰; 5% mortality (EPA nonsignificant: Peltier and Weber 1984) occurred in the other two

(Table 2). Of the 22 bioassay experiments, the only instance of mortality in a control group involved one experiment in which the control group was subjected to a high salinity: 12°C/32‰ acclimation to 3°C/31‰ test. The test resulted in mortalities of 5% for the test group and 10% for the control group. Although the 10% mortality was technically within acceptable EPA limits (Peltier and Weber 1984), results were too close and too consistent with other high-salinity experiments to be dismissed offhand. The negative effects of high salinity were also evident during acclimation. While we were able to acclimate Arctic ciscoes to a salinity of 32‰, there was a chronic background mortality of one fish every 3.6 d. This compares with a mortality rate of one fish per 24.5 d for all other lower salinity acclimation tanks combined.

Analysis of covariance (ANCOVA; Snedecor and Cochran 1980) of $\log_e$-transformed length and weight data (mortalities excluded) indicated no significant difference ($P = 0.05$) in the length–weight relationship between control and test groups in any of the 22 trials except for Test 7 (12°C/0‰ acclimation, 3°C/0‰ test). Control fish were more robust than test fish in this experiment; however, there was no mortality in either group. Control and test fish were pooled to test (ANCOVA) for differences in length–weight relationships among experiments. Results indicated three distinct groupings: Experiment 5 ($y = -10.3 + 2.7x$; $r^2 = 0.94$); Experiments 11, 14, 15, 19, and 20 ($y = -14.2 + 3.5x$; $r^2 = 0.92$); and the remaining experiments ($y = -13.1 + 3.2x$; $r^2 = 0.88$). We found no evidence that these differences affected the outcome of the experiments.

Dissolved oxygen levels for acclimation, control, and test tanks ranged from 9.2 to 11.0 mg O_2/L, depending upon salinity and temperature.

Discussion

Results of the experiments suggest that Arctic ciscoes encounter lethal salinity stress beginning at about 26‰. Salinities elicited significant mortalities at 26–28‰, but only when fish were subjected to a full salinity shock from 0‰ acclimation and a simultaneous drop in temperature from 12 to 3°C. Fish subjected to salinity shocks of 0 to 26‰ and 0 to 28‰ suffered no significant mortality (0 and 5%, respectively) when temperature was held constant at 7°C. By itself, a thermal shock of 12 to 3°C (at test salinities of 0, 15, 22, and 24‰) elicited no adverse response.

Acclimation appears to enable Arctic ciscoes to withstand higher salinities. The two highest mortality rates (60 and 80%) occurred under worst case conditions when temperature dropped from 12 to 3°C and salinities increased from 0 to 30‰ and from 0 to 31‰. For fish acclimated to intermediate conditions of 7°C and 15‰, shocks of 3°C/30‰ and 3°C/31‰ elicited mortalities of only 5 and 25%, respectively (Table 2). The ability of Arctic ciscoes to acclimate to different salinities was consistent with the physiological capabilities of anadromous and brackish-water fish. Acclimation to salinity stress involves several mechanisms, including the two principal osmoregulatory organs, the kidneys and gills. Both structures undergo adaptive morphological changes when fish are subjected to varying external conditions (Forskett et al. 1981, 1983; Hwang and Hirano 1985; Hwang and Wu 1988). The enzymatic activity of sodium–potassium adenosinetriphosphatase (Na^+, K^+ ATPase), which controls the active transport of ions in the gills and kidneys, also varies with osmotic stress (Epstein et al. 1967, 1969; Kamiya and Utida 1968; Zaugg and McLain 1970).

The absence of significant mortality at salinities of 25‰ and lower implies resistance below this threshold. Decreased stress was also evident in the behavior of the test fish themselves. During acclimation, fish generally spent their time culling or grazing along the sides of the tanks. When approached by the observer, fish would rush to the front of the tank in apparent anticipation of being fed. During two critical tests (12°C/0‰ to 3°C/22‰ and 12°C/0‰ to 3°C/24‰), fish resumed their normal activity within 15 min of transfer into their test tanks. By comparison, fish tested at salinities of 30 and 31‰ never reverted to normal activity and generally appeared disoriented. Normal behavior was less noticeable at 26 and 28‰, even when the tests resulted in no significant mortality.

Behavioral patterns also suggested that time to death was a rather useless measure of toxicity in these experiments. The most severe test (acclimation at 12°C/0‰, test at 3°C/31‰) resulted in 50% mortality at 53.25 h and 80% mortality at 96 h. Yet, many of the test fish were "environmentally dead" within several hours of the start of the test. Fish were so disoriented that they often swam into the sides of the test tank, and few individuals responded to outside visual stimuli. These fish, which are normally greenish gray in color when viewed dorsally, turned a brilliant turquoise, indicating loss of melanophore control. The colora-

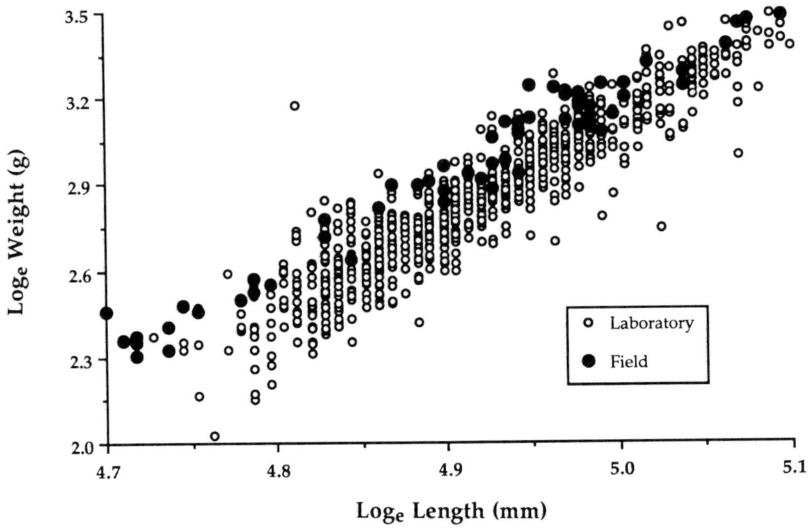

FIGURE 2.—Length–weight relationships for Arctic ciscoes in bioassay experiments and for Arctic ciscoes collected in the field from 21 to 25 July 1988.

tion change generally preceded death by as much as 36 h. It is doubtful that fish in this condition would have survived in their natural environment.

The ability of fish to osmoregulate decreases with temperature and may completely shut down in cold, saline water (Umminger 1971; Hazel and Prosser 1974). Osmoregulatory breakdown was evident in the condition of fish that died during the 26–31‰ salinity experiments. These fish appeared emaciated and weighed approximately 14% less than either the controls or the test survivors in their particular experiment; dehydration was the likely cause. In hypersaline conditions, fish offset osmotic water loss through their gills by actively drinking salt water, then actively excreting excess ions through the gills and to a lesser extent the kidneys (Evans 1979; Eddy 1982). After death, osmotic loss of water continues in the absence of physiological countermeasure, and dehydration occurs. The extent to which hypersaline osmotic stress and dehydration may have contributed to the actual mortality of the fish is unknown.

The condition (weight/unit length) of Arctic ciscoes in the bioassay experiments was generally below that of fish taken in the field, for two reasons. First, most of the test subjects were collected at the beginning of the open-water season. Because the condition of anadromous fish increased markedly during the summer feeding season (LGL 1989), test subjects were collected when their condition was naturally the poorest. Second, the feeding regimen during acclimation was insufficient for fish to gain weight. Length–weight relationships showed that experimental Arctic ciscoes were in poorer condition than fish collected in the field during early summer (Figure 2). Fish actually lost weight during acclimation. Insufficient ration was partially due to lack of precedent on the bioenergetic requirement of this species at different temperatures. However, it was an a priori decision that it would be better to underfeed rather than overfeed experimental fish so as not to "bulk up" the subjects for experimentation. The degree to which this poorer condition may have influenced test results is unknown.

Ecological Implications

The purpose of the static bioassay experiments was to obtain some idea of the tolerance levels of Arctic ciscoes to a variety of temperature and salinity regimes; however, caution must be exercised when extrapolating test results to real-world situations. Laboratory experiments effectively represented worst case scenarios. Arctic ciscoes were subjected to instantaneous environmental shocks without the slightest chance for acclimation. Fish were also prevented from escaping from their predicament, an option that would be available to them in their natural surroundings. In addition, Arctic ciscoes maintained in the laboratory may not have truly reflected the capabilities of fish in the wild. Because experimental fish in our tests were maintained on low feeding rations

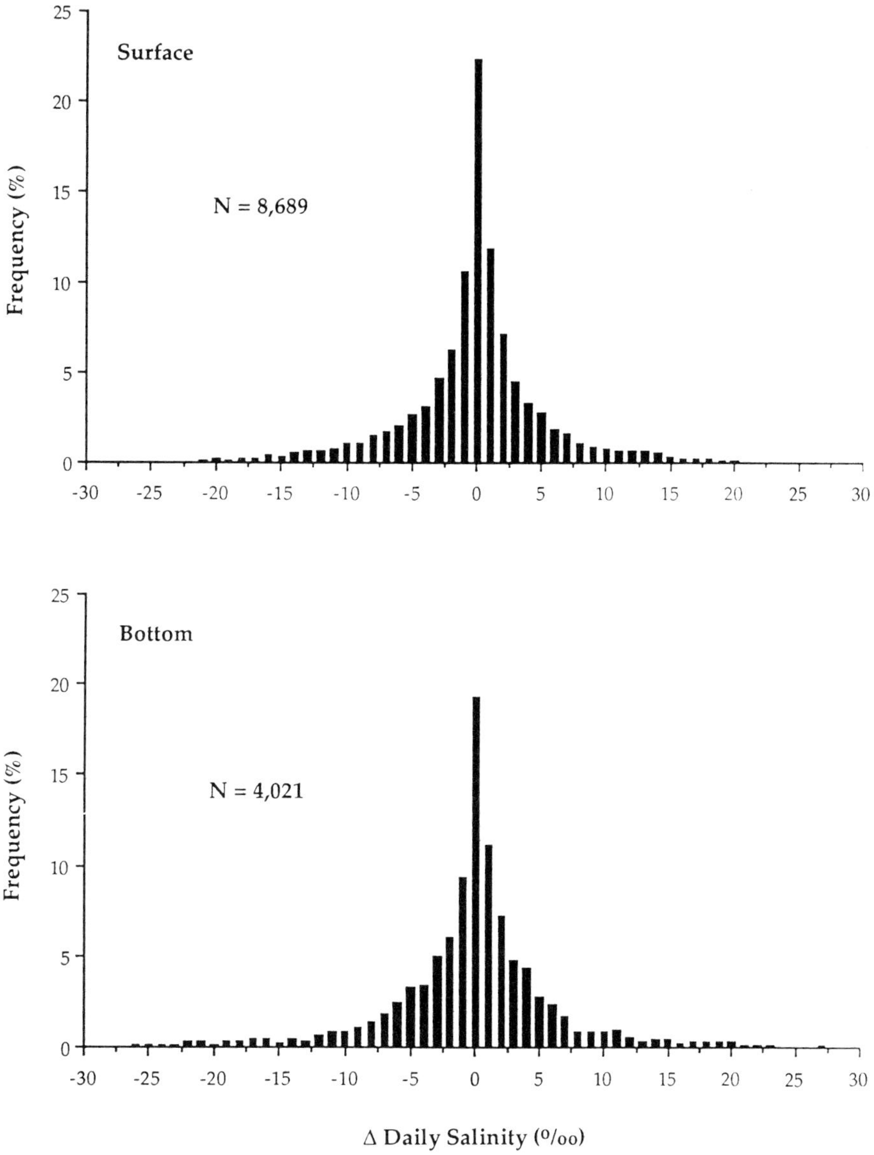

FIGURE 3.—Frequencies of daily changes in surface and bottom salinities at individual fyke net locations from 1981 to 1988. (Sources: Griffiths and Gallaway 1982; Critchlow 1983; Griffiths et al. 1983; Biosonics, Inc. 1984; Moulton et al. 1986; Cannon et al. 1987; Glass et al. 1988; Reub et al. 1988; LGL 1989.)

and were generally leaner than fish caught in the field, results may have underestimated tolerance levels of the natural population. Conversely, the culling of weaker individuals during initial capture and acclimation may have inflated the inferred resistance of the population.

While such concerns are legitimate, the results of the bioassay experiments are consistent with current knowledge of the species. Field data indicate that small (<250 mm) Arctic ciscoes generally prefer warmer, less saline waters and tend to avoid marine conditions (Griffiths and Gallaway 1982; Critchlow 1983; Griffiths et al. 1983; Biosonics, Inc. 1984; Moulton et al. 1986; Cannon et al. 1987; Glass et al 1988). Similar responses are also evident to one degree or another for the region's other two dominant anadromous fish species—least cisco and broad whitefish. The greatest room

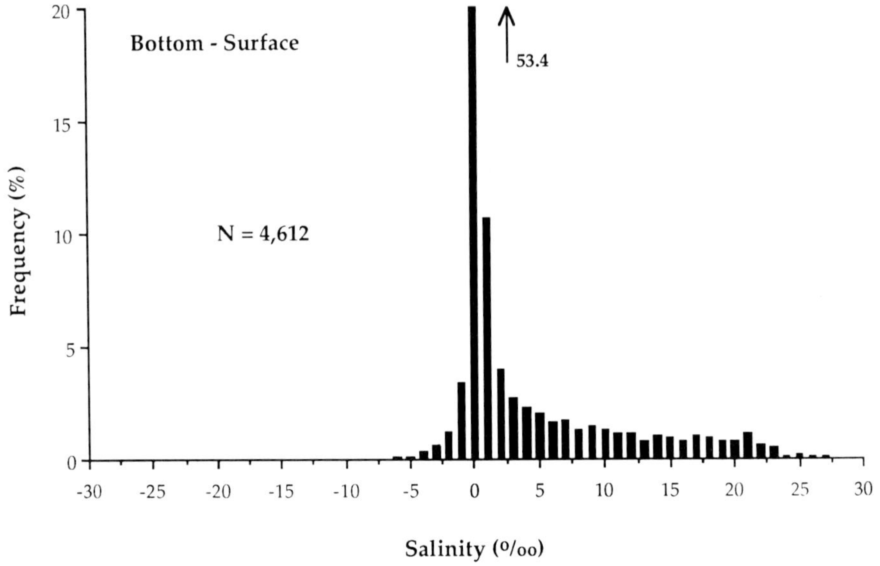

FIGURE 4.—Frequencies of daily salinity stratification differentials (bottom–surface) at individual fyke net locations from 1986 to 1988. (Sources: Glass et al. 1988; Reub et al. 1988; LGL 1989.)

for argument probably rests in how precisely the ≤25‰, 26–28‰, and ≥30‰ salinity levels represent the acceptable, transition, and high-stress levels for Arctic ciscoes. While delineation of these salinity levels can be expected to vary slightly, experiments showed that Arctic ciscoes went from an extremely stressed state at salinities of 30‰ or higher to a nonstressed state at salinities of ≤25‰ or lower. Such a response suggests that the ability of Arctic ciscoes to withstand salinity–temperature stress changes very rapidly over the upper range of the salinity spectrum. Although the exact transition point may vary, salinities in the low to mid-20s are probably not lethal to Arctic ciscoes in their natural environment. Such nonlethal salinity levels would undoubtedly increase when fish are provided any opportunity to acclimate.

Based upon the historical record, the probability that Arctic ciscoes would encounter lethal temperature–salinity conditions in the Prudhoe Bay area seems small. From 1981 to 1988, an average of 19 coastal locations (low of 7 in 1981 to a high of 32 in 1985) were sampled by fyke net during the summer of each year as part of ongoing fish monitoring programs (Figure 1). Hydrographic measurements were collected at these sites on a daily basis. While net sites were biased toward nearshore areas, they did provide an adequate spatial cross section of the Prudhoe Bay–Sagavanirktok River delta area, including zones affected by the presence of the Endicott and West Dock causeways.

Analysis of 12,710 (day-to-day changes) salinity measurements from all locations and years showed few instances of daily salinity increases greater than 20‰ (Figure 3). Daily instances of vertical stratification for 1986–1988 (the only years in which simultaneous surface and bottom salinities were recorded) revealed few cases of salinity differentials greater than 25‰ (Figure 4). Of 13,931 total hydrographic measurements from 1981 to 1988, absolute salinities of 30‰ or higher occurred only 0.7% (N = 96) of the time, while salinities of 25‰ or higher occurred with 4.4% (N = 618) frequency. Thus, acute salinity shocks of greater than 25‰, which were required to elicit detectable signs of stress in the laboratory Arctic ciscoes regardless of the accompanying thermal shock, occur with negligible frequency in the coastal environment around Prudhoe Bay. Independent physical process studies have confirmed low incidence rates of acute salinity transitions of greater than 25‰ (Savoie and Wilson 1983, 1986; Hachmeister et al. 1987). Similarly, there was a marginal incidence of nearshore salinities of 30‰ or higher, the salinities that caused low-level, chronic mortality during acclimation in the laboratory.

Had lethal conditions been more prevalent in

the environment, the possibility of natural kills might have increased because of a higher likelihood of fish entrapment in adverse conditions. The occurrence of such phenomena would be partially offset by the fishes' ability to respond to coastal shifts in local hydrography and their ability to acclimate. The low natural occurrence of lethal conditions, coupled with the avoidance and acclimation capabilities of the fish, make acute kills an unlikely prospect in the Prudhoe Bay region of Alaska. It is important to note that bioassay results in no way imply tolerance of, or preference for different salinity–temperature regimes. The fact that specific environmental conditions do not elicit severe signs of stress does not mean that fish will endure those conditions. Further, temperature and salinity conditions that enhance physiological and biochemical processes such as growth and energy conversion may be far removed from the lethal portion of the spectrum. Additional work in these latter areas will be necessary to determine potential longer-term impacts to the Arctic cisco population.

Acknowledgments

We would like to thank James Bryan, Robert Dillinger, Jean Erwin, G. F. Hubbard, and Benny Gallaway for reviewing the manuscript; William Gazey, Lynn Maritzen, and Virginia McCarter for statistical advice; and the staff of the Endicott Main Production Island for their invaluable help and assistance. Thanks are extended to Larry Dugan, Paul Fitzgerald, Tim Green, James Itta, Steve Klimowski, Dominique Rentenaar, and Cathy Taylor for their phenomenal effort in constructing the laboratory. Additional thanks to crew leaders Don Campbell, Bill Coltharp, Steve Hartz, Brent Seavey, and Kirk Waggoner, along with field technicians Mary Bingham, Greg Colonell, John Lennartson, Chris Long, Rob Morrison, Christa Morrow, David Roseneau, Aaryn Valencia, and Joanne Young. Lastly, we would like to thank field manager David Schmidt and expediter Diane Roche for putting up with us. This work was funded by the Endicott Unit Owners.

References

Biosonics, Inc. 1984. Prudhoe Bay Waterflood Project fish monitoring program, 1983. Report to U.S. Army Corps of Engineers, Anchorage, Alaska.

Cannon, T. C., B. Adams, D. Glass, and T. Nelson. 1987. Fish distribution and abundance. Pages 1–129 *in* 1985 Final report for the Endicott Environmental Monitoring Program, volume 6. Report of Envirosphere Company to U.S. Army Corps of Engineers, Anchorage, Alaska.

Critchlow, K. 1983. Fish study. Pages 1–327 *in* Prudhoe Bay Waterflood Project Environmental Monitoring Program 1982. Report of Woodward-Clyde Consultants to U.S. Army Corps of Engineers, Anchorage, Alaska.

Eddy, F. B. 1982. Osmotic and ionic regulation in captive fish with particular reference to salmonids. Comparative Biochemistry and Physiology B, Comparative Biochemistry 73:125–142.

Epstein, F. H., A. I. Katz, and G. E. Pickford. 1967. Sodium- and potassium-activated adenosine triphosphatase of gills: role in adaption of teleosts to sea water. Science (Washington, D.C.) 156:1245–1247.

Epstein, F. H., A. Manitius, E. Weinstein, A. I. Katz, and G. E. Pickford. 1969. Sodium- and potassium-activated adenosine triphosphatase in kidneys of *Fundulus heteroclitus* adapted to fresh and salt water. Yale Journal of Biology and Medicine 41: 388–393.

Evans, D. H. 1979. Fish. Pages 310–390 *in* G. M. Maloy, editor. Comparative physiology of osmoregulation in animals. Academic Press, New York.

Fechhelm, R. G., J. S. Baker, W. B. Griffiths, and D. R. Schmidt. 1989. Localized movement patterns of least cisco (*Coregonus sardinella*) and Arctic cisco (*C. autumnalis*) in the vicinity of a solid-fill causeway. Biological Papers of the University of Alaska 24:75–106.

Forskett, J. K., H. A. Bern, T. E. Marchen, and M. Conner. 1983. Chloride cell and hormonal control of teleost fish osmoregulation. Journal of Experimental Biology 106:255–281.

Forskett, J. K., C. D. Logsdon, T. Turner, T. E. Marchen, and H. A. Bern. 1981. Differentiation of the chloride extrusion mechanism during seawater adaptation of a teleost fish, the cichlid *Sarotherodon mossambicus*. Journal of Experimental Zoology 93:209–224.

Glass, D., C. Whitmus, and M. Prewitt. 1988. Fish distribution and abundance. Pages 1–188 *in* Endicott Environmental Monitoring Survey, 1986, volume 4. Report of Envirosphere Company to U.S. Army Corps of Engineers, Anchorage, Alaska.

Griffiths, W. B., and B. J. Gallaway. 1982. Prudhoe Bay Waterflood fish monitoring program 1981. Pages D1–D98 *in* Prudhoe Bay Waterflood Environmental Monitoring Program, volume 4. Report of LGL Alaska Research Associates to Woodward-Clyde Consultants and U.S. Army Corps of Engineers, Anchorage, Alaska.

Griffiths, W. B., and seven coauthors. 1983. Fish ecology. Pages 1–342 *in* B. Gallaway and R. Britch, editors. Environmental summer studies (1982) for the Endicott Development, volume 3. Report of LGL Alaska Research Associates, Inc., and Northern Technical Services to Sohio Alaska Petroleum Company, Anchorage, Alaska.

Hachmeister, L. E., K. S. Short, G. C. Schrader, K. B.

Winnick, and J. W. Johannessen. 1987. Oceanographic monitoring, 1985. Pages 1–162 *in* Endicott Environmental Monitoring Program, volume 3. Report of Envirosphere Company to U.S. Army Corps of Engineers, Anchorage, Alaska.

Hazel, J. R., and C. L. Prosser. 1974. Molecular mechanisms in temperature compensation in poikilotherms. Physiological Reviews 54:620–677.

Hwang, P. P., and R. Hirano. 1985. Effects of environmental salinity on intercellular organization and junctional structure of chloride cells in early stages of teleost development. Journal of Experimental Zoology 236:115–126.

Hwang, P. P., and S. M. Wu. 1988. Salinity effects on cytometrical parameters of the kidney in the euryhaline teleost *Oreochromis mossambicus* Peters. Journal of Fish Biology 33:89–95.

Kamiya, M., and S. Utida. 1968. Changes in activity of sodium- and potassium-activated adenosine triphosphatase in gills during the adaption of the Japanese eel to sea water. Comparative Biochemistry and Physiology 26:675–685.

LGL. 1989. The 1988 Endicott Development fish monitoring program—analysis of fyke net data. Pages 1–160 *in* Volume 1: synthesis. Report of LGL Alaska Research Associates to BP Exploration (Alaska), Inc., and North Slope Borough, Anchorage, Alaska.

Moulton, L., and seven coauthors. 1986. 1984 central Beaufort Sea fish study. Pages 1–300 *in* Waterflood Monitoring Program fish study, 1984, volume 2. Report of LGL Alaska Research Associates, Entrix, Inc., and Envirosphere Company to U.S. Army Corps of Engineers, Anchorage, Alaska.

Ostle, B., and R. E. Mensing. 1975. Statistics in research, 3rd edition. Iowa State University Press, Ames.

Peltier, W., and C. I. Weber, editors. 1984. Methods for measuring the acute toxicity of effluents to aquatic organisms, 3rd edition. U.S. Environmental Protection Agency, Cincinnati, Ohio.

Reub, G. S., J. D. Durst, and D. Glass. 1988. Fish distribution and abundance. Pages 1–47 *in* 1987 Endicott Environmental Monitoring Program. Report of Envirosphere Company to U.S. Army Corps of Engineers and Standard Alaska Production Company, Anchorage, Alaska.

Savoie, M. A., and D. E. Wilson. 1983. Physical processes monitoring program for 1982. Pages 1–162 *in* Prudhoe Bay Waterflood Environmental Monitoring Program, volume 2. Report of Kinnetic Laboratories, Inc., to U.S. Army Corps of Engineers, Anchorage, Alaska.

Savoie, M. A., and D. E. Wilson. 1986. Physical marine processes monitoring program 1984. Pages 2-1 to 2-195 *in* Prudhoe Bay Waterflood Environmental Monitoring Program, volume 1. Report of Kinnetic Laboratories, Inc., to U.S. Army Corps of Engineers, Anchorage, Alaska.

Snedecor, G. W., and W. G. Cochran. 1980. Statistical methods, 7th edition. Iowa State University Press, Ames.

Umminger, B. L. 1971. Chemical studies of cold death in the gulf killifish, *Fundulus grandis*. Comparative Biochemistry and Physiology A, Comparative Physiology 39:625–632.

Zaugg, W. S., and L. R. McLain. 1970. Adenosinetriphosphatase activity in gills of salmonids: seasonal variations and salt water influence in coho salmon, *Oncorhynchus kisutch*. Comparative Biochemistry and Physiology 35:587–596.

American Fisheries Society Symposium 11:119–131, 1991

Effects of Temperature, Salinity, and Prey Abundance on the Growth of Arctic Ciscoes and Broad Whitefish Feeding on Epibenthic Prey in Field Enclosures

K. K. English

LGL Limited, environmental research associates
9768 Second Street, Sidney, British Columbia V8L 3Y8, Canada

Abstract.—The primary objectives of this study were (1) to test the feasibility of using enclosures to study fish feeding at two different sites in coastal waters of the Beaufort Sea; (2) to obtain reliable estimates of growth and prey densities for juvenile fish that feed in nearshore waters adjacent to the Endicott Causeway and to compare these results to similar findings for fish in a control area in Prudhoe Bay; and (3) to quantify the relationships among prey abundances, feeding efficiency, and growth for juvenile Arctic cisco *Coregonus autumnalis* and broad whitefish *C. nasus*. The fish were retained in 310-m^2 mesh enclosures at two coastal sites near Prudhoe Bay, Alaska. The enclosures permitted ample water and prey circulation while retaining juvenile fish (120–180 mm fork length, 15–45 g wet weight). Prey concentrations inside the enclosures were not significantly different from those outside the enclosures. Fish in these large enclosures could be readily captured, weighed, and returned alive to the enclosures. Fish growth rates could be estimated for each enclosure group, and significant differences were detected both between sites and species. Solutions were identified for all problems encountered during the study; however, the combination of several field logistic problems compromised the quality and quantity of data obtained from one enclosure (Niakuk Arctic cisco). The highest growth rates were measured at the study site adjacent to the Endicott Causeway, where Arctic ciscoes grew at a mean rate of 0.63%/d over 6 weeks. The mean growth rate for broad whitefish reared in adjacent enclosures was only 0.08%/d. The mean growth rates for Arctic ciscoes and broad whitefish reared at the Niakuk Islands site were 0.08 and −0.01%/d, respectively. Survival rates for both species were substantially lower at the Niakuk site. Arctic cisco appeared to be six times more "efficient" than broad whitefish at consuming prey in the enclosures. Results from the Endicott enclosures suggested that the major determinants of growth for each fish species were water temperature and the biomass of key prey taxa less than 6 mm in body length. The study also provided some evidence that the nearshore environments along the lower portion of the Endicott causeway were suitable habitats for juvenile Arctic cisco, and that neither the Endicott nor Niakuk site were suitable habitats for juvenile broad whitefish.

One of the major issues associated with the oil industry's development of the coastal waters of the Beaufort Sea is the effect of solid-fill causeways on the growth and survival of arctic anadromous fish. No single experiment could provide all the data required to quantify the effect of the causeways on fish growth. This study was part of an ongoing series of field and laboratory studies and provides data on the growth of key fish species feeding on known concentrations of prey. The long-term goal of these studies was to combine information from the laboratory, fyke net sampling, and enclosure studies into a bioenergetic assessment of fish growth potential in the waters adjacent to the Endicott Causeway.

Arctic anadromous fish in the nearshore coastal zone of the Beaufort Sea may experience large changes in water temperature and salinity (up to 12°C and 28‰). In 1988, laboratory studies were conducted as part of the Endicott Fish Monitoring Program to determine the lethal nature of these changes to Arctic cisco. Results of 96-h tests suggested that the fish must experience instantaneous increases in salinity from 0 to 26‰ along with instantaneous decreases in temperature from 12 to about 2°C before any mortality occurred (Fechhelm et al. 1991, this volume). A review of hydrographic data showed that the frequency of daily changes in temperature and salinity of this magnitude was low. When such changes do occur, arctic anadromous fish appear to respond by moving to more favorable habitat (e.g., Fechhelm et al. 1983; Griffiths and Gallaway 1983; Neill et al. 1983; Craig et al. 1985; Moulton et al. 1986; Neill and Gallaway 1989). Although seldom lethal, the magnitude of the temperature–salinity fluctuations that frequently occur do have the potential to substantially affect the growth rates of most freshwater and anadromous fish species (e.g., Brett 1979). Reduced growth can lead to lower fecundity, decreased individual fitness, and ultimately a reduced population.

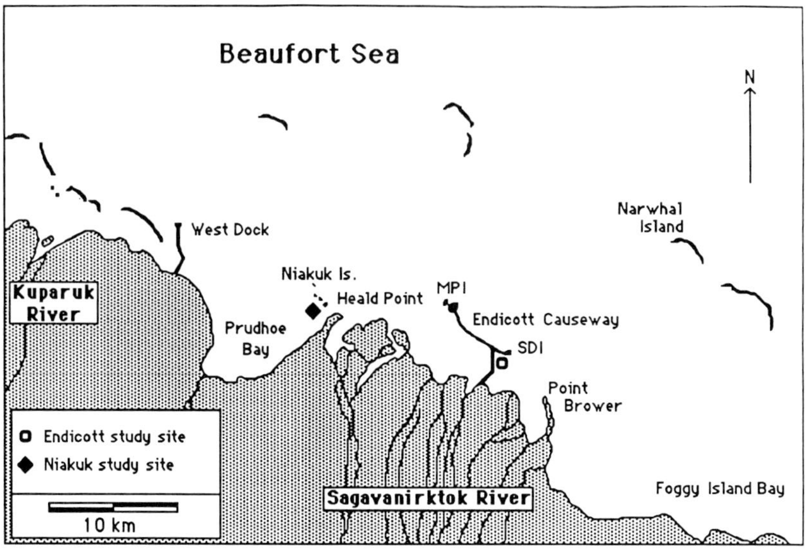

FIGURE 1.—Location of the Niakuk and Endicott enclosure sites near Prudhoe Bay in the coastal Beaufort Sea, Alaska. MPI is the Main Production Island on the Endicott Causeway, and SDI is the Satellite Drilling Island.

The relationship of these factors (temperature, salinity, and prey) to fish growth was considered conceptually in the assessment models in the Endicott Environmental Impact Statement (USACE 1984). No effect was found, because the model fish were able to mitigate much of the possible adverse effect on growth from temperature and salinity changes by avoiding extreme conditions. Ongoing distribution studies continue to strengthen the case that behavioral responses to temperature and salinity changes enable fish to exploit the most favorable parts of their environment. However, prior to this study, no data were available to confirm the presumed relationships between temperature, salinity, and prey with fish growth in the coastal Beaufort Sea. It is essential to define these relationships before any final evaluation of the effects of the Endicott Causeway on fish growth can be made (e.g., Ross 1988).

It is generally recognized that growth in relation to abiotic factors cannot be considered without including the biotic factor of food ration (Brett 1979). It would be equally difficult to extract information on the relationship between fish growth and prey abundance without accounting for the controlling effects of temperature and the masking effects of salinity. An understanding of these relationships is therefore essential to resolving issues of the causeway's effects.

This paper reports the results of a study that was conducted to determine the relationships between prey densities and fish growth for Arctic cisco *Coregonus autumnalis* and broad whitefish *C. nasus*. These species were selected because they are abundant in the nearshore areas of the Beaufort Sea and they had been the focus of most of the previous fish monitoring studies in the study area. The primary objectives of this study were (1) to test the feasibility of using enclosures to study fish feeding at two different sites in coastal waters of the Beaufort Sea; (2) to obtain reliable estimates of growth and prey densities for juvenile fish that feed in nearshore waters adjacent to the Endicott Causeway, and to compare these results to similar findings for fish in a control area in Prudhoe Bay; and (3) to quantify the relationships among prey abundances, feeding efficiency, and growth for Arctic cisco and broad whitefish.

General Methods

Arctic ciscoes and broad whitefish were studied for 6 weeks during July, August, and September 1988 in enclosures in coastal waters of the Alaskan Beaufort Sea. Three circular enclosures (two large and one small) were deployed at each of two sites, one near the satellite drilling island (SDI) at Endicott Causeway and one near the Niakuk Islands in the northeast portion of Prudhoe Bay (Figure 1). Water depth at both sites averaged about 1.5 m. Each large enclosure was 20 m in diameter (area = 310 m^2), with 2-m sides of

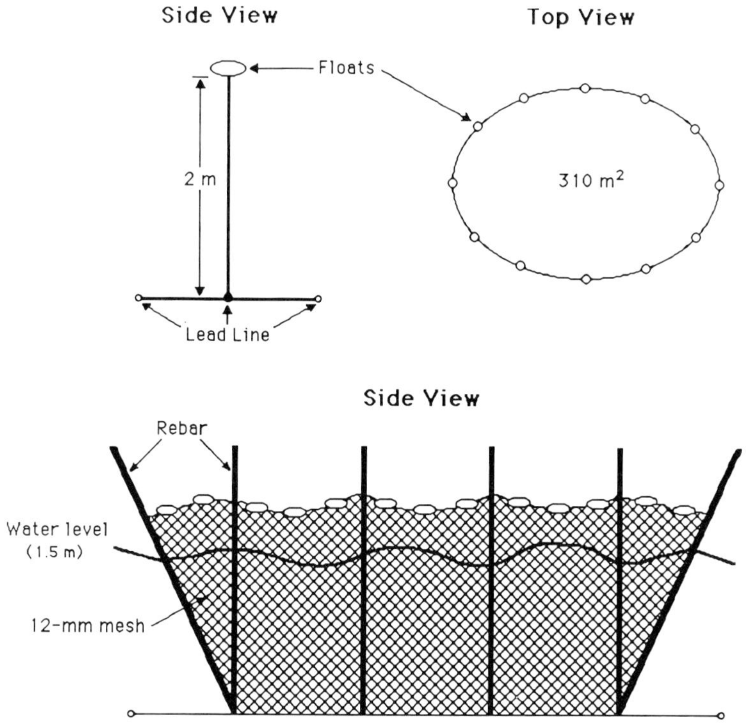

FIGURE 2.—Diagram of the fish enclosures at the Niakuk and Endicott sites.

12-mm, stretched, knotless nylon mesh (Figure 2). The smaller control enclosures were used to determine enclosure effects on prey densities. They were of similar design, but measured only 5 m in diameter (area = 20 m^2). The base of each side was attached to a 1-m-wide piece of nylon netting fitted with three lead lines. Sides of the enclosures were held in place by the lead lines, rebar, and mud that eventually covered most of the base netting. Several of the rebar supports were secured to steel pipe piles with guide wires to give additional stability to the windward, exposed sides of the enclosures. Plastic net floats kept the sides of the enclosures at or above the water surface. Shallow-water fyke nets were sewn into the sides of the large enclosures to facilitate recapture of fish for growth measurements.

Juvenile fish (120–180 mm fork length, 15–45 g wet weight) were obtained from fyke nets set near the Endicott Causeway and were transferred in coolers with air stones to the laboratory at Endicott's Main Production Island (MPI). These fish were anesthetized in 4 L of water with 2-phenoxyethanol (0.04% solution for broad whitefish, 0.05% solution for Arctic cisco), tagged, fork length measured to the nearest millimeter, and wet-weighed to the nearest tenth of a gram with a triple beam or electronic balance. To minimize stress, fish were then returned to holding tanks and held for 1–4 d before introduction to the enclosures.

While anesthetized, individual fish were marked with numbered visual implant (VI) tags and dye dot codes. The VI tags were injected under the clear tissue in the nose of Arctic ciscoes and behind the eye of broad whitefish. Up to five dye dots were placed near the midline of the back and behind the anal fins to identify the remaining fish.

At the Endicott site, 110 fish were introduced into both the Arctic cisco and broad whitefish enclosures. The Arctic ciscoes were transported to the enclosures on 28 July and the broad whitefish on 29 July 1988. At Niakuk, 114 broad whitefish were transported to the enclosure between 30 July and 4 August 1988. In total, 114 Arctic ciscoes were transported to the Niakuk site between 25 and 30 July 1988, but none were recaptured. Therefore, 105 Arctic ciscoes were stocked in the Niakuk enclosure on 21 August 1988 for the remaining 3 weeks of the experiment. The initial 114 fish were probably placed in the wrong enclosure and escaped when a portion of the netting was raised to install the fyke trap. Uncertainty associated with our ability to capture stocked fish, and difficulties installing a fyke trap delayed the restocking of the Niakuk Arctic cisco enclosure.

Fyke nets on the sides of the enclosures were opened for 24–48 h prior to weekly measurements of fish growth. Once captured, 40 fish were transferred into holding pens and then coolers with air stones for measurements on land adjacent to the study sites. The fish were anesthetized, weighed, and measured in the same manner as during original tagging. About 1–1.5 h were needed to measure the fish from each enclosure. Fish were put into a recovery cooler also equipped with air stones before reintroduction to the enclosures. During measuring, every fourth fish was sacrificed for later analysis of stomach contents. The body cavities were opened and the gastrointestinal tract removed and placed in a whirl-pac containing a label and 10% formalin. Stomachs were processed at the Institute of Marine Science, University of Alaska, Fairbanks.

Twenty-six Arctic ciscoes and 30 broad whitefish were starved in laboratory holding tanks at MPI. Temperatures and salinities in the tanks were adjusted intermittently to compare with ambient values at the study sites (usually 3–7°C and 10–15‰). Laboratory-starved fish were measured and weighed weekly in the same manner as the enclosure fish.

Measurements of prey availability were taken three to four times per week outside and inside each of the three enclosures at both study sites. Sampling gear consisted of a 1.5-m-long conical plankton drop net attached to a 0.5-m-diameter stainless steel frame at the opening and a removable sampling cup at the cod end. A pursing line ran through grommets at the bottom of the collar, up through the center of the frame, and out through a grommet near the sampling cup. A 333-μm mesh net was used with the frame. Four replicate samples of water column and epibenthic prey were taken from outside the enclosures and inside each of the large enclosures on each sampling day. Only two drop net replicates were taken from the smaller control enclosures. The time required to complete all drop net casts at each study site was 60–75 min. After a cast was made and the net pursed, the net was inverted and the top one-third rinsed three times in water adjacent to the boat. The bottom of the net was then dipped repeatedly in the water until all visible material was washed into the cod end. Material on the sides of the cod-end bucket was drained and rinsed once with water into bottles containing at least 10% formalin. Drop net samples were labeled and stored in whirl-pacs at the MPI laboratory. Each day that drop net samples were collected, a YSI meter was used to measure water temperature and salinity at 0.5-m depth intervals at the enclosure sites. Either three or four measurements were taken, depending on total water column depth.

Drop net and stomach samples were analyzed at the Institute of Marine Science, University of Alaska, Fairbanks. Sample analysis procedure consisted of first cleaning the samples and removing large animals such as fish and medusae. Some samples were split down to facilitate counting. Subsamples were rarely less than 25% of the total, except for copepods smaller than 2 mm, which were split to 1%. Each species of major taxonomic category was counted by size-class with a petri dish separator and a 2-mm square grid under a dissecting microscope. Size-classes were less than 2 mm, 2–4 mm, 4–6 mm, and greater than 6 mm (all prey) and 6–10 mm and greater than 10 mm (mysids only). After completing the count for a specific species, a weight estimate was obtained for that species. Wet weights of each species and size-class were determined with a Mettler PM200 analytical balance. All counted organisms were stored by species in containers labeled such that every species count could be traced.

Analytical Methods

Daily instantaneous growth rates for the fish reared in the enclosures and laboratory tanks were estimated with the equation

$$G = 100\ [\log_e(W_{\text{end}}) - \log_e(W_{\text{start}})]/T; \qquad (1)$$

G is the percent change in wet weight per day; W_{start} and W_{end} are the weights of each fish at the beginning and end of an interval of T days. Mean weekly growth rates were calculated by averaging the weekly growth rates for individual fish weighed at the beginning and end of a particular week. Mean changes in weight for individuals measured at the end of the study were used to estimate the mean total growth for each enclosure group.

A series of computer programs were developed to convert the numbers and weights of each prey category into mean daily, weekly, and total estimates of abundance and biomass per square meter. Mean abundance estimates were obtained by dividing the total number of prey in all samples for a selected location, sampling interval, size-group, and prey category by the number of samples taken. The mean biomass estimates were calculated by combining the abundance estimates with the mean weights for each size-group and prey category stratified by location and sampling interval. The programs and procedures for the stomach content data analyses were identical to those for the drop net analyses.

Growth estimates for 120–180-mm Arctic ciscoes and broad whitefish rearing under natural conditions in the study area were obtained from the 1985–1988 Endicott fyke net data (Gallaway et al. 1991, this volume). A modified von Bertalanffy growth curve (Parrack 1979) was fitted to each year's length-frequency data in a least-squares algorithm (Marquardt 1963). The parameters for each curve (L_{max}, L_0, and K) were used in the following equation to estimate the mean size of "wild" fish on any day between June 15 and September 15.

$$L = L_{max} - (L_{max} - L_0)e^{-Kt}; \qquad (2)$$

L is the mean length of the fish population t days from June 15; L_{max} is the mean asymptotic length; L_0 is the mean length on June 15; and K is von Bertalanffy's growth coefficient. Length–weight relationships for each group of enclosed fish were used to convert the mean lengths into the mean weights required to estimate instantaneous growths in equation (1).

Evaluation of Experimental Procedures

Enclosure Performance

All enclosures maintained their integrity throughout the 6-week study period. Enclosures at Niakuk were exposed to currents and waves driven by 50–80 km/h winds on several occasions. Enclosures at Endicott were more protected from wind, but were exposed to longshore currents and wind-driven ice floes as large as 10–20 m during the last few weeks of the study. The combination of rebar, lead lines, and mud substrate ensured that the base of each enclosure remained firmly attached to the bottom. Each enclosure was inspected upon removal and only three holes were found; the largest of these was 3 cm square. The tops of the enclosures were at or above the water surface for most of the study. However, high water levels resulting from consistent west winds caused some of the floats to be submerged on a few days. Fouling of the mesh was not a substantial problem at either site. Cleaning once per field season was sufficient to keep all nets free from large debris. The smaller control enclosure nets appeared to be somewhat more subject to plankton fouling, particularly at the Endicott site.

Fish Marking

The VI tags were easily read throughout the study, and tag loss rates were only 2.2% for Arctic cisco and 6.7% for broad whitefish. Dye dot codes proved to be a much poorer method of identifying fish. Fish retained in the laboratory starvation tank revealed errors in the application and interpretation of dye dot codes that were compounded by fish that lost some or all of their dye dots. Dye dot codes were only applied to the broad whitefish reared in the laboratory and Arctic ciscoes reared at Niakuk. Problems with dye dot interpretation did not affect our ability to differentiate individual laboratory fish, because each fish was weighed and measured each week. However, these problems did compromise our ability to estimate reliable weekly growth rates for the Arctic ciscoes at the Niakuk site.

Sampling Fish from Enclosures

Most of the fish samples were obtained from fyke nets sewn into the walls of each enclosure. At the end of the study period a small-mesh gill net was used like a beach seine to catch those fish not captured by the fyke net. The fyke net technique was an efficient method of capturing fish while minimizing stress. However, the small-mesh gill net proved to be a more consistent method of capturing all the fish from an enclosure in one pass, without damaging a single fish. For example, in a test with 100 fish, the passive fyke net method caught 55 in the first 24 h and 6 in the next 24 h. The gill net method caught the remaining 39 in the first pass.

Enclosure Effects on Epibenthic Prey Abundances

Effects of enclosures on prey abundances were analyzed by comparisons of abundances outside to those inside each of the three enclosure types at each site. We used paired, two-tailed t-tests of the replicate means from each sampling day for these analyses (Tables 1, 2). Comparisons were made for each of four prey size-classes that included all species, and for the 6–10-mm and greater than 10-mm size-classes of mysids.

No significant differences ($P > 0.05$) between prey abundances outside and inside each of the

TABLE 1.—Probabilities of equality between abundances of epibenthic prey outside versus inside enclosures at the Endicott site. All comparisons were paired, two-tailed t-tests with 23 df. Significant differences are indicated by asterisks ($P < 0.05$).

Enclosure	Prey size-class (mm)					
	<2	2–4	4–6	>6	6–10[a]	>10[a]
Arctic cisco	0.543	0.109	0.486	0.637	0.222	0.258
Broad whitefish	0.434	0.090	0.087	0.137	0.162	0.176
Control	0.081	0.044*	0.867	0.054	0.023*	0.234

[a]Mysids only.

fish enclosures were detected at the Endicott site (Table 1). This suggests that the fish were not cropping a detectable amount of the available prey as found in previous enclosure studies with juvenile chinook salmon *Oncorhynchus tshawytscha* (English 1983). In two size-classes (2–4-mm and 6–10-mm mysids), a significant difference was detected between the control and outside prey abundances at Endicott. This may have been caused by the relatively greater fouling on the smaller control enclosures, which could have obstructed some prey. No differences between prey abundances outside and inside each of the enclosures, including the control enclosure, were detected at the Niakuk site (Table 2).

Results

Substantial differences were observed in the growth and survival of the enclosed fish at each study site (Table 3). Mean growth rate for Endicott Arctic ciscoes was over seven times that of any other group, and they were the only group that maintained positive growth rates throughout the study (0.25–0.98%/d). The other enclosure groups showed little or no net growth, while the fish reared in the laboratory declined at a rate of 0.23–0.36%/d.

The survival percentages show that, with the exception of the broad whitefish at Niakuk, the majority of the fish stocked in the enclosures were recovered alive (Table 3). The remainder of the fish either died in the enclosures, escaped, or were removed by predators. Given the similarity in the construction and integrity of each enclosure and the lack of evidence for differential predation, we believe that the daily mortality rates reflected the relative effects of environmental factors (e.g., prey abundance, temperature, and salinity) on the survival of each species at each site. Consequently, conditions at the Niakuk site appear to have resulted in loss rates that were two to three times those estimated for the fish reared at the Endicott site. Weekly survival estimates for Niakuk broad whitefish indicated that the majority of the losses occurred in the third week of the study, and may have been the result of a rapid increase in salinity to a maximum of 28‰ on 16 August. The majority of the Niakuk Arctic cisco losses occurred in the fifth week of study, a period without any extreme temperature or salinity fluctuations.

These growth and survival statistics are a fairly clear indication that the conditions at the Niakuk site were not suitable for either species. In contrast, the conditions at the Endicott site were obviously adequate for Arctic cisco growth and marginal for broad whitefish growth. It should be noted that the late start of the Arctic cisco enclosure and the poor broad whitefish survival at Niakuk resulted in 40% less growth and survival data compared with data from the Endicott site. Therefore, the remaining analyses will focus on the factors affecting fish growth at the Endicott site, with limited reference to Niakuk results.

TABLE 2.—Probabilities of equality between abundances of epibenthic prey outside versus inside enclosures at the Niakuk site. All comparisons were paired, two-tailed t-tests with 23 df. No significant differences ($P < 0.05$) were detected.

Enclosure	Prey size-class (mm)					
	<2	2–4	4–6	>6	6–10[a]	>10[a]
Arctic cisco	0.310	0.969	0.217	0.425	0.654	0.450
Broad whitefish	0.356	0.472	0.871	0.724	0.773	0.077
Control	0.246	0.089	0.629	0.108	0.278	0.156

[a]Mysids only.

TABLE 3.—Summary of Arctic cisco (AC) and broad whitefish (BW) growth, survival, and sampling for the 1988 enclosure study. Laboratory fish were starved in the laboratory at the Main Production Island on the Endicott Causeway.

Statistic	Endicott		Niakuk		Laboratory	
	AC	BW	AC	BW	AC	BW
Start number	110	110	105	114	26	30
End number	22	24	36	10	13	19
Mean start weight (g)	24.6	29.1	23.1	37.9	22.4	36.6
Mean end weight (g)	32.3	30.1	23.4	37.8	20.2	32.4
Study period (d)	43	43	17	30	44	34
Growth						
Mean total (g)	7.7	1.0	0.3	−0.1	−2.2	−4.2
Mean daily (%/d)	0.63	0.08	0.08	−0.01	−0.23	−0.36
Standard error (%/d)	0.05	0.03	0.06	0.05	0.02	0.03
Survival						
Fish recovered alive[a]	62	66	59	38	23	29
Percent survival	56	60	56	33	88	97
Relative loss rate (fish/d)	1.12	1.02	2.71	2.53	0.07	0.03
Sample size						
Fish measurements	149	190	59	109	113	115
Drop net samples	336	336	322	322		
Temperature–salinity	24	24	23	23		

[a]Number of fish alive at end + fish sacrificed during study period.

Fish Growth versus Temperature and Salinity

In general, water temperatures declined and salinities increased throughout the study period. The within-week variability in salinity was generally greater than that for temperature, especially during the last 3 weeks of the study (Figure 3). Mean weekly temperatures and salinities were similar at the Endicott and Niakuk study sites. Weekly growth of Arctic ciscoes and broad whitefish showed no consistent relationship to either temperature or salinity at the Endicott site over the entire study period (Figure 3). However, the 50% decline in the mean water temperature between weeks 3 and 4 was accompanied by similar large declines in mean growth rates for both Arctic cisco (0.98 to 0.29%/d) and broad whitefish (0.41 to −0.06%/d).

Fish Growth versus Prey Abundance

An assessment of the relationship between fish growth and prey abundance requires (1) identification of the major prey species consumed by the fish, and (2) an examination of the relationships between the major prey species consumed and the abundance of those species in the environment.

In total, 62 Arctic cisco and 62 broad whitefish stomachs were examined to determine the number and biomass of each prey group consumed by the fish at the Endicott site. Table 4 lists the major prey species and biomass found and provides an index of how prey consumption differed between the two fish species. The prey taxa consumed were the same for each fish species, with the exception of the larger isopods, which were found only in broad whitefish stomachs. It is interesting to note that the biomass of prey found in the broad whitefish stomachs each week was only 15–50% of that found in the Arctic cisco stomachs, yet both fishes were exposed to similar concentrations of each prey taxa. The largest differences were for small copepods and large mysids, which were 20–30 times more abundant in the Arctic cisco stomachs. Aside from isopods, there were only two groups, *Gammaracanthus loricatus* and 6–10-mm mysids, that were more abundant in the broad whitefish than in the Arctic cisco stomachs.

Table 4 also provides an estimate of the relative feeding "efficiency" for each fish species, calculated by dividing the mean biomass per fish stomach by the mean biomass (mg/m^2) for these prey taxa. Mean prey biomass was estimated from the drop net samples that were taken on the days when the fish stomach samples were obtained. The data suggest that Arctic ciscoes were six times more efficient at exploiting the type of prey in the enclosures. These findings were consistent with the observation that broad whitefish weekly growth rates were consistently less than those for Arctic cisco. Similar results were observed for the fish at the Niakuk site; however, there was only 1 week when both fish species were reared simultaneously and stomach samples were obtained on the same day.

The stomach content data were combined with

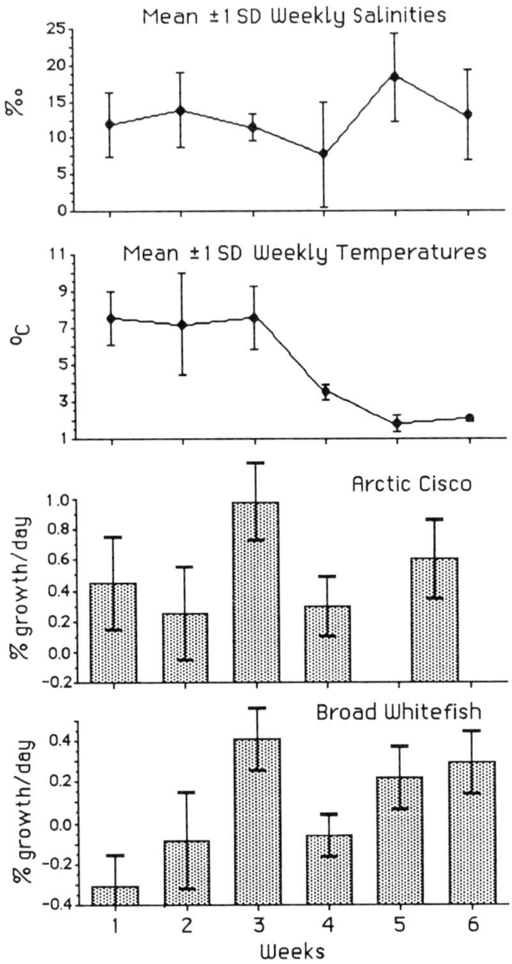

FIGURE 3.—Average weekly growth rates of Arctic ciscoes and broad whitefish at the Endicott enclosure site in relation to ambient salinity and temperature. Weekly averages of ambient temperature and salinity were calculated by taking the average of several (3–7) daily values that had been averaged over total water column depth. The origin of each growth plot is the growth rate estimated for the fish starved in the laboratory. The error bar for each growth estimate represents the 95% confidence interval (2 standard error units each side of the mean). The last growth estimate for Arctic ciscoes is the mean rate for the last 2 weeks because no fish were sampled in the fifth week.

drop net sampling data for the same days to produce an Ivlev electivity index (Ivlev 1961) for each major prey species (Figures 4, 5). Electivity indices close to zero indicate that prey consumption was roughly proportional to the prey biomass measured in the environment. Index values of 1.0 identify those prey consumed that were not found in the drop net samples (the larger copepods and *Apherusa glacialis*). Drop nets are not designed to sample the pelagic and under-ice areas where these two prey groups are commonly found. It should be noted that these prey groups were only found in the stomach samples obtained when the fish were sacrificed at the end of the study. Large negative index values identify those prey species that were not fully exploited by the fish. These indices suggest that over 85% of the prey biomass measured with drop net samples (i.e., *Monoculodes* sp., *Gammaracanthus loricatus, Mysis* sp., *Mysis relicta*) was not fully exploited by Arctic cisco. The broad whitefish consumed less total biomass, but appeared to exploit a wider variety of prey sizes and types. If these stomach content data are an accurate reflection of the basis for fish growth, there should be little or no relationship between the abundance of prey larger than 6 mm and the growth of Arctic cisco.

Figure 6 shows the weekly mean Arctic cisco growth estimates along with drop net estimates of the weekly mean biomass of the major prey taxa greater and less than 6 mm in body length. The prey less than 6 mm (primarily 1–4-mm copepods and *Pontoporeia affinis*) were 85% of the prey biomass found in the Arctic cisco stomachs in the first 4 weeks of the study. The rationale for separating the prey into these two size-groups is that the amount of time a 120–160 mm fish would require to process prey larger than 6 mm would be substantially greater than digestion time for smaller prey. This difference in digestion time, coupled with the much greater biomass of the large prey species, would make it more difficult for the fish to exploit large prey in proportion to their abundance. This is one possible explanation for the lack of a relationship between growth rate and the biomass of large prey (Figure 6). The combination of the abundance of the major prey taxa smaller than 6 mm and temperature data could account for the fluctuations in weekly growth rate for Arctic ciscoes at the Endicott study site. Abundance of small prey accounted for fluctuations in growth in the first 3 weeks when temperature was relatively high and constant. In the last 3 weeks, low temperatures appear to have limited Arctic cisco growth rates to 60% of those observed for higher temperatures and similar prey abundances. The only major inconsistency in the growth pattern of the two fish species was in the first week, when broad whitefish growth rates were proportionately much lower than in subsequent weeks.

TABLE 4.—Major prey taxa and biomass in the stomachs of 62 Arctic ciscoes (AC) and 62 broad whitefish (BW) from the Endicott enclosures. Taxon codes are used in Figures 4 and 5.

Prey size-class (mm)	and taxon, or statistic	Taxon code	Predator AC	Predator BW	Relative prey consumption (AC/BW)
Stomach contents (mg)					
<2	Copepoda	CO	2,225	71	31.3
2–4	Copepoda	CO	103	27	3.8
2–4	Isopoda	IS	12	12	1.0
2–4	*Monoculodes* sp.	MO	108	100	1.1
2–4	*Pontoporeia affinis*	PA	35	7	5.0
4–6	Copepoda	CO	330	48	6.9
4–6	Isopoda	IS	0	60	0.0
4–6	*Apherusa glacialis*	AG	42	4	10.5
4–6	*Pontoporeia affinis*	PA	60	24	2.5
6–10	Copepoda	CO	504	48	10.5
6–10	Isopoda	IS	0	54	0.0
6–10	*Apherusa glacialis*	AG	747	77	9.7
6–10	*Gammaracanthus loricatus*	GL	28	56	0.5
6–10	*Mysis* sp.	MU	70	244	0.3
>10	*Mysis relicta*	MR	308	13	23.7
Total			4,572	845	5.4
Biomass and feeding efficiency					
Mean biomass per fish stomach (mg)			73.7	13.6	
Mean biomass in environment (mg/m^2)			1,207.6	1,394.2	
Relative feeding efficiency (%)			6.1	1.0	

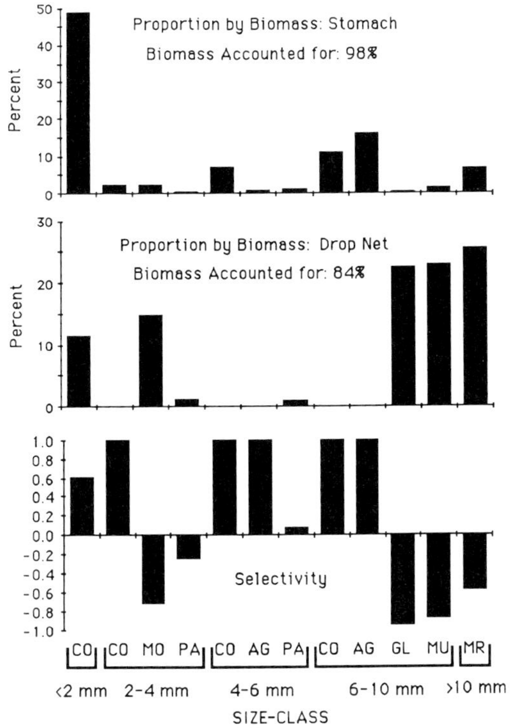

FIGURE 4.—Composition of prey consumed (stomach), prey available (drop net), and prey selectivity for Arctic ciscoes at the Endicott enclosure. Prey taxon codes along the horizontal axis are defined in Table 4.

The data from the 1988 enclosure studies were not sufficient to conduct meaningful correlation or regression analyses. Minor modifications to the 1988 study design (2–3 week extension of the study period, the use of small-mesh gill nets to sample fish, and the selection of sites that could be sampled by a single crew on a daily basis) would ensure that the data collected were adequate for a more detailed statistical analysis of the relationship between prey abundance and fish growth.

Growth Rates for Enclosure Fish versus "Wild" Populations

The results of the enclosure study were put in perspective by comparing the growth rates for the enclosed fish with the growth rates estimated for fish of similar size from the Endicott–Prudhoe Bay fyke net size-frequency data (Gallaway et al. 1991). Critical to this comparison was the ability to calculate growth rates for the wild population for intervals of the same duration and timing as those for the enclosure experiments. This was made possible by combining the interval growth estimates from the von Bertalanffy growth curves with the length–weight relationships for each fish species (Table 5). While these growth rate estimates are period specific, the data from fyke net sites close to the enclosure sites were not adequate to produce site-specific growth rates for

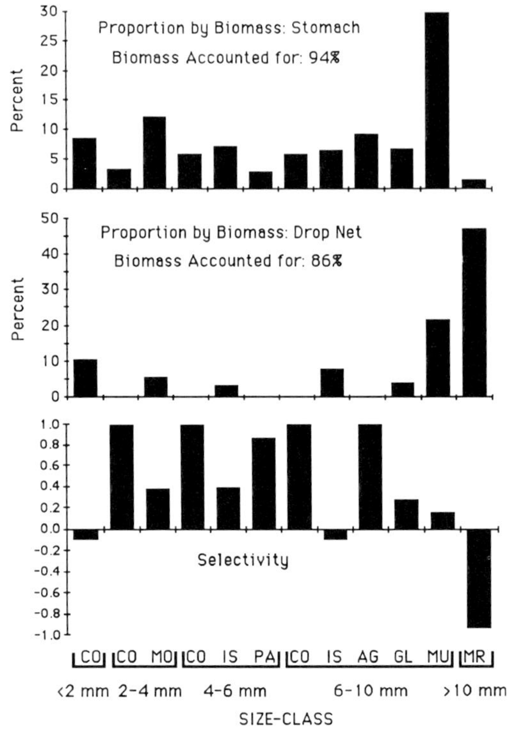

FIGURE 5.—Composition of prey consumed (stomach), prey available (drop net), and prey selectivity for broad whitefish at the Endicott enclosure. Prey taxon codes along the horizontal axis are defined in Table 4.

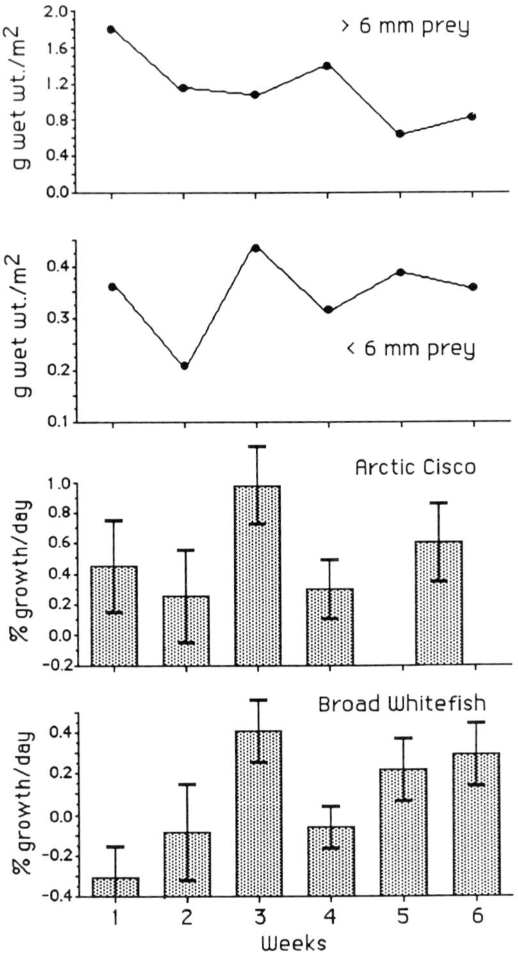

FIGURE 6.—Comparison of prey abundance and growth rates for the fish reared at the Endicott site. Prey abundance estimates are the means for the 42–98 drop net samples collected at Endicott each week. The origin of each growth plot is the growth rate estimated for the fish starved in the laboratory. The error bar for each growth estimate represents the 95% confidence interval (2 standard error units each side of the mean). The last growth estimate for Arctic ciscoes is the mean rate for the last 2 weeks because no fish were sampled in the fifth week.

wild fish. Differences in growth rates in Table 5 for the Endicott and Niakuk sites thus result from differences in study times, not study locations.

The growth rates for Arctic ciscoes enclosed at the Endicott site (0.63% wet weight/d) and wild Arctic ciscoes in the Endicott–Prudhoe Bay area in 1988 (0.67%/d) were almost identical, and similar to growth achieved by this species over similar periods in 1985 and 1986. Wild Arctic ciscoes exhibited particularly high growth rates in 1987, a year characterized by an extended period of high water temperature late in the season (Gallaway et al. 1991). The marginal growth of enclosed broad whitefish was not reflective of growth achieved by wild fish (Table 5). The conditions at the enclosure sites were obviously unfavorable for this species.

Bioenergetics of Fish Growth

A complete bioenergetic assessment is probably the only defensible way to quantify the combined effects of prey abundance, temperature, and salinity on fish growth. Unfortunately, the information required for such an assessment is not available for the fish species reared in the enclosures.

Brett (1979) summarized most of the available data on the effects that ration size, temperature, and salinity have on fish growth rates. Although there is very little data available for arctic species, the data for temperate species showed that the growth rates resulting from a particular ration size

TABLE 5.—Comparison of growth rates for 1988 enclosure study to growth rates calculated with fyke net length-frequency data for similar-size fish of each species (Gallaway et al. 1991). Growth parameters were derived from June–September length-frequency data for each year; however, growth rates were estimated for just the appropriate study periods. *K* is the von Bertalanffy growth coefficient.

Growth values	Fyke net data 1985	1986	1987	1988	Enclosure data
Arctic cisco					
Parameters					
$K \times 1{,}000$	0.8613	11.07	0.9603	25.69	
Initial length (mm)	133.8	142.9	113.9	94.3	
Maximum length (mm)	775.8	237.1	961.1	179.4	
Rates (%/d)					
Endicott (29 Jul–10 Sep)	0.73	0.55	1.14	0.67	0.63
Niakuk (23 Aug–9 Sep)	0.66	0.44	0.98	0.43	0.08
Broad whitefish					
Parameters					
$K \times 1{,}000$	15.23	14.02	5.631	9.788	
Initial length (mm)	117.3	106.5	107.3	103.2	
Maximum length (mm)	194.2	206.2	298.0	243.4	
Rates (%/d)					
Endicott (29 Jul–10 Sep)	0.71	0.92	1.21	1.13	0.08
Niakuk (30 Jul–29 Aug)	0.85	1.10	1.41	1.33	−0.01

could vary substantially over the range of temperatures and salinities observed in the enclosures. Most fish species studied showed a gradual increase in growth rate with increases in temperature to a peak, followed by an increasingly rapid decline in the growth rate as temperatures approached lethal levels (Brett et al. 1969; McCormick et al. 1972; Elliot 1976). The magnitude of the effect of temperature on growth rate increased with ration size. Therefore, the effect of temperature on Endicott Arctic cisco growth was probably greatest in the last 3 weeks, when prey abundances were fairly high and temperatures were low (1.8–3.5°C).

The effect of salinity on growth of freshwater and anadromous fish is limited to a few studies (Saunders and Henderson 1969; Otto 1971). These studies indicated that the effect of low salinities (0–15‰) on growth was relatively minor, but growth rates could be substantially reduced as salinities approached lethal levels. Fechhelm et al. (1991) found that Arctic ciscoes were fairly resilient to large temperature and salinity changes for salinities up to 25‰, but mortality rates increased substantially as salinities were increased from 25 to 32‰. The threshold at which temperature and salinity changes affect broad whitefish would probably be lower than that measured for Arctic cisco (R. G. Fechhelm, LGL Ecological Research Associates, personal communication). Although the mean weekly salinities in the enclosures were consistently less than these lethal levels, there were several days when high salinity could have had a significant impact on fish growth. Sustained high salinities of less than 48 h may have occurred but gone undetected because sampling was conducted every 2 d.

Discussion

In terms of growth, broad whitefish fared poorly at both study sites and exhibited much higher mortality rates at the exposed Niakuk site than at the Endicott site. While I cannot rule out the possibility that the broad whitefish cannot grow in this type of experimental enclosure, it is also possible that the major factor affecting broad whitefish growth was the location of the enclosures outside the "normal" rearing area for this species. Data from the 1988 fyke net sampling in nearby areas appear to support this latter possibility (B. J. Gallaway, LGL Ecological Research Associates, personal communication). At the Niakuk site, 1988 netting effort yielded a total of only 38 fish (0.5% of the total catch). At the fyke net station located at the artificial Duck II Island near the Endicott enclosure site, 492 broad whitefish (7.3% of the total catch) were taken. In contrast, sampling at a station along the shoreline at Heald Point yielded 1,984 (29.3% of the total catch) broad whitefish. Sampling along the shoreline at Point Brower on the east side of the delta yielded 701 fish (10.3% of the total catch). The growth rate of wild broad whitefish, which were mostly taken in more nearshore habitats, was over 16 times that of the enclosed fish at the selected sites.

Arctic ciscoes appeared to have grown better at the Endicott than at the Niakuk site, but problems associated with the Arctic cisco enclosure at Niakuk (e.g., only 3 weeks of data, possible errors in fish identification due to the use of dye dots) removed the possibility of any defensible explanation for the observed difference in the Arctic cisco growth rates at each site. One of the most promising findings of the study was that Arctic ciscoes in the Endicott enclosures and wild fish were estimated to have grown at about the same rate in 1988. The similarity in these growth rates does not prove that conditions in the enclosures were identical to those encountered by wild fish. The results simply show that fish can grow in these enclosures and did so at a rate similar to the only available estimate for the average growth of similar-size Arctic ciscoes in the Endicott–Prudhoe Bay area during the same rearing period.

Another significant finding of the study was the large difference in the ability of Arctic cisco and broad whitefish to feed and grow in the same environment. Arctic ciscoes grew at seven times the rate of broad whitefish and appeared to be six times more "efficient" at consuming the prey items in the enclosures. The results from the Endicott enclosures suggested that the major determinants of growth for each fish species were water temperature and the biomass of key prey species less than 6 mm in body length. These results were consistent with both the fish bioenergetics discussed above and the physical limitations of 120–160-mm fish to capture and eat different size-classes of prey. Wankowski and Thorpe (1979) investigated the role of food particle size in the growth of juvenile Atlantic salmon. Their equations relating body size to the size of prey consumed suggested that the optimum prey size for 150-mm Atlantic salmon would be 3–4 mm. Given the relatively small differences in mouth size for Arctic cisco (12 mm), broad whitefish (9 mm), and Atlantic salmon *Salmo salar* (10 mm) of the same fork length (150 mm), we would expect that the optimum prey size would be less than 6 mm for the 120–160-mm Arctic cisco and broad whitefish reared in the enclosures.

The main objectives of the 1988 enclosure study were to quantify the key determinants of juvenile Arctic cisco and broad whitefish growth in the Endicott area, and to test some of the field and laboratory techniques required to determine whether causeway development could have a significant effect on fish growth. These objectives were partially fulfilled. The enclosure studies provided a good opportunity to quantify in situ growth rates that a species could achieve given a specific range of temperatures, salinities, and prey abundances. However, the relationship between prey abundance and fish growth can be confounded by the daily fluctuations in water temperature and salinity. These factors will probably continue to cloud our understanding of fish growth in the study area until the individual effects of temperature and salinity on fish growth are assessed in a laboratory where the other factors can be held constant.

Future fish growth research in the study area should focus on two areas (1) identification of the areas, prey types, and prey abundances required to maintain broad whitefish growth rates at levels similar to those estimated from the fyke net catches, and (2) quantification of the effects of temperature and salinity on fish growth rates. Given the information from this and previous studies, suitable locations for future broad whitefish studies would be along the coast close to the Sagavanirktok River delta, where salinities are generally lower and there is likely to be a high abundance of their primary prey species, chironomids (Knutzen et al. 1986). Laboratory studies in which fish are fed fixed rations at different temperatures and salinities could also be conducted near the Endicott site. Information from these types of enclosure and laboratory studies could be combined to define critical rearing habitats for juvenile Arctic ciscoes and broad whitefish, and to determine whether the areas most affected by causeway development are important for fish growth.

Acknowledgments

I would especially like to thank Christopher Haney for his major contributions to the field work and data analysis portions of this project and Benny J. Gallaway for his editorial assistance and guidance throughout this study. I would also like to thank Dave Schmidt, Bob Fechhelm, Larry Dugan, Steve Klimowski, and Bob Rodrigues for assistance in field work and logistical support. Stephen Jewett and Kathy Omura at the Institute of Marine Science, University of Alaska Fairbanks, analyzed the drop net and stomach samples, and Stephen Jewett provided comments on a draft of this manuscript. The 1988 anadromous fish project was funded through Standard Alaska

Production Company, now BP Exploration (Alaska), Inc.

References

Brett, J. R. 1979. Environmental factors and growth. Pages 599–675 *in* W. S. Hoar, D. J. Randall, and J. R. Brett, editors. Fish physiology, volume 8. Academic Press, New York.

Brett, J. R., J. E. Shelbourn, and C. T. Shoop. 1969. Growth rate and body composition of fingerling sockeye salmon, (*Oncorhynchus nerka*) in relation to temperature and ration size. Journal of the Fisheries Research Board of Canada 26:2363–2394.

Craig, P. C., W. B. Griffiths, L. Haldorson, and H. McEldery. 1985. Distributional patterns of fishes in an Alaska arctic lagoon. Polar Biology 4:9–18.

Elliott, J. M. 1976. The energetics of feeding, metabolism and growth of brown trout (*Salmo trutta* L.) in relation to body weight, water temperatures and ration size. Journal of Animal Ecology 45:561–580.

English, K. K. 1983. Predator–prey relationships for juvenile chinook salmon, (*Oncorhynchus tshawytscha*) feeding on zooplankton in "in situ" enclosures. Canadian Journal of Fisheries and Aquatic Sciences 40:287–297.

Fechhelm, R. G., J. D. Fechhelm, C. J. Herlugson, and D. K. Beaubien. 1991. Bioassay tests of acute temperature and salinity shock to Arctic cisco. American Fisheries Society Symposium 11:109–118.

Fechhelm, R. G., W. H. Neill, and B. J. Gallaway. 1983. Temperature preference of juvenile Arctic cisco (*Coregonus autumnalis*) from the Alaskan Beaufort Sea. Biological Papers of the University of Alaska 21:24–38.

Gallaway, B. J., W. J. Gazey, J. M. Colonell, A. W. Niederoda, and C. J. Herlugson. 1991. The Endicott Development Project—preliminary assessment of impacts from the first major offshore oil development in the Alaskan arctic. American Fisheries Society Symposium 11:42–80.

Griffiths, W. B., and B. J. Gallaway. 1983. Prudhoe Bay Waterflood Project, fish monitoring program. Report of LGL Alaska Research Associates to Woodward-Clyde Consultants and U.S. Army Corps of Engineers, Anchorage, Alaska.

Ivlev, V. S. 1961. Experimental ecology of the feeding fishes. Yale University Press, New Haven, Connecticut.

Knutzen, J. A., M. S. Brancato, and S. C. Jewett. 1986. Fish food habits (stomachs). *In* 1986 Endicott Environmental Monitoring Program, Part 4. Report of Envirosphere Company to Standard Alaska Production Company, Anchorage, Alaska.

Marquardt, D. W. 1963. An algorithm for least squares estimation of non-linear parameters. SIAM (Society for Industrial and Applied Mathematics) Journal 11:431–441.

McCormick, J. H., K. E. F. Hokanson, and B. R. Jones. 1972. Effects of temperature on growth and survival of young brook trout (*Salmo gairdneri*) Richardson. Canadian Journal of Zoology 47:631–663.

Moulton, L. L., and seven co-authors. 1986. 1984 central Beaufort Sea fish study. Pages 1-1 to 6-8 *in* Prudhoe Bay Waterflood Project Environmental Monitoring Program, 1984, volume 2. Final report of Woodward-Clyde Consultants, Entrix, Inc., and LGL Ecological Research Associates to U.S. Army Corps of Engineers, Anchorage, Alaska.

Neill, W., R. G. Fechhelm, B. J. Gallaway, J. Bryan, and S. Anderson. 1983. Modeling movement and distribution of young Arctic cisco (*Coregonus autumnalis*) relative to temperature salinity regimes of the Beaufort Sea near the waterflood causeway, Prudhoe Bay, Alaska. Biological Papers of the University of Alaska. 21:39–61.

Neill, W. J., and B. J. Gallaway. 1989. "Noise" in the distributional responses of fish to environment: an exercise in deterministic modeling motivated by the Beaufort Sea experience. Biological Papers of the University of Alaska 24:123–130.

Otto, R. G. 1971. Effects of salinity of the survival and growth of pre-smolt coho salmon (*Oncorhynchus kisutch*). Journal of the Fisheries Research Board of Canada 28:343–349.

Parrack, M. L. 1979. Aspects of brown shrimp, (*Penaeus aztecus*) growth in the northern Gulf of Mexico. U.S. National Marine Fisheries Service Fishery Bulletin 76:827–836.

Ross, B. D. 1988. Causeways in the Alaskan Beaufort Sea. U.S. Environmental Protection Agency Report, EPA 910/0-88-218, Anchorage, Alaska.

Saunders, R. L., and E. B. Henderson. 1969. Survival and growth of Atlantic salmon parr in relation to salinity and diet. Fisheries Research Board of Canada Technical Report 147.

USACE (U.S. Army Corps of Engineers). 1984. Endicott Development Project final environmental impact statement. USACE, Anchorage, Alaska.

Wankowski, J. W. J., and J. E. Thorpe. 1979. The role of food particle size in the growth of juvenile Atlantic salmon. Journal of Fish Biology 14:351–370.

American Fisheries Society Symposium 11:132–144, 1991

Movement of Young-of-the-Year Arctic Ciscoes across the Beaufort Sea Coast, 1985–1988

DAVID R. SCHMIDT

Dames & Moore
5761 Silverado Way, Suite P, Anchorage, Alaska 99518, USA

WILLIAM B. GRIFFITHS

LGL Limited, environmental research associates
9768 Second Street, Sidney, British Columbia V8L 3Y8, Canada

DEBRA K. BEAUBIEN AND CHRISTOPHER J. HERLUGSON

BP Exploration (Alaska), Inc.
900 East Benson Boulevard, Anchorage, Alaska 99519, USA

Abstract.—The hypothesized migration of young-of-the-year Arctic ciscoes *Coregonus autumnalis* from Mackenzie River drainages to the coastal areas of the central Alaskan Beaufort Sea has been documented by several researchers. Studies conducted in the Prudhoe Bay area of Alaska and along the Yukon Territory coast of Canada in response to potential oil and gas development indicated that migrations during the open-water seasons of 1985–1987 were large and strongly associated with winds having a predominant easterly component. These winds generally increase the westerly flow of nearshore water, which carries young Arctic ciscoes into Alaskan waters. In 1988, winds during the open-water season differed from those of the preceding 3 years in that they were either variable or had a strong westerly component. During that season, several scientific studies were conducted along the Alaskan–Canadian Beaufort Sea coast. Catches in 1988 along the Yukon Territory coast were considerably lower those those documented in 1986. This pattern was also evident in the Prudhoe Bay area, where fewer than 500 young-of-the-year Arctic ciscoes were caught, and in the Harrison Bay area west of Prudhoe Bay, where fewer than 20 were caught. Catches of young-of-the-year were also low (<1,000 fish) at sites near Kaktovik, Alaska, east of Prudhoe Bay. Evidence is presented to suggest that at least some young-of-the-year Arctic ciscoes may migrate farther offshore than previously thought. This wind-driven migration and potential offshore migration routes suggest that the westward movements of young-of-the-year Arctic ciscoes have not been altered by the presence of gravel-fill, breached causeways in the nearshore Beaufort Sea. These conclusions may also guide management and mitigation of the effects of nearshore oil and gas developments in the Beaufort Sea.

Arctic cisco *Coregonus autumnalis* is one of the more abundant species of anadromous fish found along the Alaskan and Canadian Beaufort Sea coast (Griffiths et al. 1975, 1977; Kendel et al. 1975; Griffiths and Gallaway 1982; Lawrence et al. 1984; Moulton et al. 1986; Glass et al. 1987). The fish are a major component in both commercial and subsistence fisheries along the Alaskan Beaufort Sea coast (Craig 1989a, 1989b) and constitute a portion of the subsistence catches in the Canadian Beaufort Sea (Nelson et al. 1987). Consequently, there are concerns that oil and gas development in the nearshore Beaufort Sea may influence abundance or migration of this species.

In recent years, several solid-fill, breached causeways have been constructed to service oil and gas developments in the coastal zone near Prudhoe Bay, Alaska. The two of primary concern are the West Dock and Endicott causeways. The West Dock Causeway, located on the western perimeter of Prudhoe Bay, was constructed to service barge traffic and to support a large seawater treatment plant. The Endicott Causeway, which extends into the Sagavanirktok River delta, east of Prudhoe Bay, was constructed to provide road access and pipeline support for two offshore oil production facilities.

Although these causeways are breached, concern exists among various regulatory agencies that these structures may impede the nearshore movements of several species of fish including Arctic cisco. This species occurs in generally high abundances along the central Beaufort Sea coast and uses the nearshore zone for feeding, migration to and from natal streams, and access to overwintering areas in larger river deltas.

The life history and migratory patterns of Arctic cisco in the Alaskan Beaufort Sea have been only recently described. Documentation of large numbers of this species in this area, including the

Colville River commercial fishery, led early investigators to suggest that these fish probably spawned in the Colville River drainage (Craig and Mann 1974). Several initial surveys conducted in this drainage, however, failed to locate any spawning or ripe Arctic ciscoes, although numerous juveniles were reported (Kogl 1972; Kogl and Schell 1974; McElderry and Craig 1981).

Recent studies have also failed to find spawning Arctic ciscoes in other Alaskan North Slope rivers. The Sagavanirktok River delta has been investigated, and no evidence of an Arctic cisco spawning population was found, although young-of-the-year from the previous summer, as well as adult fish, were located in several isolated pools under the ice (McCart et al. 1972; Adams and Cannon 1987; Schmidt et al. 1989). Subsequent dissection of adult female Arctic ciscoes failed to show any evidence of spawning activity, that is, flaccid ovarian tissue or retained eggs (Schmidt et al. 1989). Subsistence fishermen from the Point Barrow region suggested that Arctic ciscoes spawned in the Meade or other larger rivers flowing into Dease Inlet and Admiralty Bay. A survey in these areas in August 1988 found large populations of least cisco *C. sardinella*, as well as broad whitefish *C. nasus* and humpback whitefish *C. pidschian*, but only three adult Arctic ciscoes were caught, and none was in spawning condition (L. Moulton, Environmental Science and Engineering, Inc., personal communication).

Gallaway et al. (1983) suggested that Alaskan Arctic ciscoes originate from the Mackenzie River and are carried passively as young-of-the-year fish by westerly currents that flow along the Alaskan Beaufort Sea coast during the open-water season. These fish then become associated with the delta areas of one or several of the larger drainages (e.g., Colville and Sagavanirktok rivers) and remain in Alaskan waters, overwintering in the delta areas and dispersing to feed along the coast each summer, until they reach the age of first spawning (between 7 and 9 years). These fish then return to the Mackenzie River to spawn.

Viewed in this context, the data from earlier research appear to support the idea that the Arctic ciscoes in Alaskan waters probably belong to populations from the Mackenzie River tributaries known to support spawning populations. These tributaries include the Peel, Arctic Red, Great Bear, Mountain, and Liard rivers (Hatfield et al. 1972; Stein et al. 1973; O'Neil et al. 1982; McLeod and O'Neil 1983; R. Dillinger, LGL Ecological Research Associates, personal communication). In these rivers, mature fish typically spawn in late fall; the young hatch in early spring and probably descend passively, similar to other whitefish species, into the lower river by late spring. Upon reaching the outer Mackenzie River delta, they then become entrained in nearshore currents and are subject to transport.

The dispersal patterns of young-of-the-year Arctic ciscoes out of the Mackenzie River and into the nearshore zone are unknown. Fraker et al. (1979) estimated that the proportions of the total discharge through major Mackenzie River distributaries were 38% into Shallow Bay on the west side of the delta, 25% into northern Mackenzie Bay, and 28% into Kugmallit Bay on the east side of the delta. If the migration of young Arctic ciscoes at this stage is passive, the initial dispersal patterns into the coastal zone might resemble the discharge patterns (B. J. Gallaway, LGL Ecological Research Associates, personal communication).

The nearshore currents of the Beaufort Sea are wind driven. Typically these currents move with or slightly to the right of the wind at about 3% of the wind speed (J. Colonell, Environmental Science and Engineering, Inc., personal communication). Although wind patterns vary considerably across the Beaufort Sea coast, Gallaway (personal communication) in a summary of a number of oceanographic investigations, showed that currents in the vicinity of the Mackenzie River delta are characterized by a persistent divergence in flow: one of the currents flows northwest in the direction of Herschel Island and the other flows to the northeast along the Tuktoyaktuk Peninsula. Under the influence of the predominant northeast winds, the nearshore currents along the Yukon and Alaskan Beaufort Sea coast flow to the west. In years when winds with a westerly component dominate, the westward-flowing current does not become established.

From the above observations, it has been suggested that a portion of the young-of-the-year out-migrants from the Mackenzie River could be subjected to a westerly flow and carried into Alaskan waters. Fawcett et al. (1986) and Cannon et al. (1987) documented this movement during the open-water season of 1985. Fechhelm and Fissel (1988) presented convincing evidence that movement into Alaskan waters is a function of the percentage of east winds occurring during a given year.

In 1988, several studies were conducted along the Beaufort Sea coast at sites extending from

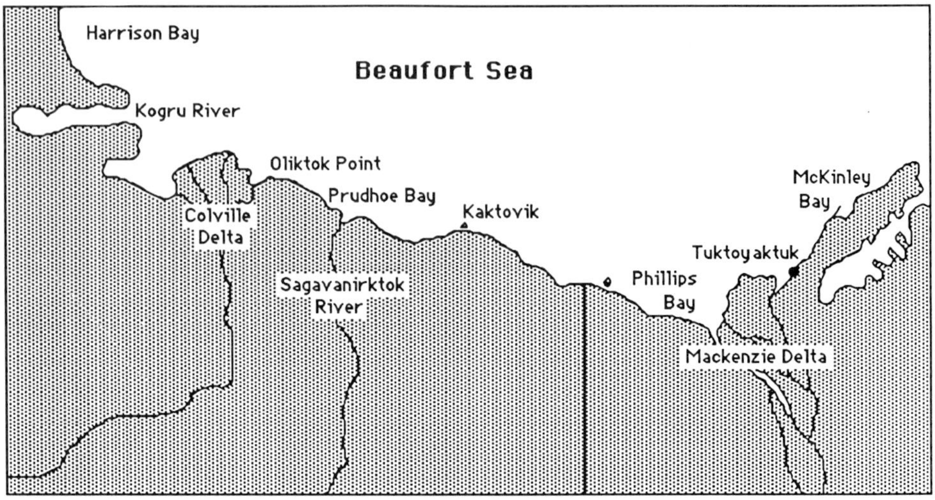

FIGURE 1.—The Beaufort Sea coast from Harrison Bay, Alaska, to McKinley Bay, Northwest Territories.

Harrison Bay, Alaska, to McKinley Bay, Northwest Territories. This extensive array of study sites provided a unique opportunity to document sequential movement of young-of-the-year Arctic ciscoes along the Beaufort Sea coast from the Mackenzie River delta into Alaskan waters. This paper presents a broad-scale look at young-of-the-year Arctic cisco migration by combining the results of the 1988 studies with those of previous years.

Methods

Fish sampling.—The fish data were collected during the open-water seasons of 1985–1988. The 1985–1987 data were collected as part of the Endicott Environmental Monitoring Program and were centered in the Prudhoe Bay area near the Endicott Causeway in the Sagavanirktok River delta (Cannon et al. 1987; Glass et al. 1987; Reub et al. 1988). In addition, during 1985, sites in Kaktovik, Alaska, (farther to the east) and sites at Oliktok Point on the eastern shore of Harrison Bay were sampled. In 1987, the Oliktok Point sites were surveyed again.

In 1988, study sites extended from the Kogru River in western Harrison Bay, Alaska, to McKinley Bay in the Northwest Territories (Figure 1). This array of sites provided extensive coverage that allowed a broad view of movement and migration along the Beaufort Sea coast. Identical fyke nets were used to sample fish at each site during all years (described in Cannon et al. 1987).

The total numbers of Arctic ciscoes less than 100 mm in length caught every 24 h at individual net sites in a given region were pooled by year. The regions and years analyzed were Harrison Bay–Oliktok Point 1985, 1987, 1988; Prudhoe Bay 1985, 1986, 1987, 1988; Kaktovik 1985, 1988; Phillips–McKinley Bay 1985, 1986, 1988. The catch data, while not strictly comparable due to the nature of passive sampling gear, provided a relative measure of the strength of the migration.

Wind data.—Meterological data from the Deadhorse, Alaska, Federal Aviation Administration station near Prudhoe Bay and from a satellite weather station at King Point, Yukon Territory, were used to determine the coastal wind patterns across the Beaufort Sea coast in 1988 and to correlate migration patterns with wind direction. In addition, wind direction and speed data collected during the Endicott Environmental Monitoring Program for the years 1985–1987 were analyzed (Hummer 1987a, 1987b, 1987c). The frequency of daily readings varied from 8 readings at King Point to 24 readings at Deadhorse. The data are presented as progressive vector diagrams, which were constructed by plotting hourly wind direction and speed vectors for the entire season and placing them successively end to end.

Results

The trends in winds for the open-water seasons of 1985–1987 are presented in progressive vector diagrams (Figure 2). For the 3 years in question, the diagrams showed a fairly consistent pattern of winds with a dominant easterly component. The

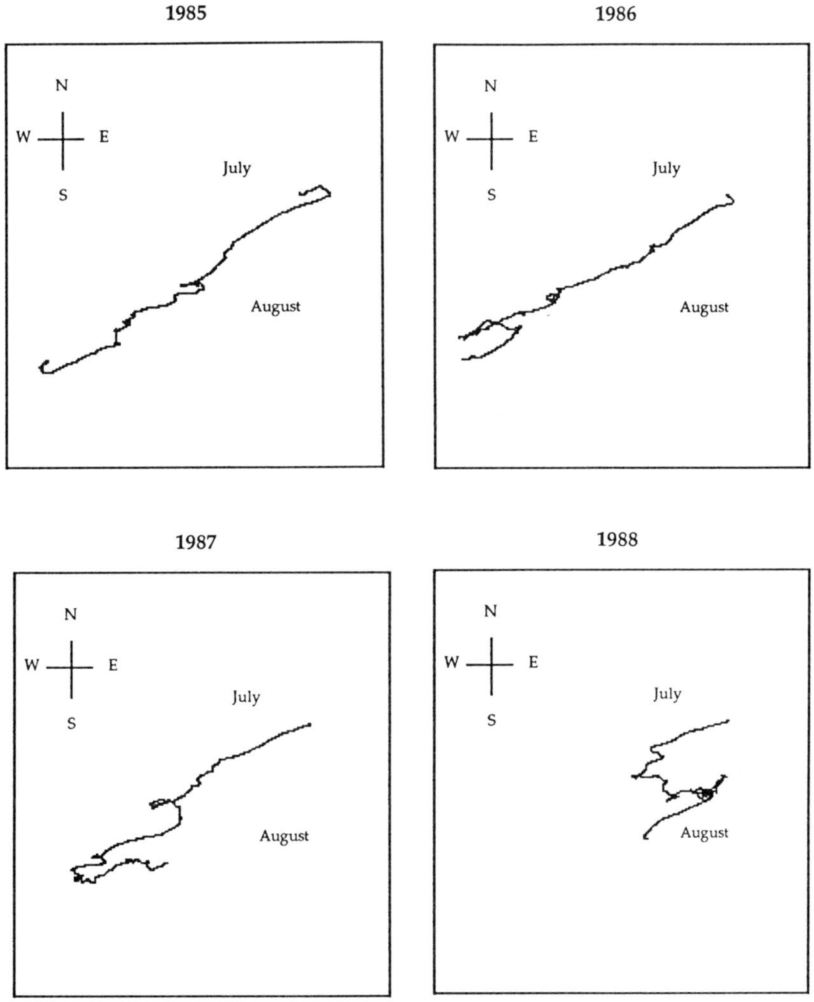

FIGURE 2.—Progressive vector diagrams for winds in the Prudhoe Bay area, 1985–1988.

two exceptions to this occurred during late August of 1986 and 1987, when there were periods of predominantly west winds.

A more detailed analysis of the hourly plots of wind direction and speed for July and August, the months of greatest migratory activity, is presented in Figures 3 and 4. The results from 1985 showed predominantly easterly winds throughout the season. A similar pattern existed for 1986 and 1987, again with the exception of late August each year, when persistent west winds occurred.

During July and August of 1988, the wind pattern in the Prudhoe Bay area differed from the previous 3 years (Figures 2, 4). The wind direction for the first half of July was similar to that of previous years, but during the latter half of the month winds with a strong westerly component predominated, resulting in a roughly balanced wind direction pattern for July. Data for August 1988 also showed a more balanced pattern in wind directions (Figure 4). The most important feature of the 1988 data was that neither month showed a persistent pattern of winds with an easterly component.

In 1988, wind data were also collected at King Point, Yukon Territory (Figure 5). The results showed that the winds had a predominantly west and/or north component. To determine the relative contribution of the various wind directions in terms of wind speed at King Point during the 1988 open-water season, wind speed was plotted against direction (Figure 5). The results showed

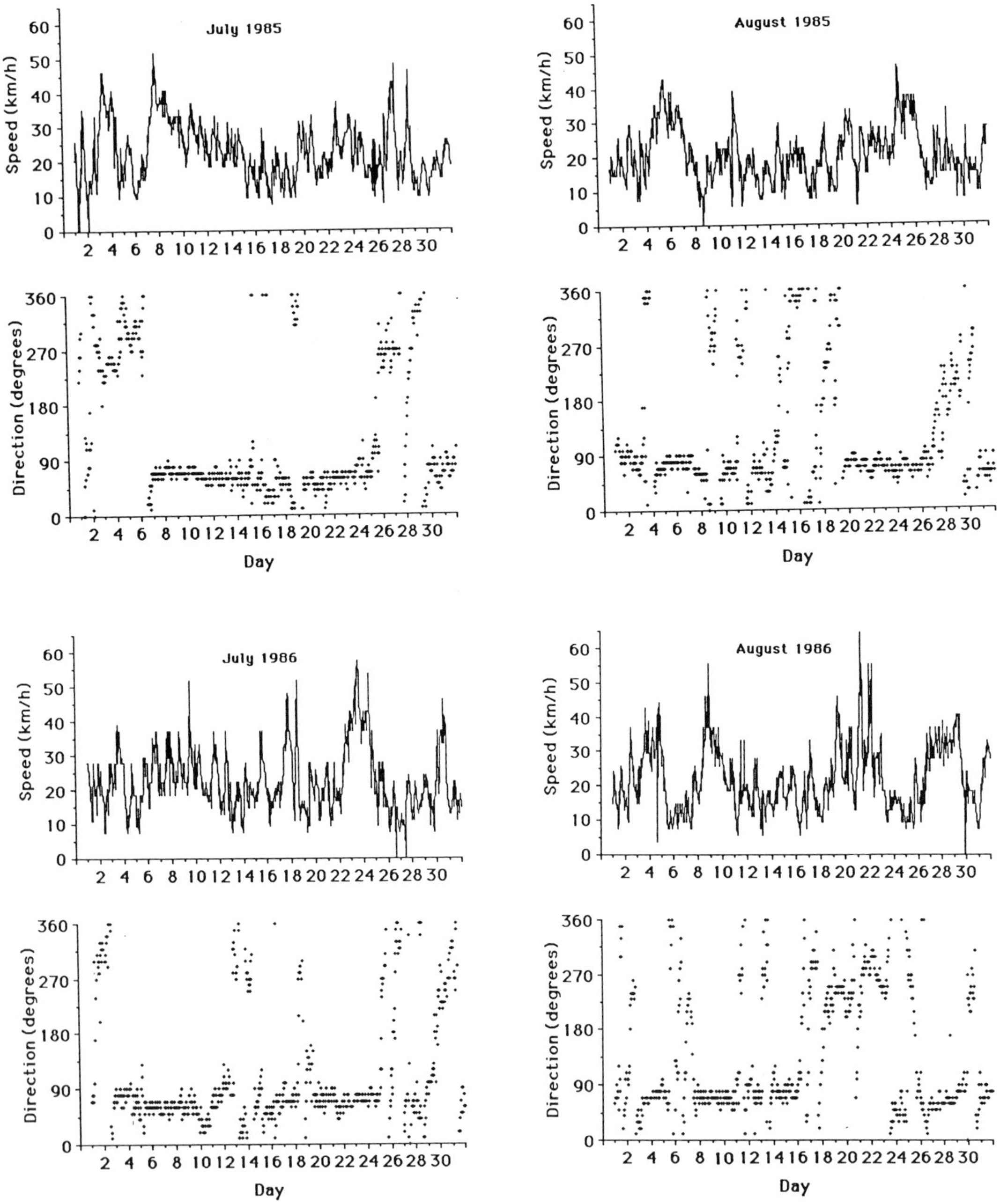

FIGURE 3.—Hourly plots of wind speed and direction for July and August 1985 and 1986 at the Deadhorse, Alaska, Federal Aviation Administration weather station.

that winds from nearly all directions occurred during the summer, but those with a westerly or northerly component (≥180°) generally had a higher speed. This was particularly true of winds out of the southwest (~225°) and out of the northwest (~315°). This wind regime would tend to produce a net easterly flow of the water mass along the Yukon coast.

In 1985–1987, there were relatively strong catches of young-of-the-year Arctic ciscoes in the Prudhoe Bay area (Figure 6). These fish typically began to appear in mid-August and extended into

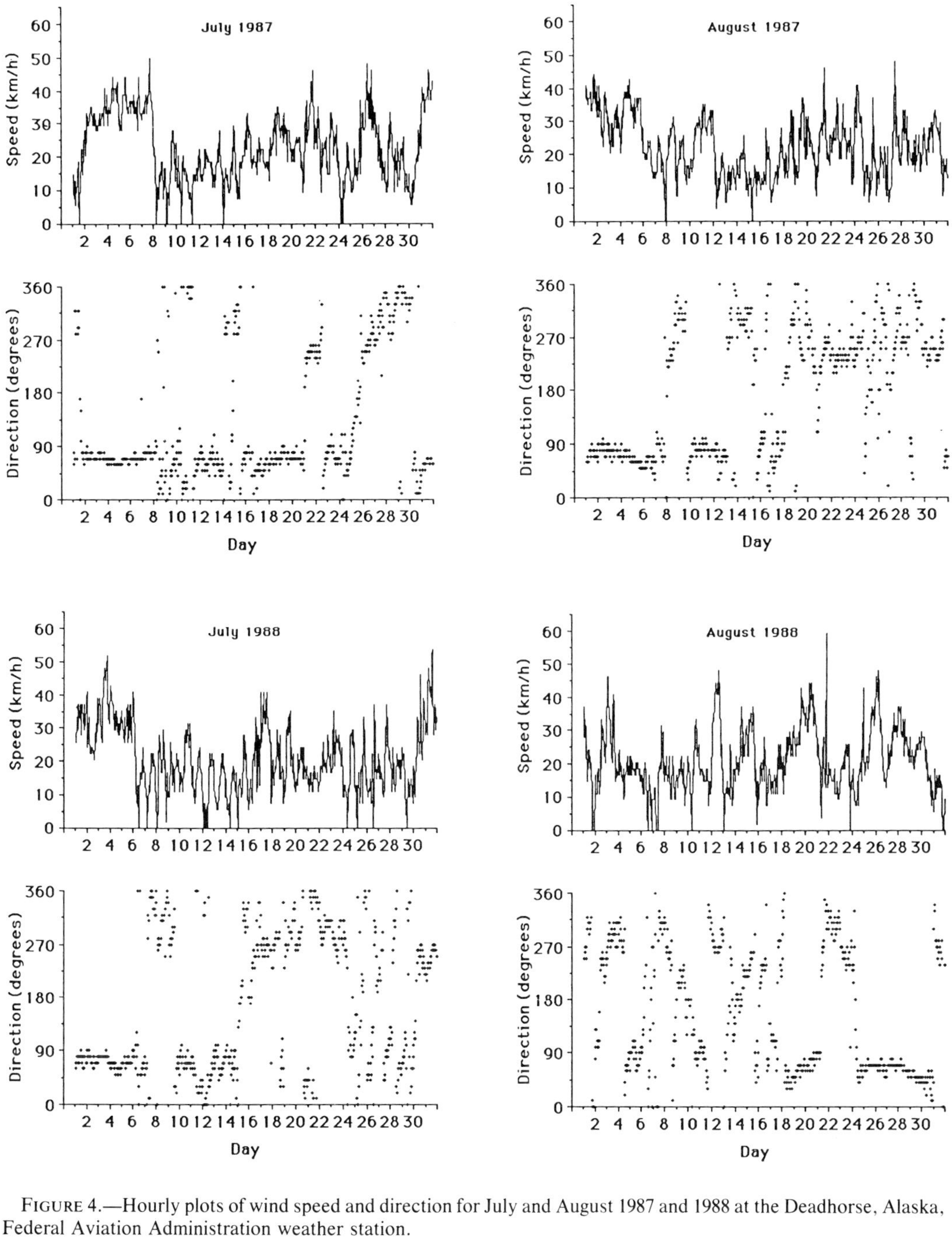

FIGURE 4.—Hourly plots of wind speed and direction for July and August 1987 and 1988 at the Deadhorse, Alaska, Federal Aviation Administration weather station.

September, suggesting that the major portion of the migration from Canada occurred in July and early August, periods typified by winds with an easterly component. In 1988, the movement of Arctic ciscoes into the same area was reduced, and as reported above, winds along the Yukon and Alaskan Beaufort Sea coast during July and early August of 1988 were predominantly from the west and northwest.

During the 1985–1988 period, while fish catches were continually monitored in the Prudhoe Bay area, limited surveys were also conducted at

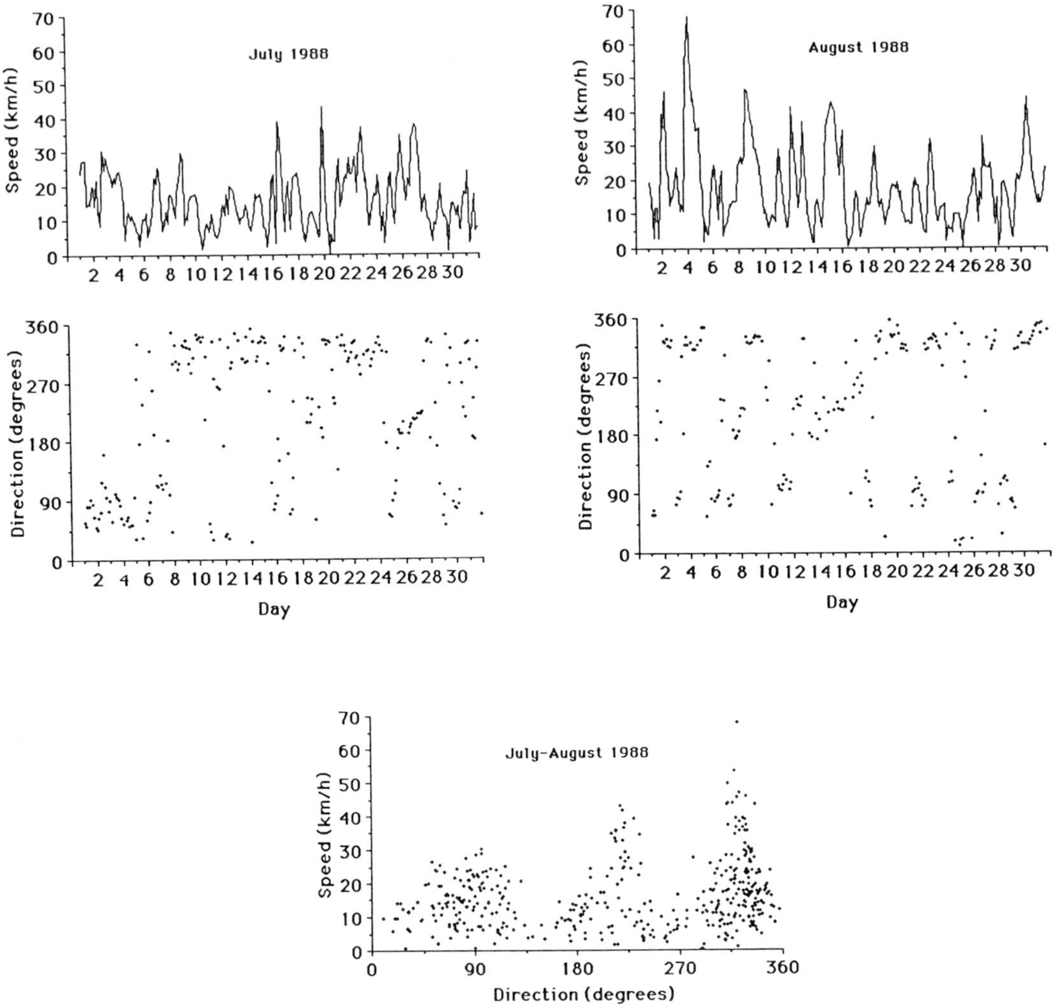

FIGURE 5.—Wind speed, direction, and speed as a function of direction for July and August 1988 at King Point, Yukon Territory. Readings were taken eight times per day.

Kaktovik, to the east of Prudhoe Bay, in 1985 (Nelson et al. 1987) and 1988 (D. Fruge, U.S. Fish and Wildlife Service, personal communication). Results indicated that Arctic cisco movement into the Kaktovik area in 1985 was almost as strong as that into the Prudhoe Bay area (Figure 6). In 1988, preliminary indications suggest that the numbers of young-of-the-year Arctic cisco were substantially lower than those in 1985 in the Kaktovik area. In 1985, 33,340 small Arctic ciscoes (<100 mm) were caught in three double fyke nets (6 traps), and in 1988, fewer than 1,000 fish of the same size were caught in seven double fyke nets (14 traps).

In the Harrison Bay area, to the west of Prudhoe Bay, similar studies were conducted during 1985, 1987, and 1988 (Cannon et al. 1987; Reub et al. 1988; LGL et al. 1989). In 1985 and 1987, the movement of Arctic ciscoes into the region was strong, even though the timing was different in each year (Figure 6). During 1988, however, very few Arctic ciscoes were captured in the Harrison Bay area (Figure 6). These results are consistent with those from corresponding years from the other locations along the Alaskan–Canadian Beaufort Sea coast, as discussed above.

The movement of young-of-the-year Arctic ciscoes is believed to start in the Mackenzie River drainages and follow the Yukon coast (Gallaway et al. 1983). In 1985, Bond and Erickson (1987)

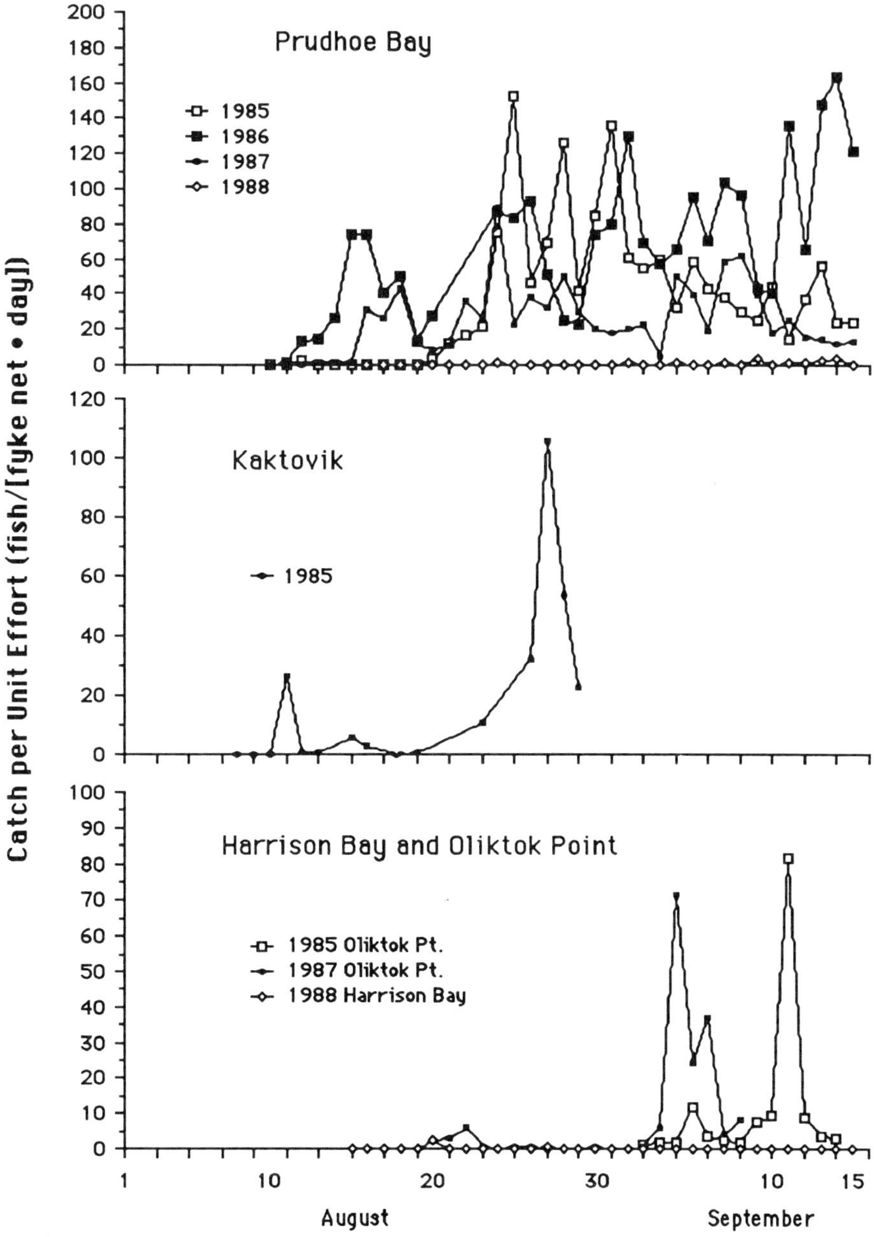

FIGURE 6.—Catch per unit effort (fish/[fyke net · d]) of young-of-the-year Arctic ciscoes recruited into the Prudhoe Bay, Kaktovik, and Harrison Bay–Oliktok Point regions, 1985–1988. Data are from Cannon et al. (1987), Glass et al. (1987), Reub et al. (1988), and this study.

conducted a summer fisheries study at Phillips Bay on the Yukon Territory coast. During this study, gill nets and seines were used, and although direct comparisons to the Alaskan fyke net studies are not possible, some trends were notable. They documented the presence of young-of-the-year Arctic ciscoes along the Yukon coast between 12 July and 21 August.

Previous estimations of the rate of migration suggested that the mid-July 1985 catches of young-of-the-year Arctic ciscoes made by Bond and Erickson (1987) probably represented migrants destined for Alaskan waters. Gallaway et al. (1983) suggested that average summer wind conditions would produce a passive drift of small Arctic ciscoes from the Mackenzie River to the Colville River, a distance of approximately 650 km, in an average of 35 d. Young-of-the-year fish began to appear in the Prudhoe Bay area by mid-August and in the Harrison Bay area 5–6 d later (Figure 6). This timing suggested that they should have been in the Phillips Bay area in early to mid-July, and this was the case (Bond and Erickson 1987).

The Phillips Bay study was repeated in 1986 during the same general time period, but sampling was conducted with fyke nets similar in design to those used in the Alaskan studies (Bond and Erickson 1989). Approximately 26,000 young-of-the-year Arctic ciscoes less than 75 mm in length were caught. As in 1985, there was a similar timing of the first pulse of young-of-the-year fish into the Phillips Bay area; in 1985 they appeared between 12 and 13 July and in 1986 between 13 and 19 July.

During the 1988 Phillips Bay study, logistical problems prevented sampling prior to 17 July; thus, the initial pulse of young-of-the-year Arctic ciscoes may have been missed. However, catches of these Arctic ciscoes were low during the entire 1988 survey period; approximately 3,200 small Arctic ciscoes (<75 mm) were captured with two double-sided fyke nets (four traps). This contrasts to a catch of 26,000 similar-sized Arctic ciscoes in 1986 with three traps (Bond and Erickson 1989). The catch of young-of-the-year Arctic ciscoes in the Phillips Bay area in 1988, a season with predominantly west to northwest winds, was much lower than in 1986, a year of predominantly east winds.

The catch of Phillips Bay young-of-the-year Arctic ciscoes is compared to that of McKinley Bay in Figure 7. During the early sampling dates catch-per-unit-effort rates were similar between the two sites. In late August, however, and continuing into September, there was a large influx of young-of-the-year Arctic ciscoes into the McKinley Bay area. It is not possible at this time to say whether 1988 was an anomalous year, or if areas to the east of the Mackenzie River delta typically accommodate more migrants than areas west of the delta. What is certain, however, is that the numbers of young-of-the-year Arctic ciscoes along the Yukon coast were lower in 1988 than in previous years.

Discussion

The timing and magnitude of the young-of-the-year Arctic cisco catch from 1985 through 1987 suggested that there was substantial movement across the Beaufort Sea coast at least as far west as the eastern perimeter of Harrison Bay during each of these years, and into the Colville River delta in 1985. Wind data from the same period indicated that these years were dominated by east winds, at least during mid-July to late August, a time when these fish have been hypothesized to travel from the Mackenzie River to the central Alaskan Beaufort Sea region.

The results of the broad-scale sampling program in 1988 indicated that the numbers of young-of-the-year Arctic ciscoes were substantially lower in the sampling region than in any of the previous 3 years. Data from Prudhoe Bay and along the Yukon Territory coast suggested that the wind patterns were also substantially different. In the Prudhoe Bay region, winds reversed direction from the east toward the west during the last half of July, while August was a month of roughly balanced east and west winds. Along the Yukon Territory coast, west winds tended to dominate and were stronger than east winds.

These results are consistent with the hypothesis of Fechhelm and Fissel (1988) that the movement pattern of young-of-the-year Arctic ciscoes is influenced by wind conditions. Years of strong movement corresponded to years of persistent winds with strong easterly components; conversely, low-catch years were associated with years dominated by winds with persistent westerly components. This suggests that young-of-the-year Arctic cisco movements into Alaskan waters will be quite variable, ranging from strong to weak. A comparison of the strength of young-of-the-year Arctic cisco movement into Alaskan waters over the previous 8 years is provided by the age data collected during the 1988 fish monitoring program along the Alaskan Beaufort Sea coast (LGL et al. 1989). Each age-group corresponded to a particular recruitment year; thus, age-0 fish corresponded to the 1988 recruitment year, while age-9 fish corresponded to the 1979 recruitment year (Figure 8). Because Arctic ciscoes reach spawning age and return to the Mackenzie drainage between ages 7 and 9, only those fish up to age 7 (i.e., years 1981–1988) were considered in this discussion. It appeared that strong migrations into Alaskan waters occurred in 1983, 1985, 1986, and 1987 and weak ones in 1981,

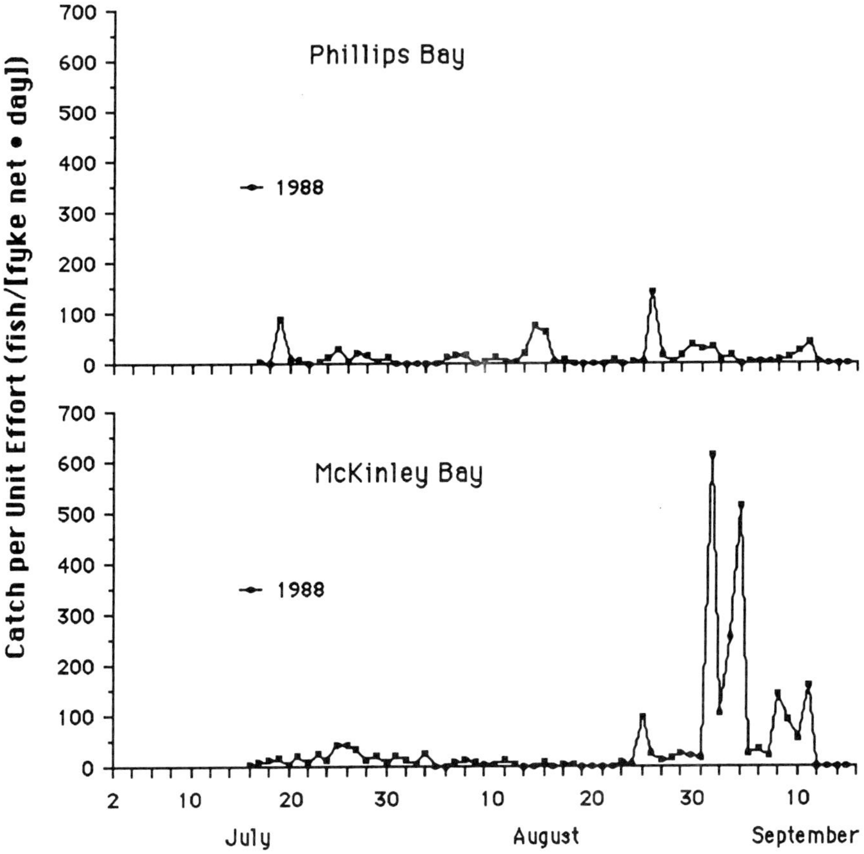

FIGURE 7.—Catch per unit effort (fish/[fyke net · d]) of young-of-the-year Arctic ciscoes in Phillips Bay, Yukon Territory, and McKinley Bay, Northwest Territories during 1988.

1982, 1984, and 1988 (Figure 8). The fish catch data from the Waterflood and Endicott monitoring programs for the corresponding years showed an identical pattern for young-of-the-year Arctic cisco movements (R. G. Fechhelm, LGL Ecological Research Associates, personal communication). Data from a permanent weather station at Barter Island, Alaska, showed that the strongest and most persistent easterly winds occurred in 1983, 1985, 1986, and 1987 during July and early August, whereas 1981 and 1984 were predominantly west-wind years, and 1982 and 1988 were weak east-wind years (Fechhelm, personal communication).

The age data from the 1988 fish monitoring program in the Canadian Beaufort Sea showed a somewhat different pattern for young-of-the-year Arctic cisco movement into Canadian waters (Figure 8). The age-class structure suggested that unlike the Alaskan situation, there was consistently strong recruitment in all years between 1978 and 1988, with the exception of 1981 and possibly 1984. The most notable difference between the Alaskan and Canadian data was that the distinct gaps in the age-class structure in the Alaskan sample were not present in the Canadian data. The number of recruits in Canadian waters from year to year is not known. Consequently, it is difficult to speculate on the relative contribution of these recruits to the migratory stock destined for Alaskan waters.

Present theory concerning the migration of young-of-the-year Arctic ciscoes suggests that the majority of these fish move along the coast within a narrow, nearshore band of warm, brackish water (Gallaway et al. 1983; Craig 1989a). At first glance, this hypothesis seems to be supported by the catch data; however, it should be noted that nearly all survey nets were situated in this nearshore water mass, so that any alternative migra-

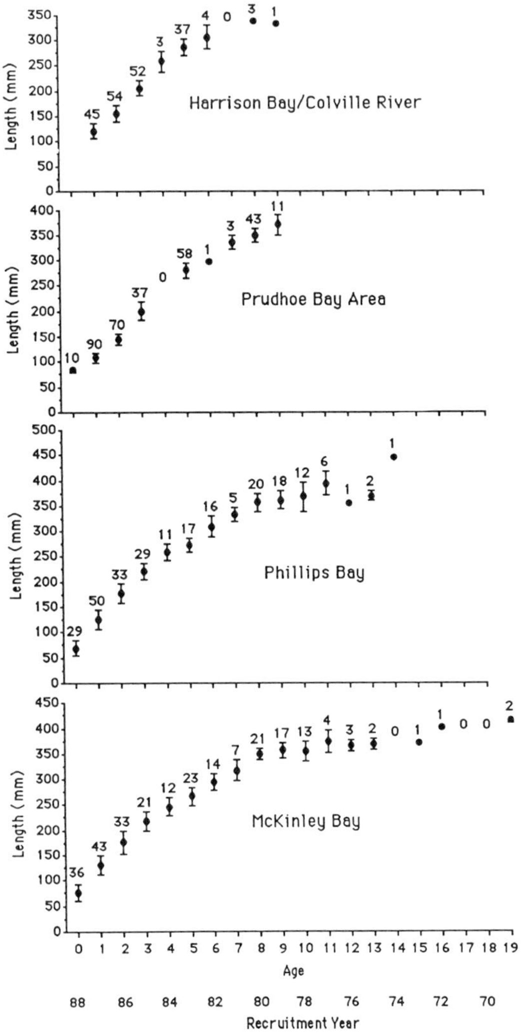

FIGURE 8.—Age–length relationships for Arctic ciscoes caught in the four study regions during the 1988 open-water season. The horizontal axis shows both age and recruitment year. The plots represent means and ranges, and numbers indicate sample sizes at each age.

tion routes would not have been detected. A recent winter survey in the Sagavanirktok River delta documented young-of-the-year Arctic ciscoes in salinities near 30‰, and 1988 laboratory studies conducted in the Prudhoe Bay area on acute temperature and salinity shock to age-1 and -2 Arctic ciscoes indicated that these smaller fish can tolerate higher salinities and lower temperatures than typically occur in the nearshore water mass (Schmidt et al. 1989; Fechhelm, personal communication).

In 1988, in an effort to document potential offshore migration, an offshore net site was established on Narwhal Island, approximately 14.5 km seaward of the Sagavanirktok River delta in waters that are typically influenced by marine conditions. No young-of-the-year Arctic ciscoes were documented at that site. In addition, tow-net sampling was conducted along two shore-to-offshore transects in the Prudhoe Bay region between 17 August and 3 September 1988 (Houghton and Whitmus 1988). Only 25 young-of-the-year Arctic ciscoes were collected out of a total of 33,944 fish. These results were not unexpected, given the poor catch of young-of-the-year Arctic ciscoes in fyke nets that season.

Glass et al. (1987) documented the migration of young-of-the-year Arctic ciscoes westward past the Prudhoe Bay area. Their net catches did not document the first large pulse of migrants in sequential order from east to west, as would be expected given a strict nearshore migration route. Instead, the migrants traversed approximately 24 km of coast around two causeways and 12 net sites in Prudhoe Bay in 1 d (LGL et al. 1989). To date, the evidence suggesting a persistent offshore migration route is minimal. Clearly, a more directed study focusing on this migratory corridor will have to be conducted before any definitive statements can be made.

This type of offshore migration of small ciscoes is not without precedent. Studies in the Yukon River delta have documented juvenile Bering cisco *C. laurettae* and least cisco (congeneric with Arctic cisco) carried within the river plume 20–30 km seaward to the delta front (Martin et al. 1989). Similarly, the migration of young-of-the-year Arctic ciscoes may encompass a broader corridor of the Beaufort Sea coast extending into offshore waters. If so, the role of nearshore causeways in inhibiting movements may have been overestimated.

The evidence strongly suggests that migration of young-of-the-year Arctic ciscoes is largely passive and wind-aided, and that the persistence of wind direction during the open-water season profoundly influences the magnitude of recruitment into Alaskan waters. Other evidence, although less extensive and more speculative, indicates that migration routes may not necessarily be confined to a narrow zone in the nearshore area. This mechanism indicates a migration success closely linked to water mass movement and independent of physical features (e.g., barrier islands, headlands, causeways), except to the extent that the features influence the water mass. Oceanographic evidence, to date, suggests that the influence of physical features such as causeways on the nearshore water mass is limited to an area extending several kilometers downstream of the structure under extreme conditions. These conclusions could improve management of fisheries during nearshore oil and gas development in the Beaufort Sea in that both nearshore and offshore water masses could be monitored, and appropriate assumptions concerning Arctic cisco recruitment could be made.

Acknowledgements

The authors would like to thank BP Exploration (Alaska) Inc. for funding the study and for providing logistic support during the field season. Doug Fruge, David Wiswar, and Larry Dugan of the U.S. Fish and Wildlife Service provided useful data on small Arctic ciscoes in the Pokok Lagoon area; William Gazey, Karl English, and Mike Blazecka assisted with data analysis; Robert Fechhelm and Lynn Maritzen supplied wind data; Randall Howard, Rob Dillinger, and Robert Pollard helped with editing; and Michele Ulofoshio helped with the manuscript production.

References

Adams, B. A., and T. C. Cannon. 1987. Overwintering study. 1985 Final report for the Endicott Environmental Monitoring Program, volume 7. Report of Envirosphere Company to U.S. Army Corps of Engineers, Anchorage, Alaska.

Bond, W. A., and R. N. Erickson. 1987. Fishery data from Phillips Bay, Yukon, 1985. Canadian Data Report of Fisheries and Aquatic Sciences 635.

Bond, W. A., and R. N. Erickson. 1989. Summer studies of the nearshore fish community at Phillips Bay, Beaufort Sea coast, Yukon. Canadian Data Report of Fisheries and Aquatic Sciences 1676.

Cannon, T. C., B. A. Adams, and T. Nelson. 1987. Fish distribution and abundance. 1985 Final report for the Endicott Environmental Monitoring Program, volume 6. Report of Envirosphere Company to U.S. Army Corps of Engineers, Anchorage, Alaska.

Craig, P. C. 1989a. An introduction to anadromous fishes in the Alaskan Arctic. Biological Papers of the University of Alaska 24:27–54.

Craig, P. C. 1989b. Subsistence fisheries at coastal villages in the Alaskan Arctic, 1970–1986. Biological Papers of the University of Alaska 24:131–152.

Craig, P. C., and G. J. Mann. 1974. Life history and distribution of Arctic cisco *Coregonus autumnalis* along the Beaufort Sea coastline in Alaska and the Yukon Territory. Arctic Gas Biological Report Series 20(4), Calgary, Alberta.

Fawcett, M., L. Moulton, and T. Carpenter. 1986. Colville River fishes: 1985 biological report. Colville River fish study, 1985 annual report. Report of Entrix, Inc., to ARCO–Alaska and North Slope Borough, Barrow, Alaska.

Fechhelm, R. G., and D. B. Fissel. 1988. Wind-aided recruitment of Canadian Arctic cisco *Coregonus autumnalis* into Alaskan waters. Canadian Journal of Fisheries and Aquatic Sciences 45:906–910.

Fraker, M. A., C. D. Gordon, J. W. McDonald, J. K. B. Ford, and G. Cambers. 1979. White whales *Delphinapterus leucas* distribution and abundance and the relationship to physical and chemical characteristics of the Mackenzie estuary. Canada Fisheries and Marine Service Technical Report 863.

Gallaway, B. J., W. Griffiths, P. Craig, W. Gazey, and J. Helmericks. 1983. An assessment of the Colville River delta stock of Arctic cisco *Coregonus autumnalis*—migrants from Canada? Biological Papers of the University of Alaska 21:4–23.

Glass, D., C. Whitmus, and M. Prewitt. 1987. Fish distribution and abundance. Endicott Environmental Monitoring Survey, 1986, volume 4. Report of Envirosphere Company to U.S. Army Corp of Engineers, Anchorage, Alaska.

Griffiths, W. B., P. C. Craig, G. L. Walder, and G. J. Mann. 1975. Fisheries investigations in a coastal region of the Beaufort Sea (Nunaluk Lagoon, Y.T.). Arctic Gas Biological Report Series 34(2), Calgary, Alberta.

Griffiths, W. B., J. DenBeste, and P. Craig. 1977. Fisheries investigations in a coastal region of the Beaufort Sea (Kaktovik Lagoon, Barter Island, Alaska). Arctic Gas Biological Report Series 40, Calgary, Alberta.

Griffiths, W. B., and B. Gallaway. 1982. Prudhoe Bay Waterflood Project, fish monitoring program, volume 4. Report of LGL Alaska Research Associates to Woodward-Clyde Consultants and U.S. Army Corps of Engineers, Anchorage, Alaska.

Hatfield, C. T., J. N. Stein, M. R. Falk, C. S. Jessop, and D. N. Shepherd. 1972. Fish resources of the Mackenzie River Valley, volumes 1 and 2. Environment Canada, Fisheries Service, Interim Report, Winnipeg, Manitoba.

Houghton, J. P., and C. J. Whitmus. 1988. Shallow neritic fish of the central Beaufort Sea. Report of Dames & Moore to Standard Alaska Production Company, Anchorage, Alaska.

Hummer, P. G. 1987a. Meteorology. 1985 Final report for the Endicott Environmental Monitoring Program, volume 2. Report of Envirosphere Company to U.S. Army Corps of Engineers, Anchorage, Alaska.

Hummer, P. G. 1987b. Meteorology. Draft report for the 1986 Endicott Environmental Monitoring Program. Report of Envirosphere Company to U.S. Army Corps of Engineers, Anchorage, Alaska.

Hummer, P. G. 1987c. Meteorology. Draft report for the 1987 Endicott Environmental Monitoring Program. Report of Envirosphere Company to U.S. Army Corps of Engineers, Anchorage, Alaska.

Kendel, R. E., R. A. Johnston, U. Lobsiger, and M.

Kozak. 1975. Fishes of the Yukon coast. Canadian Department of the Environment, Beaufort Sea Technical Report 6, Victoria, British Columbia.

Kogl, D. 1972. Monitoring and evaluation of arctic waters with emphasis on the North Slope drainages: Colville River study. Alaska Department of Fish and Game, Annual Report 12, Fairbanks.

Kogl, D., and D. Schell. 1974. Colville River delta fisheries research. Pages 483–504 *in* Environmental studies of an arctic estuarine system. U.S. Environmental Protection Agency, Ecological Research Series EPA-660/3-75-026, Washington, D.C.

Lawrence, M., G. Lacho, and S. Davies. 1984. A survey of the coastal fishes of the southeastern Beaufort Sea. Canadian Technical Report of Fisheries and Aquatic Sciences 1220.

LGL (LGL Alaska Research Associates, Inc.), Hunter/ESE Environmental Services, Inc., and W. J. Gazey Research. 1989. The 1988 Endicott Development fish monitoring program—analysis of 1985–1987 data. Report to BP Exploration (Alaska), Inc., Anchorage, Alaska.

Martin, D. J., C. J. Whitmus, L. E. Hachmeister, E. C. Volk, and S. L. Schroder. 1989. Distribution and seasonal abundance of juvenile salmon and other fishes in the Yukon River delta. Pages 123–277 *in* Environmental assessment of the Alaskan continental shelf, final report of the principal investigators, volume 63. Bureau of Land Management and National Oceanographic and Atmospheric Administration, Outer Continental Shelf Environmental Assessment Program, Boulder Colorado.

McCart, P., P. Craig, and H. Bain. 1972. Report on fisheries investigations in the Sagavanirktok River and neighboring drainages. Report of Aquatic Environments, Ltd., to Alaska Pipeline Service Co., Bellevue, Washington.

McElderry, H., and P. C. Craig. 1981. A fish survey in the lower Colville River drainage with an analysis of spawning use by Arctic and least cisco. Appendix 2, pages 657–678 *in* P. Craig and L. Haldorson. Fish. Pages 384–678 *in* Beaufort Sea barrier island–lagoon ecological process studies: final report, Simpson Lagoon. Environmental assessment of the Alaskan continental shelf, final report of the principal investigators, volume 7. Bureau of Land Management and National Oceanographic and Atmospheric Administration, Outer Continental Shelf Environmental Assessment Program, Boulder, Colorado.

McLeod, C., and J. O'Neil. 1983. Major range extensions of anadromous salmonids and first record of chinook salmon in the Mackenzie River drainage. Canadian Journal of Zoology 61:2183–2184.

Moulton, L., and seven coauthors. 1986. 1984 Central Beaufort Sea fish study. Prudhoe Bay Waterflood Project Environmental Monitoring Program 1984, volume 2. Final report of Woodward-Clyde Consultants, Entrix, Inc., and LGL Ecological Research Associates, Inc., to U.S. Army Corps of Engineers, Anchorage, Alaska.

Nelson, T., T. C. Cannon, W. R. Olmsted, and K. C. Wiley. 1987. Surveys of domestic and commercial fisheries in the central and eastern Beaufort Sea. 1985 Final report of the Endicott Environmental Monitoring Program, volume 7. Report of Envirosphere Company to U.S. Army Corps of Engineers, Anchorage, Alaska.

O'Neil, J., C. McLeod, L. Noton, L. Hildebrand, and T. Clayton. 1982. Aquatic investigations of the Liard River, British Columbia and Northwest Territories, relative to proposed hydroelectric development at Site A. Report of RL&L Environmental Services, Ltd., to British Columbia Hydroelectric and Power Authority, Vancouver.

Reub, G. S., J. D. Durst, and D. Glass. 1988. Fish distribution and abundance. Endicott Environmental Monitoring Survey, 1987. Report of Envirosphere Company to U.S. Army Corp of Engineers, Anchorage, Alaska.

Schmidt, D. R., W. B. Griffiths, and L. R. Martin. 1989. Overwintering biology of anadromous fish in the Sagavanirktok River Delta, Alaska. Biological Papers of the University of Alaska 24:55–74.

Stein, J. N., C. S. Jessop, T. R. Porter, and K. T. J. Chang-Kue. 1973. Fish resources of the Mackenzie River valley. Canada Department of Fisheries and Oceans, Interim Report II, Winnipeg, Manitoba.

American Fisheries Society Symposium 11:145–156, 1991

Predictability in the Catch of Arctic Cisco in the Colville River, Alaska

LAWRENCE L. MOULTON
5460 NE Tolo Road, Bainbridge Island, Washington 98110, USA

L. JAY FIELD
11053 38th Avenue, NE, Seattle, Washington 98125, USA

RICHARD KOVALSKY
Nuiqsut, Alaska 99789, USA

Abstract.—Arctic ciscoes *Coregonus autumnalis* caught in the Colville River during the fall gill-net fishery constitute a substantial portion of the total fish harvest in the Alaskan Beaufort Sea coastal region. Construction of causeways to support oil industry development in the nearshore region of the Beaufort Sea caused concern that adverse effects on the Arctic cisco population would reduce subsistence harvests. The Nuiqsut village fishery has been monitored for 4 years (1985–1988), and detailed records on the commercial fishery have been maintained for over 20 years. During the 4 years of monitoring in the village, fishing effort was highest in the first year and steadily declined the following years. Catch rate in the fishery is linked to the recruitment of young-of-the-year Arctic ciscoes from the Mackenzie River in Canada. From 1985 to 1987, catch rates were high because of two strong year-classes that recruited into the region around 1980. Catch rates plummeted in 1988 as the strong year-classes moved out of the fishery to return to the Mackenzie River system to spawn. Strong recruitments of young-of-the-year Arctic ciscoes into the Alaskan Beaufort Sea appear to depend on favorable winds during the summer. These recruitments translate into high catch rates in the Colville River 6 years later. The success or failure of juvenile recruitments are easily monitored with fyke nets during the summer. Abundance indices of juvenile Arctic ciscoes developed from these fyke net catches are highly correlated with catch rates in the fall fishery up to 3 years later. Future catch rates in the commercial fishery are predicted based on patterns observed from abundance indices developed from fyke net data. Strong juvenile recruitments in 1985 and 1986 are predicted to increase catches by 1991.

Since about 1974, there has been a continuing concern that causeways built to support oil industry development in the coastal region of the Alaskan Beaufort Sea are changing the summer feeding habitat of anadromous fishes in the region (USACE 1980, 1984). Studies funded by oil industry and government agencies have been conducted in the coastal region since the mid-1970s to gain basic biological, distributional, and habitat use information needed to address these concerns (Bendock 1979; Craig and Haldorson 1981; Griffiths and Gallaway 1982; Critchlow 1983; Griffiths et al. 1983; Woodward-Clyde Consultants 1983; Moulton and Fawcett 1984; Moulton et al. 1986b; Envirosphere Company 1987).

The anadromous fishes that are the subject of these studies provide an important food resource for the Inupiat communities of the Alaskan Arctic Coastal Plain and have been fished for many generations. The Arctic cisco *Coregonus autumnalis* is a highly prized food resource that is harvested near Kaktovik in the late summer and in the Colville River delta after the ice forms in the fall. In addition to subsistence harvests, a commercial fishery has been operated since the early 1950s by a nonnative family that lives in the outer portion of the Colville River delta. Prior to 1985, there was little information on the subsistence harvests of anadromous fishes, although detailed information existed on the commercial fishery (Craig and Haldorson 1981; Gallaway et al. 1983; Craig 1987).

The concerns expressed over the possible effects of causeways on the anadromous fishes of the region, especially Arctic cisco, raised concerns among local people that their traditional fishery resource would, in turn, be affected. The local government for the Arctic Coastal Plain region, the North Slope Borough, requested that information be collected to assess the fisheries in the Colville River, which were considered to be most at risk. The study reported in this paper was begun in 1985 in response to that request. The initial year of investigation included a study of fish use of the Colville River delta region and evaluation of both summer and fall fisheries (Fawcett et al. 1986; Moulton et al. 1986a). Studies in the following years (1986–1988) focused on the fall fishery for Arctic cisco and least cisco *C. sardinella*.

The evaluation of fishery harvests of Arctic cisco and least cisco in the Colville region is based on the assumption that the Arctic ciscoes are derived from spawning stocks in the Mackenzie River and that the young fish recruit into the Colville region early in life, as described by Gallaway et al. (1983). The recruitment of age-0 Arctic ciscoes into the Colville River region is thought to be aided by westerly currents generated by the predominantly easterly winds in the Beaufort Sea region. The strength of the recruitment has been correlated to the percentage of easterly winds from June to September (Fechhelm and Fissel 1988). Arctic ciscoes are assumed to return to the Mackenzie River at maturity to spawn. The anadromous least cisco population is assumed to spawn and overwinter entirely in the Colville Delta and lower Colville River.

This study of the 1988 fishery constitutes the fourth consecutive year that the fishery in the delta was subjected to a total harvest estimate. The previous 3 years of survey (1985–1987) are summarized in Moulton and Field (1988). Additional information on the fall gill-net fishery in the Colville River was developed by George and Kovalsky (1986) and George and Nageak (1986).

The objectives of the 1988 survey were to (1) continue to obtain estimates of the total effort and catch for the fall fishery in the Colville River delta; this fishery targets Arctic cisco and includes subsistence harvest by the village of Nuiqsut and commercial harvests; (2) evaluate the harvest predictions made prior to the fishing season; and (3) evaluate methods for improving harvest predictions.

Methods

The study area included the Colville River from the Itkillik River downstream to Harrison Bay (Figure 1). The same four areas of concentrated fishing effort were studied in all 4 years: (1) the Outer Colville Delta, (2) the Upper Nigliq Channel near Nuiqsut, (3) the Nanuk area of the Nigliq Channel, and (4) the Nigliq Delta. The assessment and monitoring of the fall under-ice fishery based in Nuiqsut began in early to mid-October and continued through mid-November. Fishing began later in 1988 than in the previous 3 years because a heavy snowfall early in October made the ice unsafe.

During the main fishing season, village catches were sampled daily for species composition, number of fish caught, and fork length to the nearest millimeter. Fish were examined for tags, fin clips, and dye marks applied by other researchers in the region. Whenever catch data were collected, set duration, net length, and mesh size data were also recorded so that the catch per unit effort (CPUE) could be calculated for the net set. This method chronicled the start and end dates of fishing for each net, net location, net length, and mesh size. Except for a few cases, there was a virtually complete census of village fishing effort.

The commercial fisher provided data on daily catch and effort by individual net and weekly length frequencies by species for each mesh size from the commercial fishery in the Outer Colville Delta. Length frequencies were derived from random samples of 50 fish each Monday of the fishing season.

Effort was calculated in net-days by using the start and end dates for each individually tracked net. Effort data were corrected for the various net lengths and set durations in the fishery by standardizing net length to 18 m and set duration to 24 h. Net depth was considered a fixed parameter in the calculation of effort, because most nets were of comparable depth (1.8 or 2.4 m). Within the four main sampling areas, each net was identified and monitored throughout the entire time the net was deployed. The CPUE was estimated by treating each individual sample (usually the catch from one net on a given day) as an independent sample. The total effort for each mesh size in each area and the associated CPUE estimate were calculated for each 10-d interval during the fishing season, starting October 1. Estimated catches for each mesh size by 10-d interval were then calculated and summed to provide the estimates of total catch. In some cases, CPUE was not measured because of low fishing effort. In these instances, CPUE was estimated from adjacent time periods or by extrapolating from the relative catch rate of that mesh in nearby fishing areas. In the Outer Colville Delta, complete counts of total catch were obtained.

In both the village and commercial fisheries, 76-mm mesh nets were the preferred gear. Catch rate indices used for comparisons among areas and years and evaluation of changes in length distributions were based on 76-mm mesh.

From 1984 to 1988, the catch rate of one village fisher who consistently fishes in the Main Channel of the Outer Delta region was closely monitored

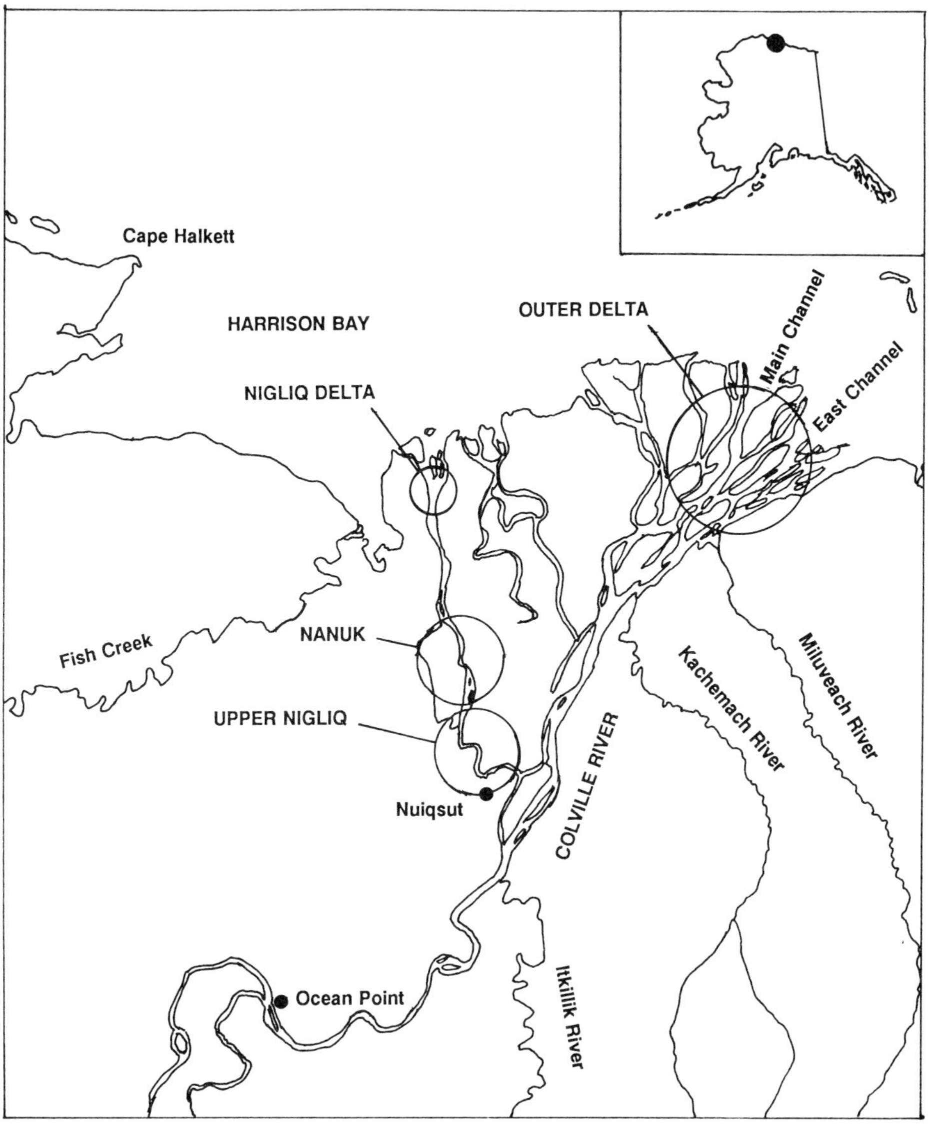

FIGURE 1.—Colville River, Alaska, study region showing major survey areas.

(data from 1984 and 1985 were provided by C. George, North Slope Borough Department of Wildlife Management). This fisher usually fished for 1 week in mid-to-late October, primarily with 76-mm mesh nets. Total counts were obtained from all sets. The catch rate index in this analysis was mean catch per day per 18 m of net for 76-mm mesh for the entire fishing period.

From 1982 to 1987, researchers obtained catch rates of Arctic cisco and least cisco from sampling with fyke nets during the summer, when these species forage in coastal waters (1982—Critchlow 1983 and Griffiths et al. 1983; 1983—Woodward-Clyde Consultants 1983 and BioSonics 1984; 1984—Moulton et al. 1986b; 1985—Envirosphere Company 1987; 1986—Glass et al. 1987; 1987—data tables in Reub et al. 1988). Three stations near Prudhoe Bay were sampled each summer (west base of West Dock, east base of West Dock, and West Beach). A fourth station (Heald Point) was sampled from 1982 to 1986. Catch rates were reported by length mode as revealed by analysis of length frequencies. These length-groups were termed "cohorts." For Arctic cisco, cohorts 1, 2,

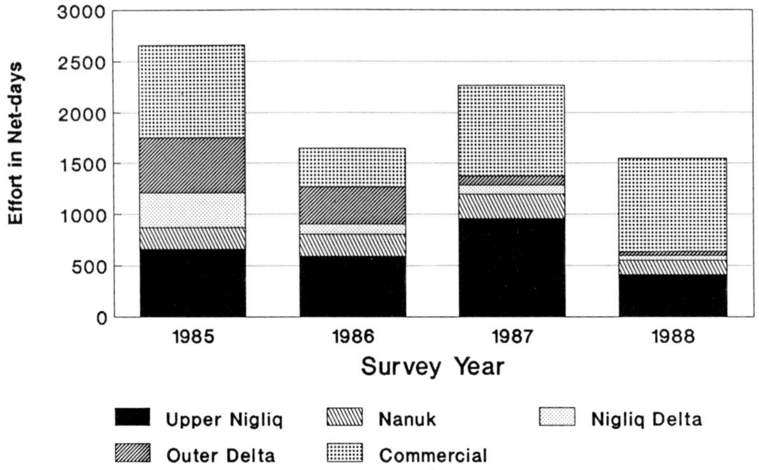

FIGURE 2.—Total effort in net-days by village fishing area and commercial fisheries in the Colville Delta, 1985–1988 (1 net-day = 18 m of net fished for 24 h).

and 3 basically represented ages 0–3 or 4, depending on the recent pattern of recruitment, and cohort 4 represented fish ages 4 or 5 and older. For least cisco, cohort 1 likely represented ages 1–4, and cohort 2 represented fish ages 5 and older. The catch rate index in this analysis was mean catch per day for the period fished between July 1 and September 1.

The catch rates observed in the commercial fishery were correlated with the other two abundance indices (i.e., village gill net and Prudhoe Bay fyke net catch rates) to compare the similarity of catch patterns over the period of record. For the comparison with fyke net data, analyses were conducted for Arctic cisco cohorts 2, 3, and 4 and least cisco cohort 2. Arctic ciscoes enter the fishery at age 6; therefore, cohort 2 was lagged 4 years relative to the commercial catch index; cohort 3 was lagged 3 years; and cohort 4 (which contained multiple age-groups) was run with both a 1-year lag and no lag. The data were not lagged for cohort-2 least ciscoes because this group was mostly large fish near harvestable size.

From 1985 to 1988, salinity was measured at net sites during the Colville River fall fishery where fishing was in progress. Measurements were taken with a YSI Model 33 salinometer at 1.0-m intervals from 1985 to 1987 and at 0.5-m intervals in 1988, starting from just under the ice to the bottom. In 1987, most of the measurements were taken near the mouth of the Main Channel of the Outer Colville Delta, whereas in 1985, 1986, and 1988, most of the measurements were from the Nigliq Channel between the village of Nuiqsut and the Nigliq Delta. Salinity and associated catch rate data were collected when and where fishing activity allowed, not through a predetermined design; thus, the information was temporally and spatially sporadic.

Results

Distribution of Fishing Effort

The number of Nuiqsut fishing groups (a family or group of families fishing cooperatively) participating in the under-ice fishery fluctuated from 30 in 1985 to 25 in 1986 and back up to 34 in 1987. In 1988, 22 fishing groups participated. Effort was highest in the four main survey areas shown in Figure 1. Additional, unquantified effort was expended on the main channel upstream from the village and at Fish Creek near the mouth of the Nigliq Channel, but villagers reported that these unmonitored harvests were targeting other species.

Village fishing effort was highest at the Upper Nigliq area (Figure 2) because of its proximity to town. Fishing effort in this area was greatest in late October to early November, then decreased by mid-November, when daylight shortened and the thickness of the ice interfered with operation of the gill nets (Figure 3).

The total effort (combined village and commercial effort) in the Colville Delta fall fishery decreased 32% from 1987 to 1988. The decrease in effort was caused by a 54% decrease in effort on

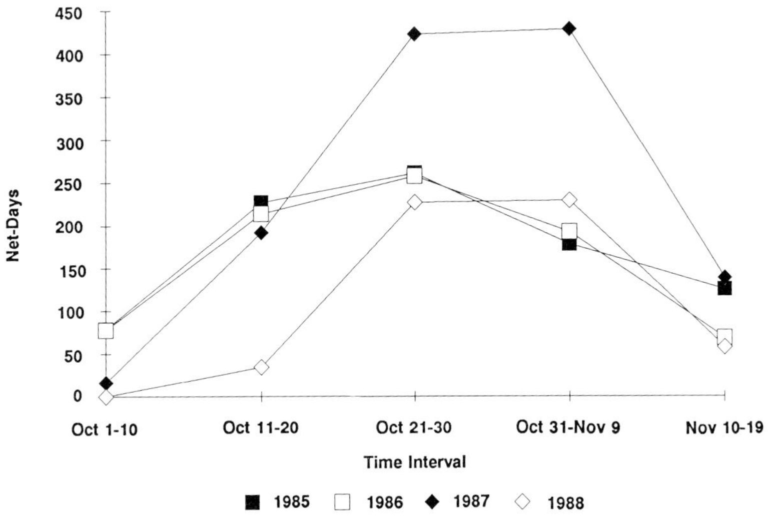

FIGURE 3.—Village fishing effort in the Upper Nigliq area by 10-d interval, 1985–1988 (1 net-day = 18 m of net fished for 24 h).

the Nigliq Channel between 1987 and 1988, and by low effort by Nuiqsut fishers in the Outer Colville Delta. The commercial fishery, which operates in the East and Main channels of the Outer Colville Delta, accounted for 34, 23, 39, and 59% of the total effort expended in the fall fishery from 1985 to 1988.

The reduction in the village effort from 1987 to 1988 resulted from (1) increased employment in the village because of two construction projects, (2) low early-season catch rates of Arctic cisco and preseason predictions that catches would be low, and (3) delayed ice formation followed by a rapid decrease to sub-zero (°F) temperatures. Effort by villagers in the Outer Colville Delta in 1988 continued to decrease and was only 12% of the effort recorded in 1985.

Variations in Salinity

In 1988, the salinity in both the Nigliq Channel and Outer Colville Delta remained low throughout the fishing season. From 1985 to 1987, salinity in the Nigliq Channel near the village reached 12–15‰ by the beginning of November. In 1988, however, salinity in this area remained near 0‰ through the season. Similarly, salinity reached over 20‰ in the Nigliq Delta in 1986 and at Nanuk in 1987, while in 1988, the highest salinity at Nigliq Delta was 6‰. Salinity also remained low in the Outer Delta, where the top 2 m of the water column below the ice remained below 1‰ for most of the season and bottom salinities did not exceed 13‰. In 1987, the freshwater layer under the ice was thin (<1 m) and salinity of much of the water column was in excess of 20‰ (Moulton and Field 1988).

Catch Composition

Arctic cisco, the target species, constituted over 90% of the total catch in the Nigliq Channel in 1988 but only 24% of the village catch in the Outer Colville Delta (Table 1). Least cisco was the dominant incidental species, followed by small broad whitefish *C. nasus* in the Nigliq Channel and humpback whitefish *C. pidschian* in both the Nigliq Channel and Outer Colville Delta. In 1988, least cisco dominated the catch in the Outer Colville Delta, representing over 50% of the catch in that area. From 1985 to 1987, least cisco never exceeded 40% of the catch. Humpback whitefish represented over 20% of the 1988 Outer Colville Delta catch and almost equaled the catch of Arctic cisco there.

In 1988, Bering cisco *C. laurettae*, a species that closely resembles Arctic cisco, represented a portion of the "Arctic cisco" catch. The species had been virtually absent in the commercial catches from the 1979–1980 season to 1986 (J. Helmericks, Colville Delta commercial fisher, personal communication). Village fishers do not distinguish between the two species. Bering ciscoes were identified in both the Nigliq Channel

TABLE 1.—Percent contribution by species to total village catch, as observed during interviews of Inupiat fishers, 1985–1988.

Species or statistic		Nigliq Channel				Outer Colville Delta			
Common name or total	Scientific name	1985	1986	1987	1988	1985	1986	1987	1988
Arctic cisco[a]	*Coregonus autumnalis*	69.5	95.9	71.8	90.6	62.1	74.6	75.1	23.6
Least cisco	*C. sardinella*	14.8	3.8	18.7	8.3	37.0	24.4	21.5	54.8
Broad whitefish	*C. nasus*	15.1	0.3	5.5	0.6	0.1	0.01		0.8
Humpback whitefish	*C. pidschian*	0.5	0.03	3.8	0.5	0.7	0.9	1.4	20.7
Rainbow smelt	*Osmerus mordax*	0.2	0.03	0.01		0.1	0.1	2.0	
Round whitefish	*Prosopium cylindraceum*		0.01						
Dolly Varden	*Salvelinus malma*			0.03			0.01		
Saffron cod	*Eleginus gracilis*			0.03					
Burbot	*Lota lota*			0.06	0.1				0.1
Fourhorn sculpin[b]	*Myoxocephalus quadricornis*								
Total observed catch (number of fish)		2,705	8,952	6,826	2,948	5,510	15,328	6,190	2,652

[a]Includes Bering cisco *C. laurettae*.
[b]Always present but not counted.

and Outer Delta fishing areas. In the Outer Delta, 39 Bering ciscoes and 376 Arctic ciscoes were identified in 10 sets that were carefully examined for species composition. Thus in 1988, Bering ciscoes represented about 10% of the Arctic–Bering cisco catch. A similar ratio was reported in the commercial fishery, where Bering ciscoes represented 10.2% of the Arctic–Bering cisco catch. All subsequent reference to Arctic cisco includes Bering cisco as well.

Comparative Catch Rates

The mean catch rates of both Arctic ciscoes and least ciscoes were higher in the Outer Colville Delta than in the Upper Nigliq area (Figure 4). In this example, the mean catch rates for each 10-d period from the commercial fishery in the East Channel of the Outer Delta were used as indices because they were the most complete record from the Outer Colville Delta over the four survey years. Within the Nigliq Channel, mean catch rates of Arctic cisco were highest near the Nigliq Delta and declined near the village (Upper Nigliq, Table 2). Least cisco mean catch rates showed the opposite trend; they were usually highest near the village and decreased downstream. Least cisco catch rates in all areas of the Nigliq Channel were well below those recorded in the Outer Delta (Table 2).

The highest mean catch rates for Arctic ciscoes during the 4-year survey period were recorded in 1986 in the Outer Delta and the Nigliq Delta (Table 2), although the catch rates reported by George and Kovalsky (1986) for village fishers on the Main Channel in 1985 were of similar magnitude. Arctic cisco catch rates were highest during the first two 10-d periods of 1986 in the Outer Colville Delta and remained high for the duration of the season (Figure 4). In 1988, catch rates approached those recorded in other years only during the first and last 10-d periods; the catch rates during the intermediate time periods were lower than in any of the previous 3 years. These catch rate patterns were interpreted as indicating that Arctic cisco abundance in the areas fished was lower during 1988 compared to the previous 3 years.

Estimated Total Catch

The total catch of Arctic cisco in the Colville region continued to decline in 1988 (Figure 5). Least cisco total catch increased in 1988 because of high catches in the commercial fishery. The majority of the village catch of Arctic cisco came from the Nigliq Channel, which annually provided 57, 56, and 78% of the village harvest from 1985 to 1987. In 1988, over 90% of the village catch was from the Nigliq Channel. As reflected by the relative catch rates and estimated effort, this high contribution resulted from intensive fishing near the village.

Size of Harvested Fish

The length frequency of Arctic ciscoes harvested in 76-mm mesh gill nets from all village and commercial harvest areas revealed two dominant size groups: 300–310 mm and 350–370 mm. The 76-mm mesh gill nets select for Arctic ciscoes between 320 and 340 mm (Moulton and Field

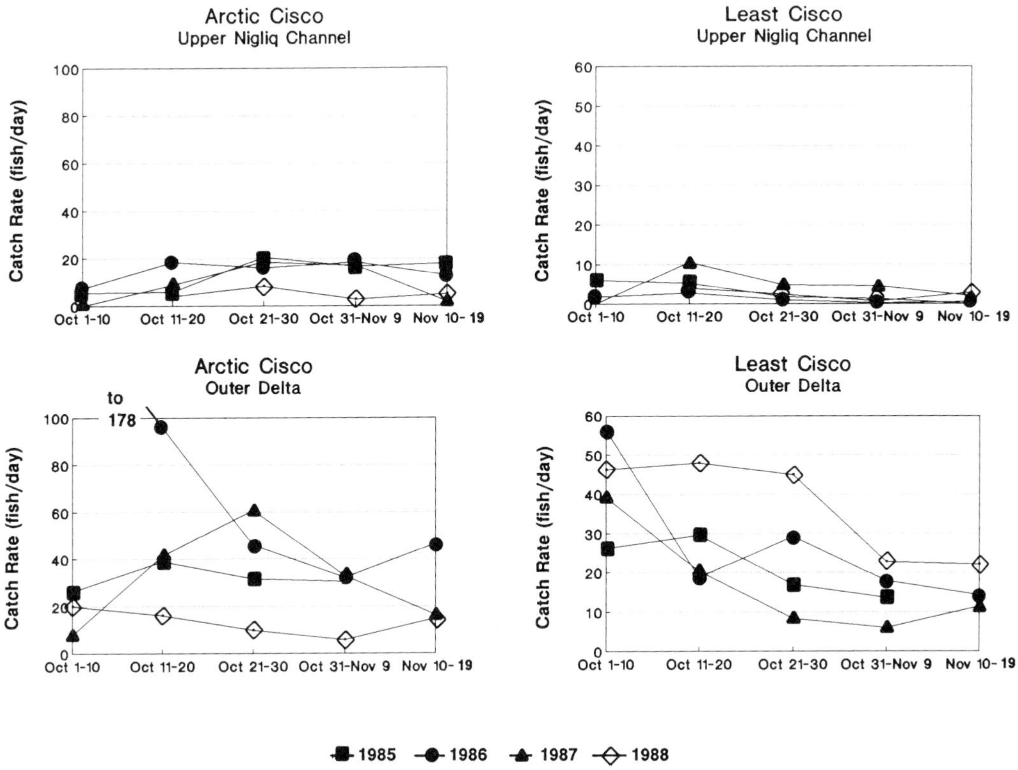

FIGURE 4.—Catch rates of Arctic cisco and least cisco in the Outer Colville Delta and Nigliq Channel by time interval, 1985–1988.

1988); thus, the frequency of fish in this size range in the population is likely even lower than that indicated in the length distribution. The large fish in the size distribution likely consisted of remnants of the abundant 1979 and 1980 year-classes, while the 300–310 mm fish were likely large members of the year-class that recruited into the region in 1983 (Woodward-Clyde Consultants 1983). The gap between 310 and 340 mm was created by the low recruitment in 1981 and 1982. The length frequency of Arctic ciscoes in the coastal region during the summer indicated that there was a group of fish from the 1983 recruitment that would become vulnerable to the 76-mm mesh gill nets in 1989 (Figure 6).

The length frequency of least ciscoes harvested in the 76-mm mesh nets in 1988 was similar to that observed in previous years (Moulton and Field

TABLE 2.—Mean catch rates of Arctic cisco and least cisco in 76-mm mesh gill nets in the Colville Delta fall fishery, 1985–1988 (fish/[d · 18 m of net]).

	Arctic cisco				Least cisco			
Fishing area and effort	1985	1986	1987	1988	1985	1986	1987	1988
Nigliq Channel (village effort)								
Upper Nigliq	12.1	17.1	11.5	6.1	3.6	1.8	5.5	1.8
Nanuk		27.9	43.0	5.1		0.9	2.8	0.5
Nigliq Delta		78.5	39.3	56.4		0.03	1.1	3.0
Outer Colville Delta								
Main Channel								
Village	76.1	62.0	47.6	19.3	47.4	18.3	15.4	57.9
Commercial	31.5	74.5			20.9	30.0		
East Channel								
Commercial	33.5	81.5	31.6	12.9	21.7	25.0	14.0	34.7

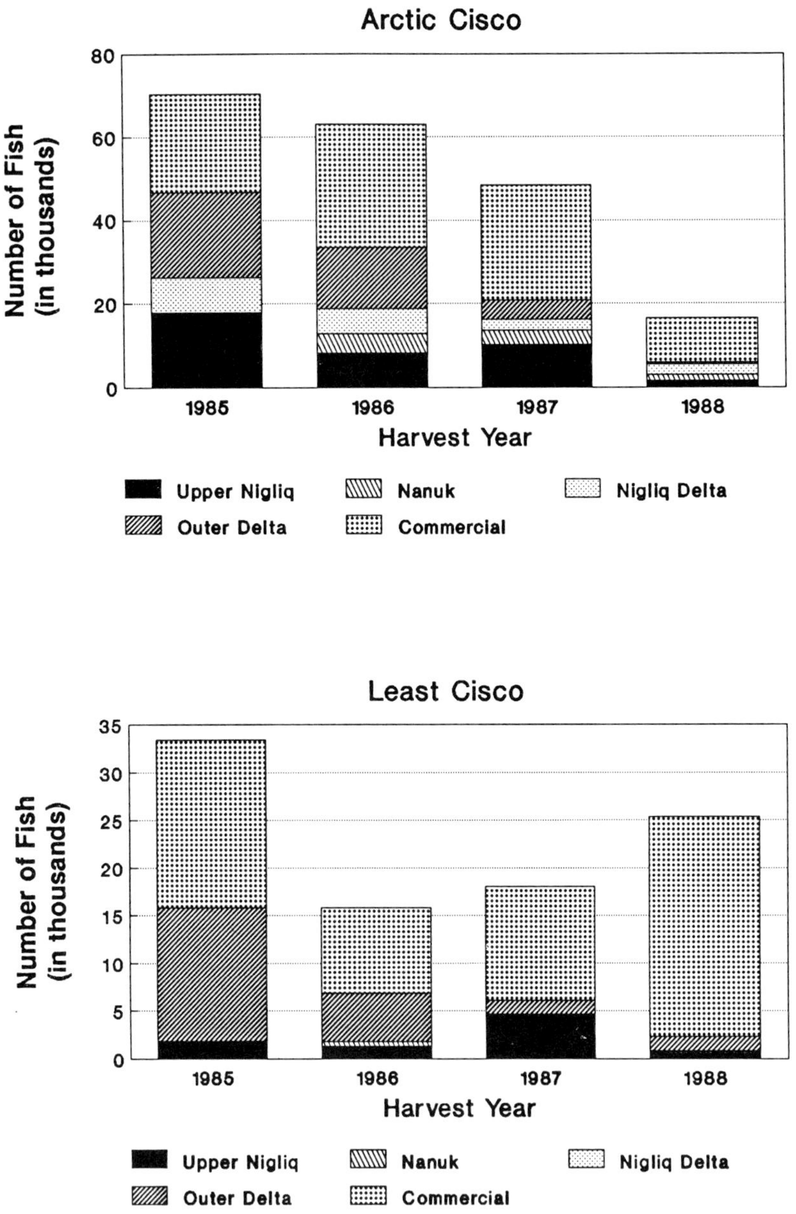

FIGURE 5.—Catch of Arctic cisco and least cisco by village fishing area and commercial fisheries in the Colville Delta, 1985–1988.

1988). The resulting distribution indicated little change in the length distribution of least ciscoes available for harvest over the four survey years. Mean catch rates for least cisco in 1988 in the Outer Delta were high (Figure 4). As will be discussed later, this pattern most likely indicated annual shifts in distribution of this species within the delta due to salinity conditions rather than changes in absolute abundance.

Evaluation of Catch Rates as Abundance Indices

The catch rates of Arctic cisco by village and commercial fishery gill nets in the Outer Colville Delta and by fyke nets in Prudhoe Bay were highly correlated (significant at 0.01 probability level; Table 3). The fyke net catch rate of cohort-3 Arctic ciscoes, with a 3-year lag, showed the

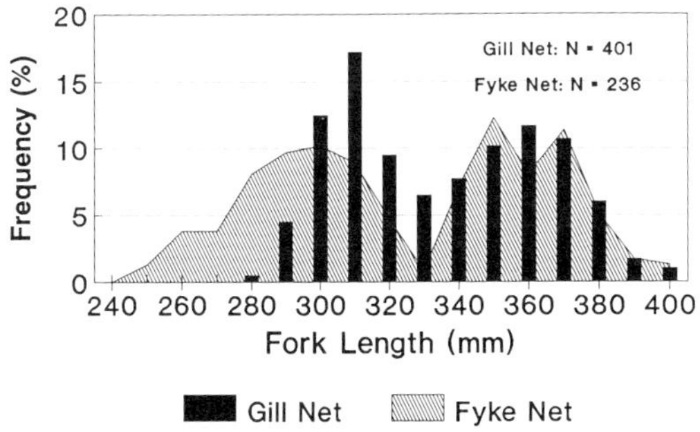

FIGURE 6.—Length frequencies of Arctic ciscoes released from fyke nets in the coastal region (after August 15) compared with those caught by 76-mm mesh gill nets in the Colville Delta fisheries, 1988. N is sample size.

highest correlation with both Outer Delta catch indices. Seven of the eight comparisons between Colville Delta and fyke net catch rates for cohort 3 were significantly correlated; the cohort-3 catch rate at West Beach was highly correlated with both Colville indices. The cohort-2 Arctic cisco correlation suffered from small sample size; only 3 years were available for analysis after the 4-year lag in the fyke net data. The cohort-4 (with no lag) Arctic cisco catch rate at the west base of West Dock were highly correlated with both Colville indices; cohort 4 (with 1-year lag) at the east base of West Dock showed a similar high correlation. The success of the correlations with the older cohorts could likely be improved by more precise definition of age-groups.

The two estimates of least cisco catch rates in the Colville Delta, unlike those of Arctic cisco, showed no consistent patterns of correlation between each other or fyke net catch rates in the Prudhoe Bay region. Only one correlation was significant (at 0.05 significance level).

Predictability in Arctic Cisco Harvest Rates

The high correlation between the mean annual catch rate of cohort-3 Arctic ciscoes at Prudhoe Bay fyke net stations and the catch rate in 76-mm mesh gill nets in the Colville Delta fishery 3 years later implies that the former may be a useful predictor of catch rate in the fishery. While catch rates at all the examined fyke net stations revealed significant correlations with commercial catch rates, the West Beach station catch rate showed the highest correlation ($r = 0.997$ between cohort 3 and commercial catch rates). The associated regression was

$$C_{(t+3)} = 22.57 + 14.458WB_t;$$

WB_t is the mean catch rate of cohort-3 Arctic ciscoes at West Beach fyke net in a given year;

TABLE 3.—Correlation coefficients between catch rates of Arctic cisco in the Colville Delta fall gill-net fishery and the abundance of Arctic cisco near Prudhoe Bay as measured by summer fyke net catches. Significant correlations are designated by asterisks for $P < 0.05$* and $P < 0.01$**.

Colville Delta fishery	Village	Prudhoe Bay cohorts 2 (4-year lag)	3 (3-year lag)	4 (1-year lag)	4 (no lag)
West base of West Dock					
Commercial	0.943**	0.943	0.976**	0.257	0.956**
	(4 df)	(2 df)	(3 df)	(5 df)	(5 df)
Village		0.907	0.914*	0.245	0.965**
		(2 df)	(3 df)	(4 df)	(3 df)
East base of West Dock					
Commercial		0.862	0.946*	0.940**	0.060
		(2 df)	(3 df)	(5 df)	(5 df)
Village		0.809	0.864	0.895*	0.078
		(2 df)	(3 df)	(4 df)	(3 df)
West Beach					
Commercial		0.853	0.997**	0.450	0.641
		(2 df)	(3 df)	(5 df)	(5 df)
Village		0.798	0.961**	0.688	0.669
		(2 df)	(3 df)	(4 df)	(3 df)
Heald Point					
Commercial		0.901	0.958*	0.624	0.274
		(2 df)	(3 df)	(4 df)	(4 df)
Village		0.855	0.880*	0.840	0.454
		(2 df)	(3 df)	(3 df)	(2 df)

and $C_{(t+3)}$ is the mean catch rate in the commercial fishery 3 years later.

In 1986 and 1987, the mean fyke net catch rates of cohort-3 Arctic ciscoes at West Beach were 0.7 and 0.85 fish/d indicating that the commercial catch rates in the Colville River fall fishery are likely to remain around 32–34 fish/d in 1989 and 1990. The utility of this harvest predictor is limited at present, however, because the small number of observations (four) we used to develop the relationship leads to wide confidence intervals (95% confidence interval: ±30 fish/d). As the time series of fyke net and gill net catch rates increases, the relationship can be tested further to evaluate the strength of the correlation.

Discussion

Accuracy of Preseason Predictions

Prior to the 1988 fishing season, it was predicted that Arctic cisco catch rates would be substantially lower than those recorded in 1987 because of the low juvenile recruitment from 1981 to 1984 and the maturation, with subsequent migration out of the region, of the 1979 and 1980 year-classes (Moulton and Field 1988). The mean catch rates did decrease, by 61% in the commercial fishery in the East Channel and by 72% in the village fishery in the Main Channel. The Arctic ciscoes that were caught were again remnants of the abundant year-classes from 1979–1980 and larger members (around 300 mm) of the weak 1983 recruitment. There were few fish between 320 and 340 mm, a size range that is highly selected for by the 76-mm meshes. The absence of fish in this size range was expected because of the weak recruitments in 1981 and 1982 (Griffiths et al. 1983; Woodward-Clyde Consultants 1983; Moulton et al. 1986b).

Village effort continued to decrease in 1988 and was only 36% of that recorded in 1985. The change was particularly striking in the Outer Delta, where effort was only 7% of that recorded in 1985. Commercial effort has been relatively stable, except in 1986 when a high catch rate allowed a high total catch with reduced effort.

Effect of Salinity on 1988 Catches

In 1988, as in 1987, the distribution of Arctic cisco and least cisco appeared to be affected by the salinity of the water in the Outer Colville Delta during the fishing season. Catch rates of least cisco in the commercial fishery were at an all-time high, and catch rates in the Main Channel were the highest in 4 years of record. The population estimate of least cisco based on 1988 tag returns did not reveal any unusual changes in abundance. Similarly, catch rates of humpback whitefish were also unusually high. The adult populations of both species are composed of many age-groups, which reduces the chance for drastic changes in abundance. In addition, there was no unusual increase in abundance by any particular size-group that may have indicated a large recruitment into the population. The only unusual phenomenon was the low salinity throughout the delta into late November. In 1987, increases in salinity were observed to cause least ciscoes to move upstream, while decreases in salinity allowed them to return downstream. The low salinities in 1988 likely caused the distribution of both least cisco and humpback whitefish to be farther downstream than normal, resulting in greater susceptibility to catch by fishers.

The lower salinity in 1988 likely contributed, in part, to the low catch of Arctic cisco. It is likely that Arctic ciscoes, which normally show highest catch rates in salinities in excess of 20‰ (Moulton and Field 1988) were farther downstream in the delta during the main fishing season. Arctic cisco catch rates in the commercial fishery increased at the end of the season while least cisco catch rates decreased, indicating that salinity may have been increasing at that time. Salinity measurements are not available for that period.

These responses to salinity in the delta region likely caused some of the variation in catch rates in the past years as well. Inspection of the daily catch patterns from the commercial fishery revealed that 1982 and 1983 were probably also characterized by low salinity through much of the fishing season.

Catch Rates as Indicators of Abundance

The historical catch rates of Arctic cisco and least cisco from the commercial fishery have been used as indices of abundance in this and other studies (Craig and Haldorson 1981; Gallaway et al. 1983, 1989). Some question remains, however, as to how accurately these indices reflect abundance. It is possible that competing fishers and annual changes in distribution of fish within the delta may cause substantial fluctuation in the commercial catch rate independent of fish abundance. The data gathered during the 4 years of study, when combined with catch indices from other studies in the region, allow evaluation of these indices as useful indicators of abundance.

The data from the harvests in the Colville Delta region and from the summer coastal studies indicate that for least cisco, the changes in catch rate caused by distributional changes resulting from random environmental factors override variations in population size, while for Arctic cisco, the changes in population size override the random distribution component. The catch rates of least cisco from the various summer and fall samples demonstrated little correlation amongst each other while Arctic cisco catch rates tended to be highly correlated. The catch patterns for least cisco thus provide little information on relative changes in population size. The fyke net and gill net catch rates of least cisco reflected changes in distribution caused by wind direction and salinity distribution, both in the coastal region during the summer and in the delta after ice formation. Arctic ciscoes also responded to changing wind and salinity, but the changes in overall abundance in the region caused by the sporadic recruitment of juveniles have been large enough to obscure the environmental effects.

Predictions for 1989

The recent (1985–1987) high catch rates for Arctic cisco were a result of a high recruitment of young in 1980. Because this group of fish has grown out of the fishery and recruitment was low from 1981 to 1984, catches declined in 1988 and will remain low until the 1985 year-class reaches a harvestable size. Catches will likely increase moderately in 1991 and should remain stable for several years after that because of the three consecutive years (1985–1987) of moderate-to-strong juvenile recruitment. The high catch rates that were observed in 1986 are not likely to be repeated in the foreseeable future because juvenile abundance is much lower than that recorded in the early 1980s.

The 1989 harvest rates will likely be similar to those observed in 1988 and will consist almost entirely of the 1983 year-class, which does not appear strong enough to significantly increase the catch rates above those recorded in 1988. If the catches of cohort-3 Arctic ciscoes at West Beach fyke net in 1986 and 1987 are used as predictors of abundance, then the catches in the Colville Delta will remain low through 1990 and will not increase until 1991.

Acknowledgments

The 1988 portion of the study was funded by the Endicott Unit Owners and managed by BP Exploration (Alaska) as part of the 1988 fish monitoring program required by the North Slope Borough. The study was administered by Chris Herlugson. The 1985–1987 portions of the study were funded by ARCO Alaska, Inc. and administered by Scott Robertson and Robert Newell. Field and data entry support were provided by Nuiqsut residents. Jim Helmericks, Colville Delta commercial fisherman, supplied logistical support, tag returns, catch data, and a variety of other information related to the fish and fishery in the delta.

References

Bendock, T. N. 1979. Beaufort Sea estuarine fishery study. Pages 670–729 *in* Environmental assessment of the Alaskan continental shelf. Final report of the principal investigators, volume 4. Bureau of Land Management and National Oceanic and Atmospheric Administration, Outer Continental Shelf Environmental Assessment Program, Boulder, Colorado.

BioSonics, Inc. 1984. Prudhoe Bay Waterflood Project fish monitoring program, 1983. Prudhoe Bay Waterflood Environmental Monitoring Program, volume 2. Report to U.S. Army Corps of Engineers, Anchorage, Alaska.

Craig, P. C. 1987. Subsistence fisheries at coastal villages in the Alaskan Arctic, 1970–1986. U.S. Minerals Management Service, Alaska Socioeconomic Studies Program, Technical Report 129, Anchorage, Alaska.

Craig, P. C., and L. Haldorson. 1981. Fish. Pages 384–678 *in* Beaufort Sea barrier island–lagoon ecological processes studies: final report, Simpson Lagoon. Environmental assessment of the Alaskan continental shelf, final report of the principal investigators, volume 7. Bureau of Land Management and National Oceanic and Atmospheric Administration, Outer Continental Shelf Environmental Assessment Program, Boulder, Colorado.

Critchlow, K. R. 1983. Prudhoe Bay Waterflood Project fish monitoring program, 1982. Pages 5-1 to 5-219 *in* Prudhoe Bay Waterflood Environmental Monitoring Program, 1982, volume 3. Report of Woodward-Clyde Consultants to U.S. Army Corps of Engineers, Anchorage, Alaska.

Envirosphere Company. 1987. 1985 Final report for the Endicott Environmental Monitoring Program. Report to U.S. Army Corps of Engineers, Anchorage, Alaska.

Fawcett, M. H., L. L. Moulton, and T. A. Carpenter. 1986. Colville River fishes: 1985 biological report. Colville River fish study, 1985 annual report. Report of Entrix, Inc., to ARCO–Alaska, Anchorage, Alaska.

Fechhelm, R. G., and D. B. Fissel. 1988. Wind-aided recruitment of Canadian Arctic cisco (*Coregonus autumnalis*) into Alaskan waters. Canadian Journal of Fisheries and Aquatic Sciences 45:906–910.

Gallaway, B. J., W. J. Gazey, and L. L. Moulton. 1989.

Population trends for the Arctic cisco (*Coregonus autumnalis*) in the Colville River of Alaska as reflected by the commercial fishery. Biological Papers of the University of Alaska 24:153–165.

Gallaway, B. J., W. B. Griffiths, P. C. Craig, W. J. Gazey, and J. W. Helmericks. 1983. An assessment of the Colville River delta stock of Arctic cisco—migrants from Canada? Biological Papers of the University of Alaska 21:4–23.

George, J. C., and R. Kovalsky. 1986. Observations on the Kupigruak Channel (Colville River) subsistence fishery, October 1985. North Slope Borough, Department of Wildlife Management, Barrow, Alaska.

George, J. C., and B. P. Nageak. 1986. Observations on the Colville River subsistence fishery at Nuiqsut, Alaska. North Slope Borough, Department of Wildlife Management, Barrow, Alaska.

Glass, D., C. Whitmus, and M. Prewitt. 1987. Fish distribution and abundance. 1986 Endicott Environmental Monitoring Program, volume 4. Report of Envirosphere Company to U.S. Army Corps of Engineers, Anchorage, Alaska.

Griffiths, W., and B. J. Gallaway. 1982. Prudhoe Bay Waterflood Project fish monitoring program, 1981. Report of LGL Ecological Research Associates to U.S. Army Corps of Engineers, Anchorage, Alaska.

Griffiths, W. B., and seven coauthors. 1983. Fish ecology. Pages 1–323 *in* B. J. Gallaway and R. P. Britch, editors. Environmental summer studies (1982) for the Endicott Development, volume 3. Report of LGL Alaska Research Associates to Sohio Alaska Petroleum Company, Anchorage, Alaska.

Moulton, L. L., and M. H. Fawcett. 1984. Oliktok Point fish studies—1983. Report of Woodward-Clyde Consultants to Kuparuk River Unit, Anchorage, Alaska.

Moulton, L. L., and L. J. Field. 1988. Assessment of the Colville River fall fishery 1985–1987. Final Report of Environmental Sciences and Engineering to ARCO–Alaska, Inc., Anchorage, Alaska.

Moulton, L. L., L. J. Field, and S. Brotherton. 1986a. Assessment of the Colville River fishery in 1985. Colville River fish study, 1985 annual report. Report of Entrix, Inc., to ARCO–Alaska, Inc., Anchorage, Alaska.

Moulton, L. L., and seven coauthors. 1986b. 1984 central Beaufort Sea fish study. Pages 1–300 *in* Prudhoe Bay Waterflood Project Environmental Monitoring Program, 1984, volume 2. Final report of Woodward-Clyde Consultants, Entrix, Inc., and LGL Ecological Research Associates to U.S. Army Corps of Engineers, Anchorage, Alaska.

Reub, G. S., J. D. Durst, and D. Glass. 1988. Fish distribution and abundance. Pages 1–47 *in* 1987 Endicott Environmental Monitoring Program. Report of Envirosphere Company to U.S. Army Corps of Engineers, Anchorage, Alaska.

USACE (U.S. Army Corps of Engineers). 1980. Final environmental impact statement—Prudhoe Bay Oil Field Waterflood Project. USACE, Anchorage, Alaska.

USACE (U.S. Army Corps of Engineers). 1984. Final environmental impact statement—Endicott Development Project. USACE, Anchorage, Alaska.

Woodward-Clyde Consultants. 1983. Lisburne Development Area: 1983 environmental studies. Report to ARCO–Alaska, Inc., Anchorage, Alaska.

American Fisheries Society Symposium 11:157–162, 1991

Habitat Use Patterns of Juvenile Arctic Cod in the Coastal Beaufort Sea near Prudhoe Bay, Alaska

T. C. Cannon, D. R. Glass, and C. M. Prewitt

Ebasco Environmental
10900 NE 8th Street, Bellevue, Washington 98004, USA

Abstract.—Round-haul seining was conducted within and adjacent to Prudhoe Bay in 1986. Physical variables (temperature, salinity, and depth) were measured with each seine haul and catch-frequency histograms were prepared for juvenile Arctic cod *Boreogadus saida*. Juvenile Arctic cod had a broad range of salinity use (0–32‰). Peak frequencies of occurrence were in salinity-stratified water: 10–20‰ at the surface and 14–22‰ at the bottom. Their temperature use peaked in the midrange between 4 and 10°C. Because Arctic cod are an important species in the arctic food web, their use of relatively warm, moderate-salinity coastal waters as nursery areas indicates that such habitat may be important for much of the fish productivity of the coastal Beaufort Sea. Thus, coastal oil and gas development such as gravel-fill causeways that alter the physical habitat of coastal waters could affect regional fish productivity.

Arctic cod *Boreogadus saida* are an important component of the Arctic Ocean ecosystem (Quast 1974; Lowry and Frost 1981; Bradstreet et al. 1986). They are a significant fraction of the food consumed by marine mammals such as the seals, whales, and walruses (Bradstreet et al. 1986); seabirds such as terns, gulls, murres, and kittiwakes (Bradstreet et al. 1986); and fishes such as fourhorn sculpin *Myoxocephalus quadricornis* and Arctic char *Salvelinus alpinus* (Craig and Haldorson 1981; Cannon et al. 1987a). Craig et al. (1982) cite examples of the common occurrence of Arctic cod strandings on beaches, where the fish become food for polar bears *Ursus maritimus*, Arctic foxes *Alopex lagopus*, and humans and their dogs. Arctic cod are generally described as a marine species, although they are common to coastal brackish waters of the central Beaufort Sea (Bendock 1979; Craig and Haldorson 1981). Their occurrence in nearshore waters of the Beaufort Sea represents a potentially important source of prey and competition for anadromous fishes that feed in coastal waters during the summer. Oil and gas developments such as gravel-fill causeways in coastal waters affect the physical habitat along the coast (Hachmeister et al. 1991, this volume), which could alter the distribution, abundance, and production of Arctic cod in the region of the development. The present study of Arctic cod habitat use was conducted in the summer of 1986 as part of the Endicott Development Environmental Monitoring Program (USACE 1986) within and adjacent to Prudhoe Bay (Figure 1) to determine habitat use criteria of arctic fishes. A companion study was conducted on marine invertebrates (Cannon et al. 1991, this volume).

Methods

Samples of fish were collected with seines over a wide variety of habitat types and conditions in the Prudhoe Bay area (Figure 1). Juvenile Arctic cod were the only species captured in sufficient numbers to determine habitat use patterns. Seining was conducted with two different boat–seine combinations. For water depths greater than 1.2 m, a 91- × 3-m seine with a graduated (4–8-cm stretch) mesh lead and a 6-mm mesh bunt was used. The net was constructed with considerable "hang-in" to allow for pursing during retrieval. The net was deployed from a seine drum mounted on a 6.7-m Munson aluminum boat fitted with a fairlead roller and hydraulic drum collection device. The boat was powered by twin 70-horsepower outboard motors. The seine was deployed by running the boat in reverse to create a round seine set. The lead end was tied to the rear rail of the boat on closing the set, and the seine was then retrieved onto the drum. A white 6-mm mesh curtain, about 3 × 3 m, was draped from the side of the boat during retrieval to prevent fish from escaping under the boat. The net was retrieved slowly so as not to sink the floats. At the end of retrieval, the lead line was pulled aboard and the bunt collected. Net retrieval in high winds was facilitated by using a 5-m Avon inflatable skiff to keep the Munson boat downwind of the seine. A smaller seine set from a 5-m Avon inflatable skiff was used for shallow-water sampling (generally less than 1.2 m deep). The seine was 45.7 m long

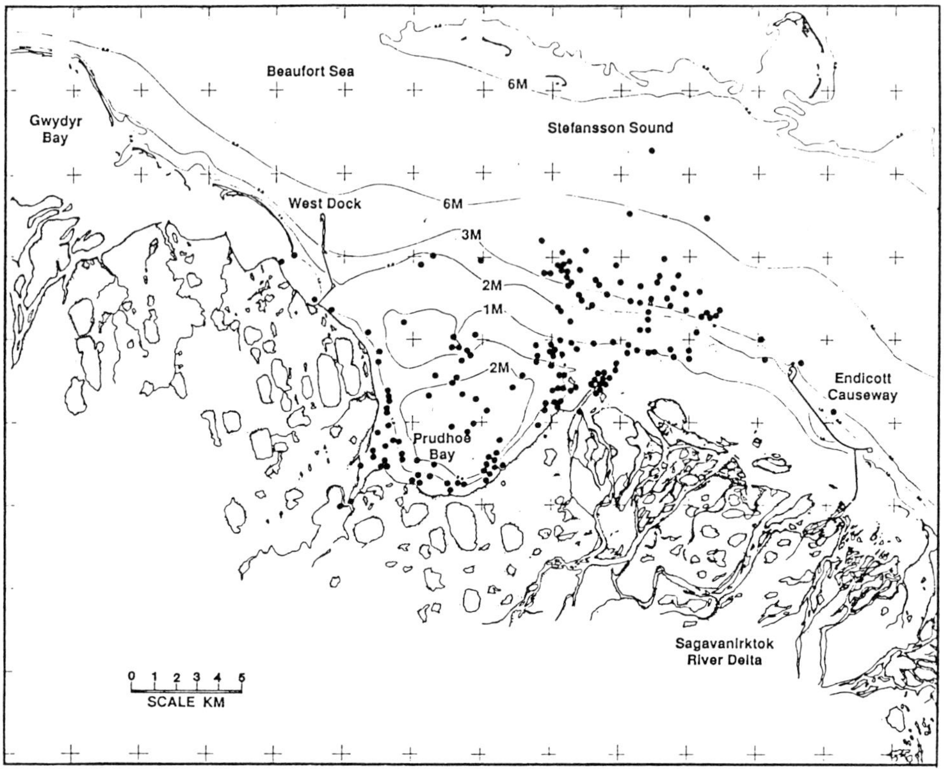

FIGURE 1.—Study area along the central Alaskan Beaufort Sea coast indicating locations of seine sets in summer 1986.

by 1.8 m deep, with 33.5-m, 6-mm mesh wings, and a 12-m long bag of 4.5-mm mesh web. The net was deployed from the skiff in a circular set and retrieved by hand. The acceptability of each seine set was evaluated by the following criteria: (1) seine set (full circular), (2) seine retrieval (retention of circular shape, floats stayed afloat, and retrieval was uninterrupted), and (3) final haul onboard (lead-line and final pursing were coordinated). Catches from hauls not meeting these criteria were not recorded.

We chose salinity, water temperature, and water depth to represent major habitat components. Salinity and temperature were measured with a YSI model 33 meter, which was calibrated weekly. Depth was measured with a calibrated sounding line.

We sampled in Prudhoe Bay and the adjacent west side of the Sagavanirktok River delta, where past studies indicated a gradient of habitat from freshwater to marine conditions was normally available (Bendock 1979; Moulton et al. 1986; Cannon et al. 1987b). The selected sampling area provided an uninterrupted transect leading from the shoreline to approximately 4-m depth, and often crossing salinity and temperature gradients from 0 to 30‰ and 0 to 10°C, respectively. Midsummer was selected for the study period because it provided the widest range of habitat conditions for sampling. Early summer was not included because of the preponderance of low-salinity water, while in late summer high-salinity water often dominates Prudhoe Bay and adjoining areas. Daily seining locations were chosen such that a wide range of salinities could be sampled, given acceptable weather and ice conditions. Daily sampling designs were planned based on observations from the previous day. Salinity was chosen as the primary variable driving the sampling design, because water temperature and depth were often correlated with salinity. Sampling sites were chosen randomly within the targeted areas of known salinity.

The approach we chose to develop fish habitat suitability relationships was based on concepts described by Baldridge and Amos (1981). The method used frequency of catch data and associated measurements of habitat variables to derive expressions of relative use by the fish for values of the variables within measured ranges. Capture frequency was plotted against the measured range of each habitat variable to depict a univariate habitat relationship.

The primary objective of this analysis was to determine relationships between fish catch (by species) and the measured habitat conditions (salinity, temperature, and depth). Before these relationships were developed, however, we tested whether catch and habitat condition data met study design criteria (i.e., sampling was distributed across the available habitat conditions, and most catches were nonzero) and assumptions (i.e., the two seine types sampled with a similar effectiveness). Evaluations included a review of the effort distribution for each habitat variable, the frequency distribution of the catches, and the catch patterns exhibited for each seine type in the data set.

Effects of effort were assessed by comparing effort versus variable (e.g., temperature and salinity) histograms to determine how seining effort was distributed across the available range of the variables. Because effort was not evenly distributed across the variable ranges, catch was adjusted (weighted) by the ratio between effort in a variable category and the effort in the category with maximum effort.

The longer seine was expected to capture more fish because of its greater surface and enclosure areas. Because the set surface area of the large (91-m) seine when deployed was approximately three times greater than that of the 46-m seine, the catch in the 46-m seine catches was multiplied by three to standardize them to the 91-m seine catches. Comparison of the catch patterns of each seine set versus habitat variables indicated similar patterns; thus, utilization histograms were developed from the combined data set of the two nets.

Utilization histograms were not adjusted for availability of salinity, temperature, and depth data because availability varied considerably among the sampling dates. In addition, availability could not be quantified for each survey day because all available habitats were not sampled during each survey day. We attempted to sample habitat conditions in proportion to their estimated availability during a period of several survey days.

Results

We began the seine survey on 14 July, and ended it on 17 August 1986. During that time period, 283 acceptable seine sets were recorded. The seining was concentrated, as planned, in Prudhoe Bay and the western side of the Sagavanirktok River delta (Figure 1). During the first 2 weeks of the study, seining was restricted to Prudhoe Bay because ice dominated the shoal along the north side of the bay and on the western side of the Sagavanirktok River delta. After the ice cleared, effort was directed to the bay, delta, and offshore areas.

Distribution of Seining Effort

Effort was distributed across the range of habitat conditions (Figure 2), roughly in proportion to the conditions in the survey area. Because sampling locations were chosen on the basis of salinity, the distributions of effort for water temperature and depth were less uniform than for salinity.

Habitat Use Patterns

Arctic cod were most abundant at intermediate salinities, water temperatures, and depths (Figure 3). Catch was greatest in areas with surface salinities between 10 and 20‰, and with bottom salinities between 14 and 22‰. Catch was greatest in waters with surface and bottom temperatures of 4–10°C and in water depths of 1–2 m.

Discussion

A factor to consider in interpretation of these habitat use data is the possible confounding of effects of the habitat variables and fish utilization. From the survey data it is obvious that salinity, water temperature, and depth are not independent variables. Thus, univariate associations may not represent true habitat preferences. Rather, one factor may determine habitat choice, while another may be only indirectly related to habitat preference because of its interdependence with the first factor. In this study, depth, salinity, and water temperature were related; thus, the specific univariate relationships of these variables with fish catch that we described above may not reflect the independent responses of the fish to this range of conditions. Only a carefully designed experiment that varied the level of each factor while keeping the others constant could determine the real effect of each habitat variable on fish habitat use or preference. In a field study such as this, it is difficult to obtain all possible conditions, as they may not naturally exist. Only in a laboratory study can these conditions be sufficiently controlled; however, laboratory studies cannot dupli-

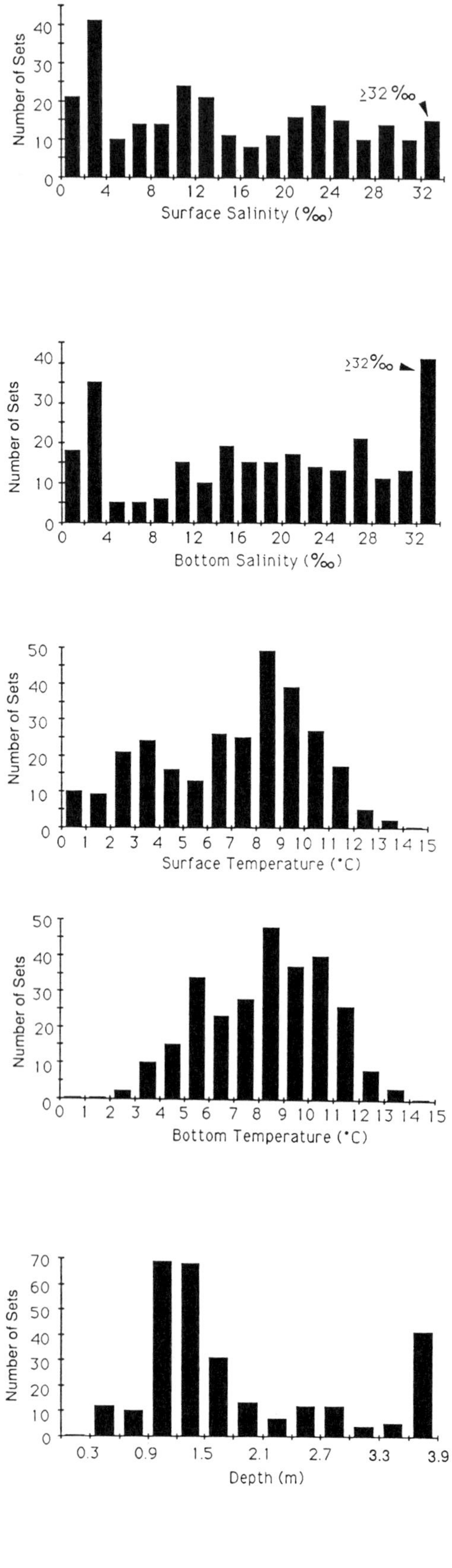

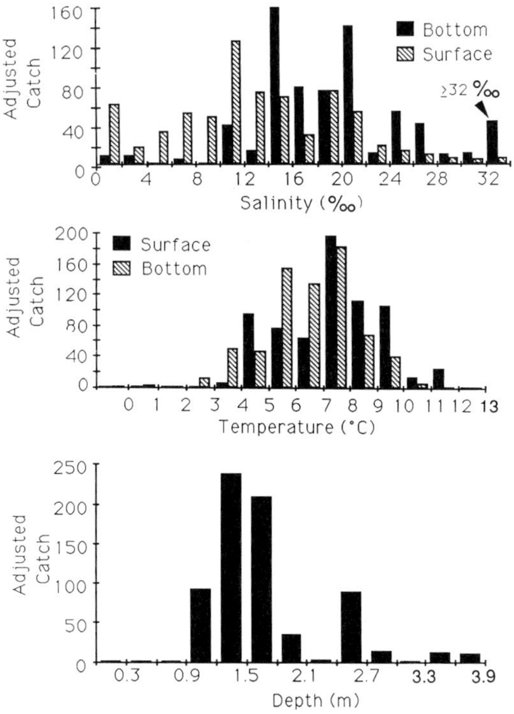

FIGURE 3.—Catch of Arctic cod by surface and bottom salinity, surface and bottom water temperature, and sample depth. Catch values are adjusted for effort in each habitat variable and for net size.

cate true natural conditions, and thus may not produce representative or natural behavioral response in the fish.

Our study found Arctic cod to be concentrated in warmer, lower-salinity water than previously reported in the Prudhoe Bay area. Craig et al. (1982) described the preferred habitat of juvenile Arctic cod in coastal waters during the summer as water of 0–6°C and 20–30‰. Cannon et al. (1987b) found juvenile Arctic cod most abundant in 2–4°C and 15–25‰ water (surface measurements). The difference between our findings and those of these other studies may be due to sampling methodology and habitat availability. Their studies used fyke nets, which tend to capture fish that are moving. The fyke nets were sampled for 24 h, during which time habitat conditions may have changed. Availability, especially the lack of warm, intermediate-salinity water, also may have

←

FIGURE 2.—Number of seine sets by surface salinity, bottom salinity, surface water temperature, bottom water temperature, and water depth.

influenced the cod distribution in the previous studies. The fyke nets also were confined to shallow water about 1 m in depth, whereas we used haul seines to sample depths up to 4 m and deeper.

Our findings are consistent with other reports of Arctic cod distribution. Moulton and Tarbox (1987) found Arctic cod concentrated along the interface between warm brackish water and cold marine waters, a pattern indicated in our study (Figure 3). Rass (1968) described the preferred water temperature of Arctic cod fry as 5–7°C or higher. Our study supports the findings of Rass, as we found Arctic cod to be most abundant at or near these temperatures (Figure 3).

The differences in the conclusions between the studies are important. If juvenile Arctic cod do prefer warmer, slightly lower-salinity water than indicated by the fyke net studies, then such habitat may be an important factor in the portions of the arctic food web that depend on Arctic cod. The importance of warmer, lower-salinity water is well established for anadromous chars, ciscoes, and whitefishes in the Beaufort Sea (Craig 1984). Our results indicate that such waters are also essential for Arctic cod, which account for much of the fish productivity of the coastal Beaufort Sea. Coastal development such as gravel-fill causeways that alter the distribution and amount of warm brackish water along the coast of the Beaufort Sea (Cannon and Hachmeister 1987; Hachmeister et al. 1991) could alter Arctic cod production, which could affect production of other fish and birds that feed upon the Arctic cod. Causeway-induced change of nearshore, low-salinity (1–15‰) habitats of Arctic char, Arctic and least cisco, and broad whitefish to intermediate salinity could encourage Arctic cod to invade these areas at the expense of these other fishes if competition for food and space is intensified. The juvenile Arctic cod would also be more vulnerable to predators in the shallow water. Overall effects would depend on the extent of habitat change induced by causeways.

References

Baldrige, J. E., and D. Amos. 1981. A technique for determining fish habitat suitability criteria: a comparison between habitat utilization and availability. Pages 251–258 *in* N. B. Armantrout, editor. Acquisition and utilization of aquatic habitat inventory information. American Fisheries Society, Western Division, Bethesda, Maryland.

Bendock, T. 1979. Beaufort Sea estuarine fishery study. Pages 670–729 *in* Environmental assessment of the Alaskan continental shelf, volume 4. Final report of the principal investigators. Bureau of Land Management and National Oceanic and Atmospheric Administration, Outer Continental Shelf Environmental Assessment Program, Boulder, Colorado.

Bradstreet, M. S. W., and six coauthors. 1986. Aspects of the biology of Arctic cod (*Boreogadus saida*) and its importance in arctic marine food chains. Canadian Technical Report of Fisheries and Aquatic Sciences 1491.

Cannon, T. C., M. S. Brancato, and S. C. Jewett. 1987a. Fish food habits. *In* 1985 Final report of the Endicott Environmental Monitoring Program, volume 7. Report of Envirosphere Company to U.S. Army Corps of Engineers, Anchorage, Alaska.

Cannon, T. C., D. R. Glass, B. Adams, and T. Nelson. 1987b. Fish distribution and abundance. Pages 1–129 *in* 1985 Final report of the Endicott Environmental Monitoring Program, volume 6. Report of Envirosphere Company to U.S. Army Corps of Engineers, Anchorage, Alaska.

Cannon, T. C., and L. E. Hachmeister. 1987. Integration and assessment. *In* 1985 Final report of the Endicott Environmental Monitoring Program, volume 1. Report of Envirosphere Company to U.S. Army Corps of Engineers, Anchorage, Alaska.

Cannon, T. C., J. A. Knutzen, and D. R. Glass. 1991. Habitat use by epibenthic and planktonic invertebrates along the coast of the central Beaufort Sea, Alaska. American Fisheries Society Symposium 11:163–172.

Craig, P. C. 1984. Fish use of coastal waters of the Alaskan Beaufort Sea: a review. Transactions of the American Fisheries Society 113:265–282.

Craig, P. C., W. B. Griffiths, L. Haldorson, and H. McElderry. 1982. Ecological studies of Arctic cod (*Boreogadus saida*) in Beaufort Sea coast waters, Alaska. Canadian Journal of Fisheries and Aquatic Sciences 39:395–406.

Craig, P. C., and L. Haldorson. 1981. Fish. Pages 384–678 *in* Beaufort Sea barrier island–lagoon ecological process studies: final report, Simpson Lagoon. Environmental assessment of the Alaskan continental shelf, final report of the principal investigators, volume 7. Bureau of Land Management and National Oceanic and Atmospheric Administration, Outer Continental Shelf Environmental Assessment Program, Boulder, Colorado.

Hachmeister, L. E., D. R. Glass, and T. C. Cannon. 1991. Effects of solid-fill gravel causeways on the coastal central Beaufort Sea environment. American Fisheries Society Symposium 11:81–96.

Lowry, L. F., and K. J. Frost. 1981. Distribution, growth, and foods of Arctic cod (*Boreogadus saida*) in the Bering, Chukchi, and Beaufort seas. Canadian Field-Naturalist 95(2):186–191.

Moulton, L. L., and seven coauthors. 1986. 1984 central Beaufort Sea fish study. Pages 1–300 *in* Prudhoe Bay Waterflood Environmental Monitoring Program, 1984, volume 2. Final report of Woodward-Clyde Consultants, Entrix, Inc., and LGL Alaska Research Associates to U.S. Army Corps of Engineers, Anchorage, Alaska.

Moulton, L., and K. Tarbox. 1987. Analysis of Arctic cod movements in the Beaufort Sea nearshore region 1978–1979. Arctic 40:43–49.

Quast, J. C. 1974. Density distribution of juvenile Arctic cod, *Boreogadus saida*, in the eastern Chukchi Sea in the fall of 1970. U.S. National Marine Fisheries Service Fishery Bulletin 72:1094–1105.

Rass, T. S. 1968. Spawning and development of polar cod. Rapports et Procès-Verbaux des Réunions, Conseil International pour l'Exploration de la Mer 158:135–137.

USACE (U.S. Army Corps of Engineers). 1986. Five-year plan, Endicott Development Project Environmental Monitoring Program. USACE, Anchorage, Alaska.

American Fisheries Society Symposium 11:163–172, 1991

Habitat Use by Epibenthic and Planktonic Invertebrates along the Coast of the Beaufort Sea, Alaska

T. C. Cannon, J. A. Knutzen, and D. R. Glass

Ebasco Environmental

10900 NE 8th Street, Bellevue, Washington 98004, USA

Abstract.—Habitat response patterns of mobile macroinvertebrate fish prey were derived from drop net survey data collected in and adjacent to Prudhoe Bay along the central Beaufort Sea coast of Alaska from 14 July to 17 August 1986. Density of *Mysis relicta* increased with water temperature, reaching a peak in relatively warm water (9–11°C). Peak average density (10–20 animals/m^2) occurred at intermediate salinities of 12–19‰ and intermediate water depths of 2–3 m. Small, unidentified mysids, probably young *M. relicta*, showed a similar distribution pattern, but were slightly more dense in shallower (<1 m), lower-salinity (3–10‰), and warmer (11–12°C) water. *Mysis litoralis* were less abundant (<5/m^2) and concentrated at depths greater than 3 m in cold (<4°C), high-salinity (>28‰) water. The density of the amphipod *Pontoporeia affinis* reached a peak (>1,000/m^2) in the warmest (12°C), lowest-salinity (0–2‰), shallowest (<1 m) water sampled. Chironomids were also most abundant (10–100/m^2) in this water. Chironomids and *P. affinis* densities were lowest at depths in excess of 2.5 m in cold (<4°C), high-salinity (>25‰) water. Copepod density was consistently higher (>1,000/m^2) in the deeper, colder, higher-salinity water. Due to the high degree of intercorrelation among water temperature, salinity, and depth, it was not possible to separate the effect of each of these parameters on invertebrate distributions. Principal components analysis results suggested that *M. relicta* and copepods were more abundant within the warmest water available, within a given salinity range. Because key fish prey are most abundant in shallow, warmer, estuarine waters, developments such as gravel-fill causeways that alter the temperature and salinity gradients in coastal habitats could affect the distribution and availability of important fish prey. Thus, causeways could directly or indirectly reduce the productivity of coastal waters of the Beaufort Sea.

Offshore oil and gas development along the Beaufort Sea coast of Alaska may affect the summer feeding habitat of anadromous fish. To determine the potential effects of physical habitat changes from development on important invertebrate fish prey, we developed habitat suitability curves for seven epibenthic and planktonic invertebrate taxa that make up most of the diet of anadromous fish in the central Beaufort Sea. Changes in physical habitat conditions from coastal developments such as gravel-fill causeways can now be translated into effects on the habitat of fish prey.

Earlier attempts to relate the distributions of invertebrate fish prey along the coast of the central Beaufort Sea to their aquatic habitat conditions were limited by very high spatial and temporal variability in invertebrate species composition, abundance, and habitat conditions (Craig and Haldorson 1981; Feder and McGee 1982; Moulton et al. 1986; Cannon et al. 1987a). High correlations among the various habitat variables (water temperature, salinity, turbidity, and water depth) made it difficult to separate the effects of individual habitat variables (Cannon et al. 1987a).

In the summer of 1986, we collected samples of epibenthic and planktonic invertebrates as part of a fish habitat study in the Prudhoe Bay area (Cannon et al. 1991, this volume). The survey was conducted over a relatively small area of the coast during midsummer conditions that provided a wide range of summer habitats (Figure 1).

Methods

Epibenthic and macroplanktonic invertebrates were sampled with a 0.5-m, 333-μm mesh vertical plankton net, commonly termed a drop net, with a pursing collar that allowed the net to be closed after it was deployed. Two net samples were collected at each sampling location. Samples were washed from the net into a jar and preserved in 10% formalin. Surface and bottom temperature and salinity, depth, and Secchi depth were measured with each sample. In the laboratory, samples were rinsed through a 0.3-mm mesh sieve, placed in a 50% solution of isopropanol, and sorted by major taxa groups. Amphipods were identified to genus and species to the degree possible. Larger mysids (>10 mm) were identified

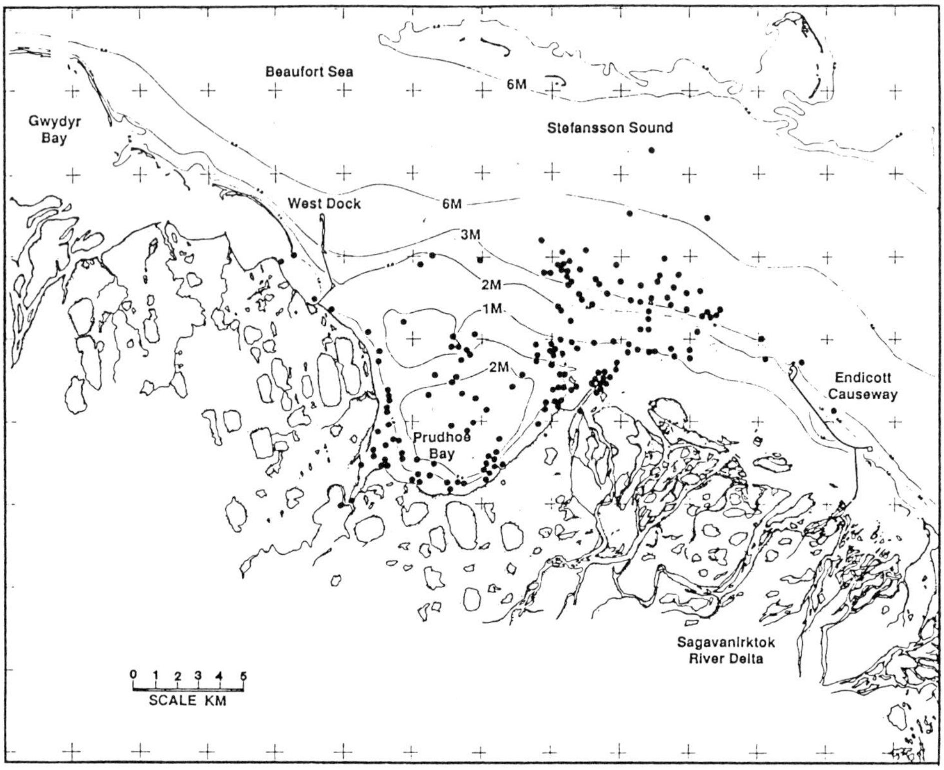

FIGURE 1.—Study area along the central Alaskan Beaufort Sea coast indicating locations sampled with drop nets in summer 1986.

to species, while smaller mysids were classified as small mysids. Other taxonomic groups were identified to class or order. Samples were collected daily from 14 July to 17 August 1986. We attempted to collect samples across the entire range of habitat conditions (depth, water temperature, and salinity) available in the study area each day.

Univariate histograms were developed for temperature, salinity, and depth. Secchi depth data were not included because the variable was confounded with depth. The dependent variable was mean density. The number of prey organisms in a taxon in each drop net at a given sample site was determined. The numbers from two replicates were summed, and the total was divided by 0.392 (each drop net sampled 0.196 m^2) to provide the average density (number of animals/m^2). A principal components analysis (PCA) was conducted to identify independent variables among the highly intercorrelated habitat variables. A relative water clarity variable was derived by dividing Secchi depth by water depth for use in the PCA.

Results

Water Conditions

In total, 274 drop nets were collected from 14 July to 15 August 1986. Effort was distributed over most of the range of habitat conditions (Table 1). There was a lack of effort in low-to-moderate salinity (0–20‰), low-temperature (<4°C) water, and in warm (>10°C), high-salinity (>20‰) water. Such habitat rarely occurred within the study area during the summer sampling period.

The principal components analysis of the water data indicated a high degree of intercorrelation among the water variables (Table 2). Waters of greater depth were less turbid, colder, and higher in salinity, as exemplified by factor 1, which accounted for 67% of the variability in the data. Factor 2 represented conditions in which water temperature and salinity were both relatively high

TABLE 1.—Distribution of samples by water temperature and salinity for near-surface and near-bottom conditions. Temperature values refer to the lowest value in the interval (e.g., 0°C refers to the range 0.0–0.9°C).

Salinity (‰)	Temperature (°C)													
	<0	0	1	2	3	4	5	6	7	8	9	10	11	12
Surface														
≥32		2	13	7	3	1	7	4	3	2				
30–32			3	6	3	1		2						
28–30	1	2	2	4	1	1		1		1				
26–28		3	2	4	4	2	2	1	1		2	1		
24–26			1	1	2			4	4	1				
22–24			1				1	1	6	3	2			
20–22				1	1		2	1	6	7		1		
18–20						3	1	4	1	2	3	1		
16–18					1		5		3	2	3	1		
14–16					1	1		2	4	4	2	2		
12–14		1		1		1	1	1		1	1	3		
10–12		1		1	1	1	3	1	4	1				
8–10						1		2	1		2			
6–8							1		3		1			
4–6								1	3	1				
2–4						1	1	1	11	9	8	2	1	1
0–2						1	2	1		5	3	3	2	1
Bottom														
≥32					3	4	2	4	2					
30–32				1	1	3	3	1	1	1				
28–30				1	4	6		2			1	1		
26–28			1			4	1	1			3			
24–26				1	2	3	2	3	1	1	2			
22–24				1	1		1	3	9	3		1		
20–22				3		2	1	2	2	6		2		
18–20				1		2	2	2	3	3		1		
16–18						2			4	2				
14–16					2		1		2	2	4			
12–14				2		2	1	3	1	2	2	7		
10–12			1	1	2	1	3	3	4	2	2	2	3	
8–10						2	1	1	1	1	3	2		
6–8				1		1	1	1	4	1	1	3	1	
4–6							1	3	4	1	1			
2–4						1	1	2	10	8	15	2	1	1
0–2						1	3		1	4	6	2	3	1

and explained 16% of the variability in the data. Factors 3 and 4 explained only 10 and 5%, respectively, of the variance in the data set. For this reason, and because these factors were poorly related to the invertebrate groups (Table 3), only factors 1 and 2 are included in the following discussion by invertebrate group.

TABLE 2.—Correlations of water variables and components (factors) derived from principal components analysis for midsummer 1986 drop net sampling in Prudhoe Bay. (Percent of variance explained by each component is shown in parentheses.)

Habitat variable	Factor 1 (67.1)	2 (15.5)	3 (9.7)	4 (5.2)
Bottom temperature	−0.87	0.45	0.08	−0.06
Surface temperature	−0.76	0.62	0.14	0.09
Bottom salinity	0.83	0.39	−0.32	−0.02
Surface salinity	0.84	0.40	−0.26	−0.09
Water clarity	0.77	0.12	0.53	−0.32
Depth	0.84	0.13	0.31	0.43

Mysis relicta

Adult *Mysis relicta* were most abundant in waters of intermediate temperature, salinity, and depth (Figure 2). Density peaked in samples for which bottom salinities were 12–20‰. Density gradually decreased below and above this level, reaching a minimum in fresh water. Density increased with increasing bottom temperatures to 11°C then fell to its lowest level at 12°C. Waters of 12°C or higher were encountered in only two samples, and these had salinities less than 3‰. Highest densities occurred at depths between 2.0 and 2.8 m. Density was generally lower in both shallow and deep water and was lowest in water less than 0.4 m deep. Density was highest in the midrange of factor 1, indicating an affinity for waters of intermediate temperature, salinity, and

TABLE 3.—Correlations of the natural logarithms of the invertebrate density and water principal components (factors) for midsummer 1986 drop net data from Prudhoe Bay. Asterisks (*) denote significance levels: * for $P \leq 0.01$ and ** for $P \leq 0.001$. Sample size was 278.

Invertebrate group	Factor 1	Factor 2	Factor 3	Factor 4
Copepods	0.57**	0.41**	0.01	0.07
Chironomids	−0.39**	−0.12	0.10	−0.08
Isopods	−0.24**	0.02	−0.08	0.02
Gammaracanthus loricatus (amphipod)	−0.23**	0.02	−0.07	0.03
Monoculodes sp. (amphipod)	−0.23**	0.08	−0.21**	0.08
Mysis relicta	−0.01	0.26**	0.01	−0.02
Mysis litoralis	0.52**	−0.09	−0.07	−0.07
Small mysids	0.07	0.09	−0.06	−0.11
Onisimus spp. (amphipod)	−0.20**	−0.09	−0.17*	−0.17*
Pontoporeia affinis (amphipod)	−0.64**	−0.12	−0.02	−0.04

depth (Figure 2). *Mysis relicta* also showed a significant positive relationship with factor 2, which reflected a relatively high abundance in warmer, moderate-to-high salinity water.

Mysis litoralis

Adult *Mysis litoralis* were collected in deeper, colder, higher-salinity waters (Figure 3). Density was greatest in salinities above 28‰ and water temperatures less than 4°C. Highest densities were in areas deeper than 2.8 m; few of these mysids were present in samples taken in depths less than 2.4 m. Highest densities occurred in samples with highest factor 1 and lower factor 2 scores.

Small Mysids

Small mysids peaked in abundance in intermediate-to-high salinity waters and in intermediate and deeper waters (Figure 4). The two peaks in salinity and water depth may indicate the presence of two species, *Mysis relicta* and *Mysis litoralis*, which peaked in density at these conditions. Except for the peaks at 8–15‰ and above 28‰, density varied little over the range of salinity sampled, except density was very low in fresh water (0‰). Small mysids were generally most abundant at the middle range of factor 1 (Figure 4), suggesting an affinity for intermediate depth, salinity, and temperature, as indicated in the single parameter plots described above. The peak at the lowest value came from two samples with high densities taken in less than 1-m depth and in 11–13°C and 2–4‰ waters. The slight positive but not significant relationship with factor 2 (Table 3, Figure 4) reflected higher abundance in the warmest, moderate-to-high salinity waters.

Pontoporeia affinis

The amphipod *Pontoporeia affinis* was most abundant in warm, low-salinity, shallow waters (Figure 5). Density was high at salinities under 4‰ and highest in fresh water. Peak density occurred at 12–13°C. There was a strong positive relationship between density and water temperature, and a strong negative relationship between density and salinity. Very few *P. affinis* were found in waters less than 4°C. Density was highest in the shallowest water sampled (<0.8 m). Few were present in depths greater than 2.4 m. *Pontoporeia affinis* were significantly and negatively related to factor 1, reflecting their affinity for warm, shallow, low-salinity water (Table 3). A slight negative relationship with factor 2 indicated a stronger affinity for low-salinity water than for warmer water.

Onisimus spp.

The amphipods classified as *Onisimus* spp. were abundant in most habitats, but peaked in shallow, warm, low-salinity waters (Figure 6). Small peaks at intermediate depths, salinities and water temperatures may indicate the presence of more than one species with different habitat use patterns. *Onisimus* spp. were found in all salinity categories, but were most abundant in 4–24‰. Density was highest at temperatures from 6 to 12°C. Distribution by depth was bimodal, with peaks in water shallower than 0.8 m, and between 1.6 and 2.4 m. Few were found in waters with depths greater than 2.4 m. Low densities were observed in samples with higher factor 1 scores, indicating that *Onisimus* spp. avoid deep, high-salinity, cold waters. There was no apparent relationship between *Onisimus* spp. density and

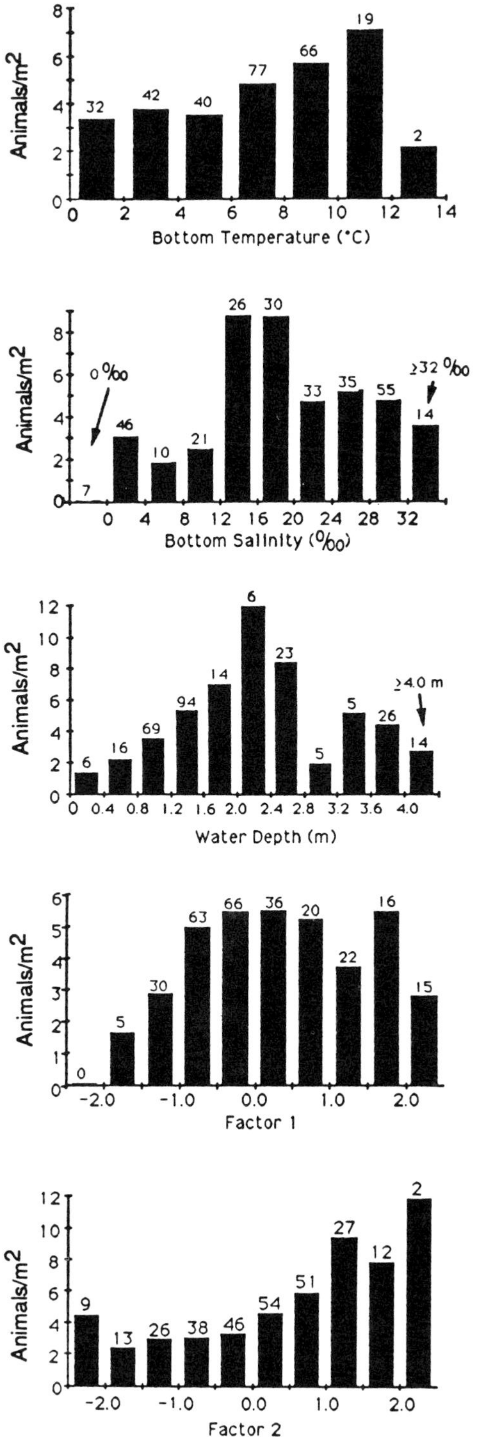

FIGURE 2.—Mean density of *Mysis relicta* versus water temperature, salinity, water depth, and principal components analysis factors 1 and 2 in midsummer 1986 in the Prudhoe Bay area. Numbers above bars indicate sample size.

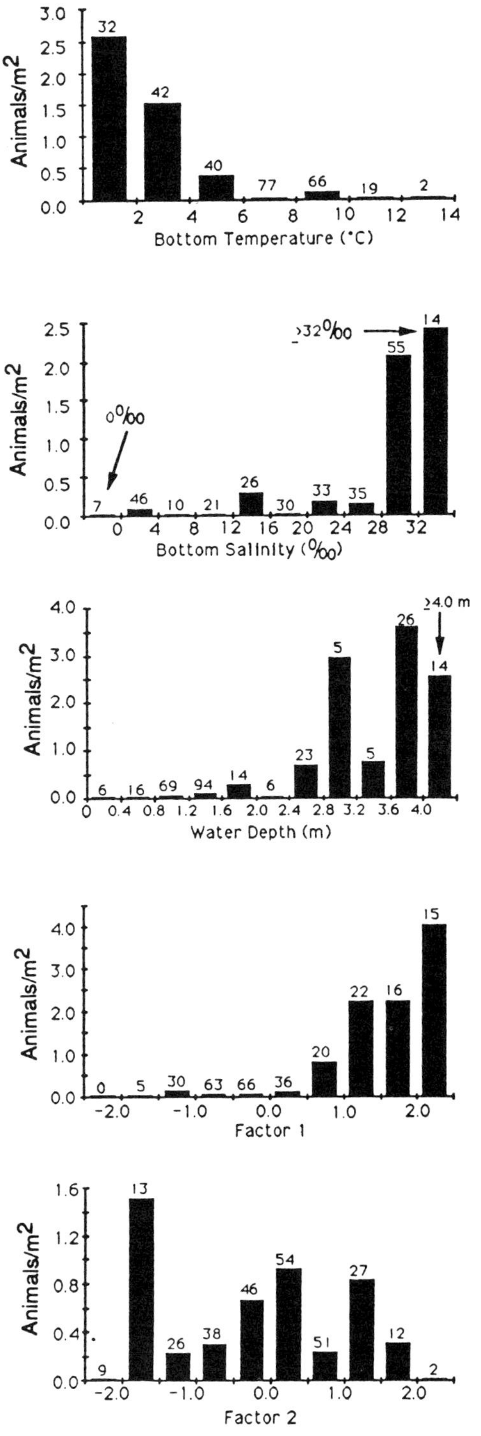

FIGURE 3.—Mean density of *Mysis litoralis* versus water temperature, salinity, water depth, and principal components analysis factors 1 and 2 in midsummer 1986 in the Prudhoe Bay area. Numbers above bars indicate sample size.

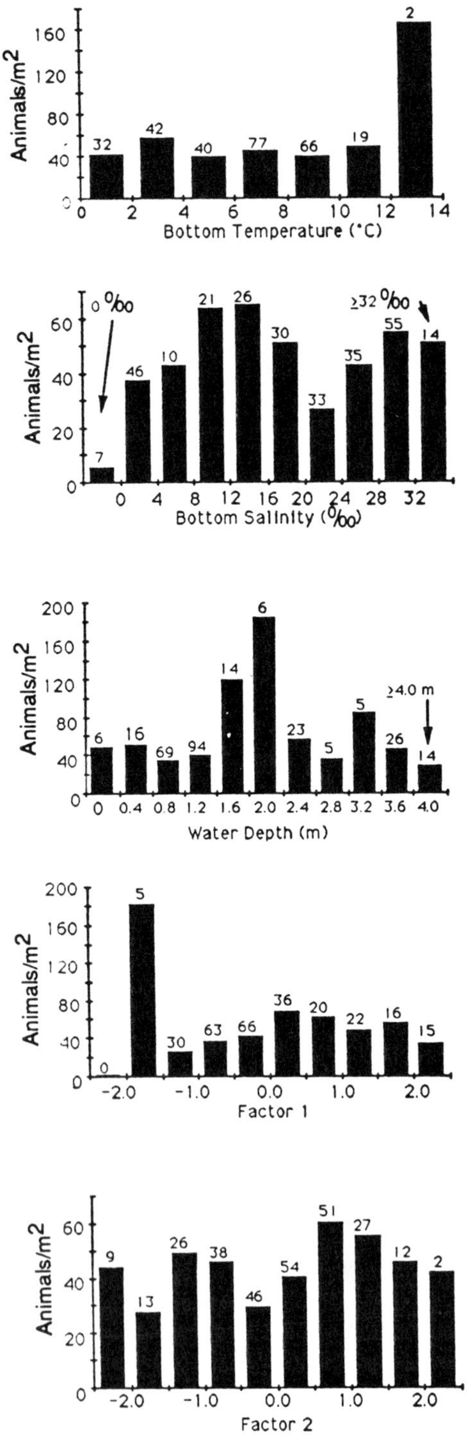

FIGURE 4.—Mean density of small mysids versus water temperature, salinity, water depth, and principal components analysis factors 1 and 2 in midsummer 1986 in the Prudhoe Bay area. Numbers above bars indicate sample size.

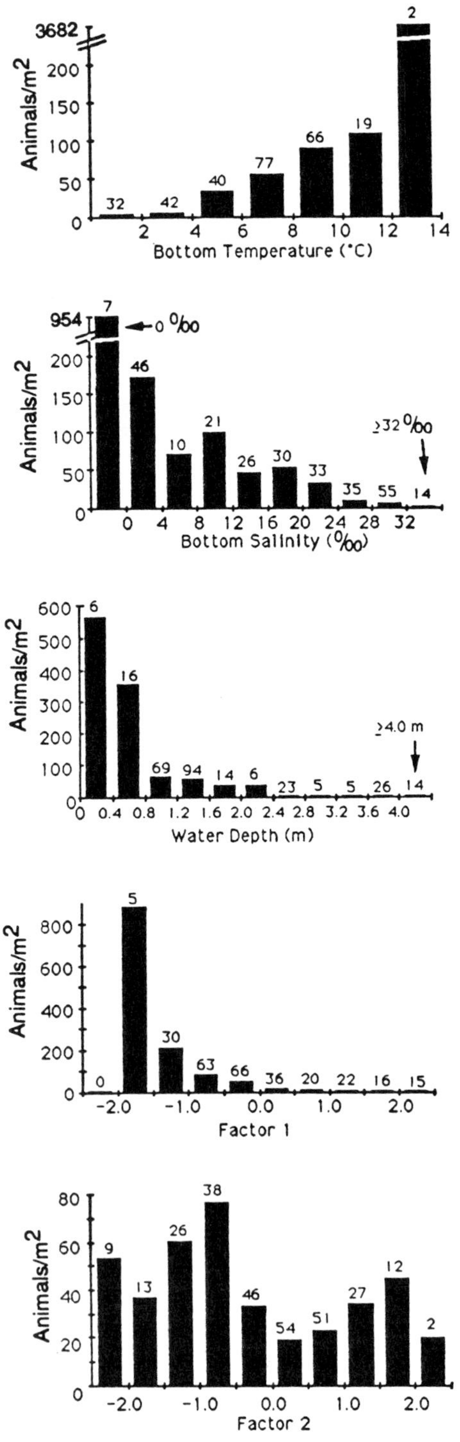

FIGURE 5.—Mean density of *Pontoporeia affinis* versus water temperature, salinity, water depth, and principal components analysis factors 1 and 2 in midsummer 1986 in the Prudhoe Bay area. Numbers above bars indicate sample size.

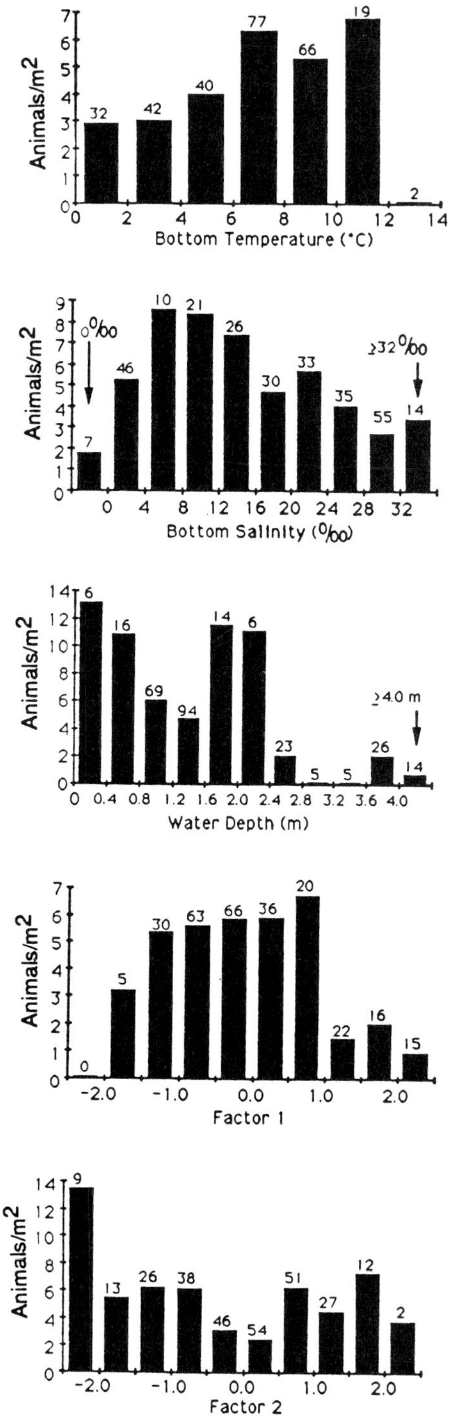

FIGURE 6.—Mean density of *Onisimus* spp. versus water temperature, salinity, water depth, and principal components analysis factors 1 and 2 in midsummer 1986 in the Prudhoe Bay area. Numbers above bars indicate sample size.

factor 2, although peak density did occur at the lowest factor 2 levels (cool, low-salinity water).

Other Amphipods

Gammaracanthus loricatus and *Monoculodes* sp. had patterns very similar to *Onisimus* spp. They were also negatively related to factor 1 (Table 3), reflecting their affinity for shallow, warmer, lower-salinity waters.

Copepods

Copepods were most abundant in deeper, colder, higher-salinity waters (Figure 7). Density was highest at surface salinities greater than 24‰. Lowest densities typically occurred when surface and bottom waters were less than 4‰. The distribution of copepod densities over surface temperature suggests a bimodal relationship. The distribution over bottom temperature is similar, but the bimodal pattern is weak. This bimodal pattern may reflect the presence of multiple species with different life history requirements. Densities were low at depths less than 2.4 m, but rose sharply, reaching a peak at depths over 3.6 m. Density peaked at the higher levels of factor 1 and factor 2. The significant positive relationship with factor 1 (Table 3) reflects their higher abundance in deeper, cooler, more saline waters. The positive association with factor 2 indicates an affinity for the warmest, most saline waters.

Chironomids

Chironomids collected were insect midge larvae and, unlike the other groups, were of freshwater origin, from the lower rivers and river deltas. Densities were highest in shallow, warm, low-salinity waters (Figure 8). Highest mean density occurred at salinities less than 8‰. Peak density occurred above 8°C and occurred in the waters less than 0.4 m deep. Density decreased substantially in samples taken in depths greater than 0.8 m. Densities were significantly and negatively related to factor 1, reflecting their affinity for shallow, warm, low-salinity water. Densities were negatively but not significantly related to factor 2 (Table 3), indicating that distribution of chironomids is more controlled by salinity than water temperature, as would be the case with a freshwater organism.

Discussion

There appear to be at least three major habitat types used by the seven major taxa surveyed. One

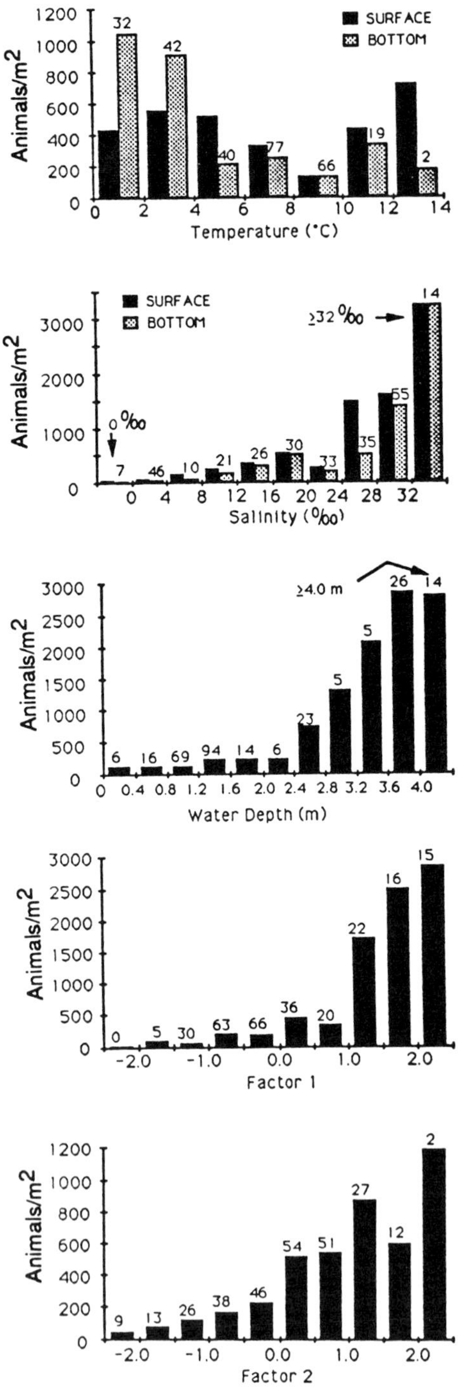

FIGURE 7.—Mean density of copepods versus water temperature, salinity, water depth, and principal components analysis factors 1 and 2 in midsummer 1986 in the Prudhoe Bay area. Numbers above bars indicate sample size.

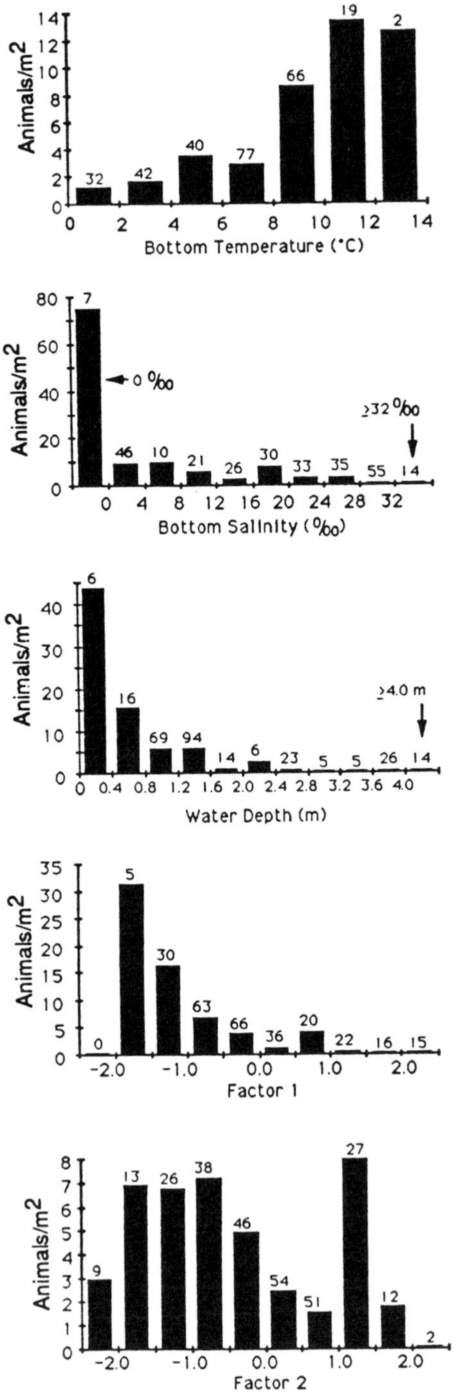

FIGURE 8.—Mean density of chironomids versus water temperature, salinity, water depth, and principal components analysis factors 1 and 2 in midsummer 1986 in the Prudhoe Bay area. Numbers above bars indicate sample size.

habitat type is warmer, fresher, shallower water. *Pontoporeia affinis* and chironomids were most abundant in this habitat. A second habitat type is characterized by deeper, colder, higher-salinity water. Copepods and *Mysis litoralis* were most common in such waters. Intermediate conditions of temperature, salinity, and depth were typical of the third habitat type. *Mysis relicta*, small mysids, and the amphipods *Onisimus* spp., *Gammaracanthus loricatus*, and *Monoculodes* sp. were generally found in the intermediate habitat.

Survey results indicate that the important prey of anadromous fish were most abundant in low-to-intermediate salinity water, which indicates that these invertebrates may depend on coastal, warm, brackish (estuarine) waters for an important part of their life cycle. The two most important prey, *Mysis relicta* and *Pontoporeia affinis* (Cannon et al. 1987b; Knutzen et al. 1990), are described by Martinez and Bergersen (1989) as glaciomarine relicts with a circumpolar distribution in the northern hemisphere that also dominate the epifauna of many remnant Pleistocene proglacial lakes in North America (e.g., Great Lakes). We believe that these species exploit the coastal environment, keying in on warm, low-salinity water with abundant prey. Because they are found in deeper waters of lakes, water depth is not likely an important determinant of habitat use. Even the major copepods, *Acartia clausi* and *Pseudocalanus* spp., are marine, neritic copepods typical of and abundant in temperate estuaries (English and Horner 1977).

The principal components analysis was designed to show primary and secondary features in the data set. The primary feature in the habitat, factor 1 which explained 67% of variance, was the negative relationship between water temperature and salinity and depth. Factor 2 of the principal components analysis explained only 16% of the data set variance, but provided information regarding prey response to a habitat gradient in which water temperature and salinity were positively related. Factor 2 appeared to represent the occurrence of warm, high-salinity water that was encountered in Prudhoe Bay at times during the sampling period. Marine water that has stood for a time in shallow bays and lagoons may warm considerably and therefore differ considerably from the normal negative gradient between water temperature and salinity in coastal waters. *Mysis relicta* and copepods densities showed a positive response to factor 2, indicating an affinity for warmed coastal waters with moderate-to-high salinity. Chironomids and *Pontoporeia affinis* showed a negative relationship with factor 2, indicating a greater affinity for low-salinity water of the river plumes, regardless of temperature.

Four improvements to the sampling design would further enhance one's ability to define the habitat requirements of the invertebrates. (1) Even though sampling covered only a 1-month period from mid-July to mid-August, seasonal difference was apparent between the early and late samples. Stratified sampling by time period would better account for this variability. (2) Stratified sampling by area would also account for more variability. Differences in distribution due to location were apparent. Areal differences in the marine invertebrate communities were apparent in regional drop net surveys (Griffiths and Dillinger 1981; Cannon et al. 1987a). (3) Stratified sampling by depth, salinity, and temperature would provide more samples in certain types of habitats that otherwise were not well represented. For example, few samples were obtained in cold, low-salinity water typical of ice melt waters, or in shallow, warm, high-salinity water typical of lagoons in late summer. Sampling in proportion to the amounts of each habitat type available or in proportion to the variance of each type could improve the design. (4) Including additional habitat parameters could improve one's ability to define the specific habitat requirements of the invertebrates. For example, sediment characteristics (such as grain size and organic content) are considered an important habitat feature of epibenthic and benthic invertebrates. Griffiths and Dillinger (1981) noticed that epifauna were more prevalent in slight depressions where light organic detritus had collected. During processing of drop net samples in this survey, we noticed what appeared to be higher concentrations of invertebrates in samples with more detritus. Current speed and wave action are two other parameters that could be considered.

Improvements in the survey design mentioned above will improve our knowledge of the optimal habitat conditions for key invertebrate fish prey. However, lack of control over the primary habitat parameters hinders our ability to separate the individual effects or importance of water temperature, salinity, water clarity, and depth. Carefully designed field experiments and surveys may provide further insight; on the other hand, controlled laboratory studies offer the possibility of separating the effects of the individual variables.

Despite the above-mentioned weaknesses in the

survey data and design, the survey results indicated that many important invertebrate prey of coastal Beaufort Sea fishes tend to associate with warm, low-to-moderate salinity water during the summer open-water season. Oil and gas developments that affect the quantity and quality of such coastal estuarine habitat may affect the distribution and abundance of these invertebrates, and possibly the overall productivity of the waters. Gravel-fill causeways affect coastal water temperature and salinity distribution patterns in the central Beaufort Sea (Cannon and Hachmeister 1987; Hachmeister et al. 1991, this volume), and thus would likely affect the distribution and abundance of the key invertebrate fish prey. The habitat suitability plots developed from this study could be used to estimate the effect of specific habitat changes induced by development such as causeways on regional epibenthic and planktonic invertebrate communities.

References

Cannon, T. C., M. S. Brancato, and S. Jewett. 1987a. Fish prey survey. *In* 1985 Final report for the Endicott Environmental Monitoring Program, volume 7. Report of Envirosphere Company to U.S. Army Corps of Engineers, Anchorage, Alaska.

Cannon, T. C., M. S. Brancato, and S. Jewett. 1987b. Fish food habits. *In* 1985 Final report for the Endicott Environmental Monitoring Program, volume 7. Report of Envirosphere Company to U.S. Army Corps of Engineers, Anchorage, Alaska.

Cannon, T. C., and L. E. Hachmeister. 1987. Integration and assessment. *In* 1985 Final report for the Endicott Environmental Monitoring Program, volume 1. Report of Envirosphere Company to U.S. Army Corps of Engineers, Anchorage, Alaska.

Cannon, T. C., D. R. Glass, and C. M. Prewitt. 1991. Habitat use patterns of juvenile Arctic cod in the coastal Beaufort Sea near Prudhoe Bay, Alaska. American Fisheries Society Symposium 11:157–162.

Craig, P. C., and L. Haldorson. 1981. Fish. Pages 384–678 *in* Beaufort Sea barrier island–lagoon ecological process studies: final report, Simpson Lagoon. Environmental assessment of the Alaskan outer continental shelf program, volume 7. Bureau of Land Management and National Oceanic and Atmospheric Administration, Outer Continental Shelf Environmental Assessment Program, Boulder, Colorado.

English, T., and R. Horner. 1977. Beaufort Sea plankton studies. Pages 275–627 *in* Environmental assessment of the Alaskan outer continental shelf program, volume 9. Bureau of Land Management and National Oceanic and Atmospheric Administration, Outer Continental Shelf Environmental Assessment Program, Boulder, Colorado.

Feder, H. M., and S. G. McGee. 1982. Epifaunal monitoring program. *In* 1981 Prudhoe Bay Waterflood Environmental Monitoring Program, volume 3. Report of University of Alaska, Fairbanks, Institute of Marine Science, to U.S. Army Corps of Engineers, Anchorage, Alaska.

Griffiths, W. B., and R. Dillinger. 1981. Invertebrates. Pages 1–198 *in* Beaufort Sea barrier island–lagoon ecological process studies: final report, Simpson Lagoon. Environmental assessment of the Alaskan continental shelf, volume 8. Bureau of Land Management and National Oceanic and Atmospheric Administration, Outer Continental Shelf Environmental Assessment Program, Boulder, Colorado.

Hachmeister, L. E., D. R. Glass, and T. C. Cannon. 1991. Effects of solid-fill gravel causeways on the coastal central Beaufort Sea environment. American Fisheries Society Symposium 11:81–96.

Knutzen, J. A., M. S. Brancato, and S. C. Jewett. 1990. Fish food habits. *In* 1986 Final report for the Endicott Environmental Monitoring Program, volume 7. Report of Envirosphere Company to U.S. Army Corps of Engineers, Anchorage, Alaska.

Martinez, P. J., and E. P. Bergersen. 1989. Proposed biological management of *Mysis relicta* in Colorado Lakes and reservoirs. North American Journal of Fisheries Management 9:1–11.

Moulton, L. L., and seven coauthors. 1986. 1984 central Beaufort Sea fish study. Pages 1–300 *in* Prudhoe Bay Waterflood Environmental Monitoring Program, 1984, volume 2. Final report of Woodward-Clyde Consultants, Entrix, Inc., and LGL Alaska Research Associates to U.S. Army Corps of Engineers, Anchorage, Alaska.